Formulating Natural

handbook of plants and their phytochemistry

by

Anthony C. Dweck FLS FRSC FRSPH

2nd Edition
Fully revised and updated
2017

The first edition was published by Allured Publishing who sadly stopped publishing and the copyright restored to the author.

This book is dedicated to the Linda Dweck who is the most patient and understanding person in my life and has been a constant source of wonderment and encouragement in all my years as a consultant.

About the author

Anthony C. Dweck BSc, CChem, CSci, FRSC, FLS, FRSPH.

Dweck has worked for Smith & Nephew, S.C. Johnson, Marks & Spencer, and Peter Black (now PB Beauty). In 1998 he formed Dweck Data in order to devote more time to the study of botanicals and their chemistry.

Past member of Council (1984-1986) Society of Cosmetic Scientists, Past President Society of Cosmetic Scientists (1996-1997), Past President Society of Cosmetic Scientists (2001 – 2002), Honorary Member of Society Cosmetic Scientists (awarded 2004), Technical Editor of the Personal Care Magazine (Asia Pacific since 1999) (Europe since 2008), Associate Editor International Journal of Cosmetic Science (2001-2003), Moderator and creator of the Formulators' Discussion Group (1998-2005), Honorary Organiser SCS Spring Conference 100% Natural in 2007, Member of the Scientific Advisory Committee of the CTPA (1992-1998). Member of the Advisory Board of Cosmetics & Toiletries Magazine (1997-1998), Member of the Advisory Board of International Society of Cosmetic Dermatology (since 2003), Member of the Editorial Scientific Advisory Panel of SPC Magazine (1997-2001), Member of the Editorial Scientific Advisory Board of SOFW Journal (1988 - 1999), Member of the LCLN (Ingredient Nomenclature) of the C.T.P.A. (1994-1998), Member of the IFSCC Monograph Review Committee (1997 - 1999), External Examiner for Society of Cosmetic Scientists (since 1991), Referee (reviewer) for International Journal of Cosmetic Science (since 1992), Chairman Sponsorship Committee of IFSCC Congress in 2002, Edinburgh (1998-2000), IFSCC listed conference speaker. Chairman of the SCS 50th Anniversary Book Committee, 1998. Joint Organiser of the Post Graduate Course in Cosmetic Science (1998, 2000, 2001, 2003, 2005, 2007), Council of Europe - Botanical Task Force (Committee of Experts) (appointed 1998). Consultant and expert witness (listed) to the Trading Standards Office

Author of over a hundred articles and papers on various aspects of the Cosmetic and Toiletry industry and numerous book chapters, Anthony is also a frequent lecturer on his favourite topic of botanicals/medicinal plants and has presented over 80 papers at conferences all over the world. He was a regular organiser of the conference programme for PCIA (Personal Care Ingredients Asia) and the honorary organiser for the SCS Spring Symposium 13-15th May 2007 "the 100% Natural Conference" at Staverton Park, Northants.

Anthony has a unique library of over 650 volumes on medicinal plants and their uses as well as a unique computerised data base on over 80,000 species of plants. He also maintains a comprehensive computer data base of supplier technical data sheets.

INDEX

Formulating Natural Cosmetics

Chapter 1

Fixed Oils

The choice of fixed oils today is huge and the beauty of them is that they all have different blends of fatty acid carbon chain lengths, all have totally different skin feel and differing rates of application. It was once thought that by examining the composition of the various fatty acids and looking at the profile that it would be possible to predict the characteristics of any particular oil. This was not the case.

The list below is not exhaustive, we believe that there are still fixed oils to be found. However, finding these oils commercially is not always assured and some of the more exotic may require a great deal of research to find the supplier.

A

Actinidia chinensis. Kiwi Seed Oil fixed oil. Iodine value: 185. Average carbon number: 17.6. Average molecular weight: 278.3. C16:0: 6%; C18:1: 11%; C18:2: 15%; C18:3: 67%. Traditional use: An oil rich in vitamin E, the kiwi, or Chinese gooseberry, grown in New Zealand, is a significant source of skin nutrition and protection. Folklore: Seeds of *A. chinensis* from the Yangtze valley were taken to New Zealand early in the 20th century, and commercial cultivation began in the 1930s. The second largest producer is France.

Actinodaphne hookeri. Pisa Oil fixed oil. Sp.gr.: 0.925. Saponification value: 256. Iodine value: 9.13. Average carbon number: 12.24. Average molecular weight: 203.59. C12:0: 96%; C18:1: 4%; C18:2: 0%.

Adansonia digitata. Baobab Oil fixed oil. Saponification value: 205. Iodine value: 80. Average carbon number: 17.379. Average molecular weight: 273.6. C14:0: 1.5%; C16:0: 30.1%; C18:0: 2%; C18:1: 31.8%; C18:2: 24.8%; C18:3: 8%; C20:0: 1%; C20:1: 0.4000%; C22:0: 0.2%; C22:1: 0.1%. Traditional use: An African tree steeped in mystery, legend and religious significance. The fruit provides an exquisite oil that is a rich and substantive moisturiser. The traditional method of oil extraction is by pounding the seeds. The oil thus produced is used as a rub to relieve aches, pains and rheumatism, but more especially, to treat skin complaints such as eczema and psoriasis.

Aleurites fordii. Tung Oil fixed oil. Melting Point: -2.5°C. Iodine value: 168. Average carbon number: 17.82. Average molecular weight: 276.6. C14:0: 4%; C16:0: 1%; C18:1: 7%; C18:2: 7%; C18:3: 1%; C18:3 (Conj n=5): 80%.

Aleurites moluccana. Kukui Nut Oil fixed oil. Sp.gr.: 0.925. Saponification value: 190. Iodine value: 165. Average carbon number: 17.8. Average molecular weight: 278. C14:0: 5%; C18:1: 28.6%; C18:2: 42.6%; C18:3: 23.8%. Traditional use: To the Hawaiians this tree is a symbol and a legend. The oil is a panacea for delicate, sensitive or dry skin, and is pure and gentle enough to use on a baby's delicate skin. Traditional use: The oil is mainly used and traditionally for acne, eczema and psoriasis. As a moisturiser moisturizer and emollient it is highly respected, an excellent penetrating oil that, when spread onto the skin, softens it and soothes sunburns and irritations. Folklore: Hundreds of years ago the early Hawaiians discovered the magical Kukui Nut. When set to flame, it lighted the soft Polynesian nights. When smoothed on the skin, it soothed and lessened sunburns and irritations. New born babies were bathed is in this easily-absorbed oil.

Aleurites triloba. Candlenut Oil. Pacific Islands, Asia, Australia. Traditional use: Seeds are source of drying oil used in soap and wood preservatives. Folklore: In Polynesia the seeds are strung on sticks and burnt as candles.

Amaranthus cruentus. Amaranthus Oil. Amaranth Oil fixed oil. Average carbon number: 15.64. Average molecular weight: 255.8. C16:0: 19%; C18:0: 3%; C18:1: 22%; C18:2: 45%.

Amaranthus cruentus. Amaranthus Oil (Critical CO_2). Amaranth Oil fixed oil. Average carbon number: 15.619. Average molecular weight: 255.8. C16:1: 2%; C18:1: 30%; C18:2: 45%; C30: 6%.

Anacardium occidentale. Cashew Nut Oil fixed oil. Sp.gr.: 0.96397. Saponification value: 168. Iodine value: 44. Average carbon number: 18.18. Average molecular weight: 279.69. C16:0: 10.6%; C18:0: 6.40%; C18:1: 74.297%; C18:2: 10.8%. Traditional use: An oil rich in proteins, this lubricious material can be used wherever a skin nourishing effect is needed. There is an acrid oil in the skin that is used to remove warts and indolent ulcers. This is not found in the kernel oil.

Annona muricata. Guanabana Seed Oil or Soursop Oil. Saponification value: 228, Iodine value: 111. Found in Mexico, Brazil and Caribbean islands, widely used in the traditional medicine of these countries for its anti-bacterial, antiparasitic, soothing and anti-fungal properties. Soursop or graviola (*Annona muricata* Linn.) seeds were analysed for proximate composition and were found to contain 22.10% pale-yellow oil and 21.43% protein. Physical and chemical characteristics of the seed-kernel oil were determined. Results show that the oil is bland in taste and possesses an acid value of 0.93 and an acetyl value of 66.77. The oil consists of 28.07% saturated and 71.93% unsaturated fatty acids.

Arachis hypogaea. Arachis, Peanut Oil, Ground nuts, Monkey nuts or Earth nuts. It is a fixed oil. Tropical Africa and in India, Brazil, southern USA United States and Australia. Melting Point: 3°C. Sp.gr.: 0.919. Saponification value: 190. Iodine value: 98. Average carbon number: 18.07. Average molecular weight: 283.5. C16:0: 7%; C18:0: 5%; C18:1: 59%; C18:2: 23%; C20:0: 3%; C22:0: 1%; C24:0: 2%. Traditional use: A traditional oil for use in sunscreen preparations and after-sun oils. It is substantive and protective to the harshest of external conditions. The oil is listed as a skin conditioning agent - occlusive; solvent. Folklore: In 1840 Jaubert, a French colonist of Cape Verde, suggested its importation into Marseilles as a seed oil. It was first mentioned by Fernandez de Oviedo y Valdes who lived in Haiti from 1513 to 1525, who reported that the Indians widely cultivated the *mani* - a name for Arachis still used in South America and Cuba. South American native, it is widely cultivated, especially India, Africa, China and America. Nut oil allergy: there are very few people affected by ingestion of peanuts and there is little evidence to suggest that topical application of peanut oil would elicit a reaction, potentially fatal or otherwise. Additionally, the allergen involved has not been accurately identified, nor the quantities required to cause an effect. Furthermore, the degree of refinement of peanut oils will more than likely remove the offending allergen.

Argania sideroxylon. Argane Oil fixed oil. Saponification value: 195. Iodine value: 98.5. Average carbon number: 17.3. Average molecular weight: 278.6. C16:0: 12.5%; C18:0: 6%; C18:1: 46%; C18:2: 32.5%; C18:3: 0.1%; C20:0: 0.4000%; C20:1: 0.450%. Traditional use: Rich in natural sterols, this oil from Morocco is used by the local women to keep their skin soft, smooth and protected. Chemistry: Four sterols have been found in argan oil. The two major are named spinasterol and schottenol (44% and 48% respectively), the two minor (stigmasta-8,22-dien-3b-ol (22-E, 24-S) and stigmasta-7,24-28-dien-3b-ol (24-Z)) have been both isolated in 4%

yield. Interestingly, no D-5 sterols that are frequently encountered in vegetable oil have been isolated so far.

Argania spinosa. Argane Oil fixed oil. Saponification value: 195. Iodine value: 98.5. Average carbon number: 17.75. Average molecular weight: 278.6. C16:0: 12.6%; C18:0: 6.78%; C18:1: 43.3%; C18:2: 37.2%. Traditional use: Rich in natural sterols, this oil from Morocco is used by the local women to keep their skin soft, smooth and protected. Argan oil and preparations including argan oil have been used in the traditional Moroccan medicine for centuries to cure skin diseases.

Astrocaryum murumuru. Murumuru Butter fixed oil. Saponification value: 245. Iodine value: 25. Average carbon number: 13.89. Average molecular weight: 226.47. C8: 1%; C10: 0.9000%; C12:0: 44%; C14:0: 28%; C16:0: 9%; C18:0: 3%; C18:1: 10%; C18:2: 4%; C20:0: 0.1%. Traditional use: A very soft butter from the heart of the Brazilian rain forest. This butter is a substantive and protective material that makes a perfect, if not better, replacement for cocoa butter, illipe butter and others.

Astrocaryum tucuma. Tucuma Butter fixed oil. Saponification value: 231.4. Iodine value: 12.5. Average carbon number: 13.49. Average molecular weight: 221.47. C8: 1.92%; C10: 1.95%; C12:0: 50.157%; C14:0: 24.44%; C16:0: 6.21%; C18:0: 2.33%; C18:1: 8.35%; C18:2: 4.16%; C20:0: 0.1%; C24:0: 5.98E-2%.

Attalea cohune. Cohune Oil fixed oil. Melting Point: 28°C. Saponification value: 254. Iodine value: 13. Average carbon number: 13.109. Average molecular weight: 215.6. C8: 7.5%; C10: 6.5%; C12:0: 46.5%; C14:0: 16%; C16:0: 9.5%; C18:0: 3%; C18:1: 10%; C18:2: 1%. Traditional use: The cohune palm is a valuable source of oil and was an important tree in the Mayan culture. The seeds of the cohune palm yield cohune oil which that is used extensively as a lubricant, for cooking, soap-making and lamp oil. It has a deliciously smooth almost butter-like quality that leaves an exceptional emolliency on the skin.

Avena sativa. Oat Oil fixed oil. Sp.gr.: 0.910. Average carbon number: 17.879. Average molecular weight: 275.5. C16:0: 24%; C18:1: 34%; C18:2: 44%. Traditional use: the The useful part of oat is the endosperm, which contains xanthophylls, phytosterols, terpene alcohol, tocopherols and hydrocarbons. Oats are the cereal with the highest lipid content (about 7 8%), most of which is located in the endosperm. It has good skin affinity and has a moisturising moisturizing action which that is particularly visible in skins dehydrated by excessive exposure to solar radiation and sea water. It is used in protective children's preparations and emulsions and pastes for problem skin conditions, especially in cases of dryness and itching.

Azadirachta Indica. Neem Oil fixed oil. Sp.gr.: 0.932. Saponification value: 185. Iodine value: 89. Average carbon number: 17.32. Average molecular weight: 278.3. C16:0: 18%; C18:0: 15%; C18:1: 50%; C18:2: 13%; C20:0: 2%.

B

Bassia latifolia. Illipe Butter or Mowrah Butter fixed oil. Melting Point: 36°C. Sp.gr.: 0.890. Saponification value: 185. Iodine value: 32. Average carbon number: 16.809. Average molecular weight: 276.19. C16:0: 23.5%; C18:0: 22%; C18:1: 35.5%; C18:2: 15%. Traditional use: Another of the African butters, used by natives for the treatment of dry skin and especially for the care and protection of babies' babies' delicate skin. Mahuwa oil has emollient properties and is used in skin disease, rheumatism and headache. It is also a laxative and considered useful in habitual constipation, piles and haemorrhoids and as an emetic. Tribal peoples also used it as an illuminant and hair fixer.

Bassia latifolia. Mowrah Butter fixed oil. Melting Point: 39ºC. Sp.gr.: 0.9010. Saponification value: 291. Iodine value: 62. Average carbon number: 17.35. Average molecular weight: 276.3. C16:0: 23.5%; C18:0: 21%; C18:1: 38.5%; C18:2: 16%. Traditional use: [Syn. Bassia longifolia] The butter is extracted from the seed kernels and further processed and refined to obtain a yellowish butter which that has a very mild odour. The butter is a solid at room temperature, but melts readily on contact with the skin. It will help prevent drying of the skin and so help reduce the formation of wrinkles. Helps soften and smooth the skin. Medicinally, Bassia oil is used as an emollient application to the skin.

Bertholletia excelsa. Brazil Nut Oil fixed oil. Sp.gr.: 0.917. Saponification value: 193. Iodine value: 98. Average carbon number: 17.719. Average molecular weight: 278.19. C12:0: 0.2%; C14:0: 0.2%; C16:0: 13%; C16:1: 0.2%; C18:0: 11%; C18:1: 39.297%; C18:2: 36.1%. Traditional use: An oil rich in proteins and vitamins A and E, this precious gift from the South American rain forest comes from the fruit (nut) of a tree that can grow for a thousand years. Brazil nuts, or para nuts, produce an oil that is a rich emollient and a moisturiser moisturizer not too dissimilar to sesame oil.

Bixa orellana. Urucum Oil, Annatto Oil fixed oil. Average carbon number: 20.559. C16:0: 15.4%; C16:1: 0.149%; C18:0: 25.6%; C18:1: 44.8%; C18:2: 28.6%; C18:3: 1.3%. Annato oil is emollient and topically soothing. Its high content of polyunsaturated fatty acids make it readily and completely absorbed by the skin. The polyunsaturated fatty acids do not hinder the oxygenation and natural secretion process of the skin, which would cause dilatation of the skin pores, formation of blackhead and accumulation of fats.

Bombax malabaricum. Kapok Oil fixed oil. Sp.gr.: 0.926. Saponification value: 193. Iodine value: 98. Average carbon number: 14.949. Average molecular weight: 241.19. C16:0: 22%; C18 Epoxy: 5.5%; C18:1: 21%; C18:2: 37%.

Borago officinalis. Borage (Starflower) Oil fixed oil. Melting Point: 29ºC. Saponification value: 185. Iodine value: 147. Average carbon number: 17.8. Average molecular weight: 278.3. C16:0: 14%; C18:0: 5%; C18:1: 17%; C18:2: 36%; C18:3: 24%; C20:1: 2%; C20:2: 2%. Traditional use: Another rich plant source of GLA. Excellent moisturiser in skin care.

Brassica alba. Mustard Seed Oil fixed oil. Average carbon number: 18.62. Average molecular weight: 310.3. C18:1: 23%; C18:2: 9%; C18:3: 10%; C20:1: 8%; C22:1: 43%.

Brassica campestris. Rapeseed Oil (low erucic acid). Canola Oil fixed oil. Sp.gr.: 0.914. Saponification value: 173. Iodine value: 107. Average carbon number: 17.94. Average molecular weight: 283.1. C16:0: 3%; C18:0: 77%; C18:1: 14%; C18:2: 6%.

Brassica napus. Rapeseed Oil (high erucic acid) fixed oil. Melting Point: -10ºC. Iodine value: 100. Average carbon number: 20.57. Average molecular weight: 312.6. C16:0: 2%; C18:0: 1%; C18:1: 21%; C18:2: 20%; C18:3: 2%; C20:0: 1%; C22:1: 53%; C24:0: 2%.

Brassica nigra. Mustard Seed Oil fixed oil. Iodine value: 103. Average carbon number: 19.82. Average molecular weight: 321.69. C16:1: 2%; C18:0: 0.1%; C18:1: 1.2%; C18:2: 23%; C18:3: 13.5%; C20:0: 10%; C20:1: 0.696%; C22:0: 10%; C22:1: 0.299%; C24:0: 34%; C24:1: 0.598%. Traditional use: This cold cold-pressed oil is traditionally used in Indian head massage to produce a stimulating effect on the tissues. It increases body heat and helps to relieve pain and stiff muscles. Mustard seed oil has the aroma of cabbage. While it is used as cooking oil in India, it is not recommended for consumption because of the high level of erucic acid. It has a low level of saturated fat content, of 5%. It is primarily monounsaturated fat at 67%, about half of which is erucic acid. Polyunsaturated fats are 21%. Mustard seed oil has application in personal care products like massage oils and muscle rubs. Oil from the hulls of seeds promotes the growth of hair according to some sources. Mustard seed oil is also used in massage therapy.

Brassica oleracea italica. Broccoli Seed Oil. Sp.gr.: 0.914. Iodine value: 105. Average carbon number: 20.6. Average molecular weight: 307.4. C16:0: 3.5%; C18:0: 0.8%; C18:1: 15%; C18:2: 15%; C18:3: 11.2%; C20:0: 0.6%; C20:1: 5.3%; C22:0: 0.9%; C22:1: 50.1%; C22:2: 2.1%. Broccoli Seed Oil is normally cold pressed from the seeds and yields an extra virgin, unrefined, odourless, pale green oil (although some suppliers do not always achieve this colour). Broccoli Seed Oil provides excellent oxidative stability and non-greasy moisturizing properties and it possesses an impressive fatty acid profile that makes it a valuable emollient in cosmetics and personal care. The oil is very high content in erucic acid and rich in vitamins K and A. These lubricating properties, non-greasy feel and rapid absorption make it an excellent emollient ideal for use on skin and hair.

Butea frondosa [Syn. *B. monosperma*]. Palash Seed Oil (Dhak or Palas oil) fixed oil. Melting Point: 45°C. Saponification value: 183. Iodine value: 75. Average carbon number: 18.23. Average molecular weight: 287.24. C16:0: 27%; C18:0: 8%; C18:1: 27%; C18:2: 16.5%; C20:0: 2.5%; C20:1: 2.700%; C22:0: 12%; C24:0: 4%. The seeds from the "Flame of the Forest" tree contain a yellow fixed oil called moodooga oil or kino-tree oil, small quantities of resin and large quantities of a water soluble albuminoid. Fresh seeds contain proteolytic and lipolytic enzymes. The seeds are beneficial in the treatment of certain skin disease. The seeds, ground and then mixed with lemon juice is painted onto dhobi's itch, an eczema-type of skin disorder, characterized by itching. The crushed seeds are used for killing maggots in wounds and sores.

Butyrospermum parkii. Shea Butter fixed oil. Saponification value: 180. Iodine value: 3. Average carbon number: 17.6. Average molecular weight: 278.1. C16:0: 20%; C18:0: 45%; C18:1: 33%; C18:2: 2%. Traditional use: This rich buttery oil from central Africa is used for the protection and care of skin cracked and dehydrated by the elements. Beurre de karite is an elegant addition to products crafted for the smoothing and replenishment of dry skins. Shea butter is a suitable base for topical medicines. Its application relieves rheumatic and joint pains and heals wounds, swellings, dermatitis, bruises and other skin problems. It is used traditionally to relieve inflammation of the nostrils.

Butyrospermum parkii. Shea Butter or Karite Butter fixed oil. Melting Point: 38°C. Sp.gr.: 0.915. Saponification value: 180. Iodine value: 65. Average carbon number: 18.42. Average molecular weight: 281.6. C16:0: 6%; C18:0: 40.5%; C18:1: 50.5%; C18:2: 6%. Traditional use: The high proportion of unsaponifiable matter, consisting of 60-70% triterpene alcohols, gives shea butter creams good penetrative properties that are particularly useful in cosmetics. Allantoin, another unsaponifiable compound, is responsible for the anti-inflammatory and healing effect on the skin. Traditional use: Shea Butter has natural antioxidant properties and is said to contain a small quantity of allantoin that is renowned renowned for its healing qualities. It is also said to protect the skin against external aggressions, and sun rays. It is an effective skin emollient and skin smoother. Folklore: In Africa, the fat (shea butter) is used as ointment for rheumatic pains and boils. A decoction from the bark is used to facilitate child delivery and ease labour pains. The leaf extract is dispensed for headaches and as an eye bath.

C

Calendula officinalis. Marigold Oil fixed oil. Iodine value: 132. Average carbon number: 17.28. Average molecular weight: 279.1. C18:2: 34%; C18:3 (n-3): 62%. Traditional use: An oil that would be in any herbalist's arsenal for the care of bruised or damaged skin. It is especially good for chapped and roughened skin. Also traditionally used for the care of varicose veins. The Commission E approved the internal and topical use of calendula flower for inflammation of the oral and pharyngeal mucosa. It was also approved externally for poorly healing wounds. Specifically, herbal infusions, tinctures, and ointments are used to respond to skin and mucous membrane inflammations such as pharyngitis, dermatitis, leg ulcers, bruises, boils, and rashes.

Calodendrum capense. Cape Chestnut Oil, Yangu Seed Oil fixed oil. Sp.gr.: 0.876. Saponification value: 192. Iodine value: 77. Average carbon number: 17.78. Average molecular weight: 279.6. C16:0: 11.1%; C18:0: 7.200%; C18:1: 75.9%; C18:2: 5.78%. Traditional use: It is also known as Wild Chestnut and not related to any other chestnut. Seeds of this African tree are crushed and boiled to obtain oil that is suitable for making soap. The Xhosa tribe believes that the seeds have magic properties, and the hunters (in order to improve their luck when hunting) tied them around their wrists to bring skill and good luck. Around Mount Kenya, the Kikuyu tribe traditionally used the oil as a skin balm to treat troublesome skin conditions.

Calophylum inophylum [Syn: *Calophylum tacamahaca*] Tamanu Oil, Undi Oil, Punnah Oil fixed oil. Melting Point: 17ºC. Sp.gr.: 0.94395. Saponification value: 193. Iodine value: 93. Average carbon number: 17.57. Average molecular weight: 278. C16:0: 13.4%; C18:0: 16%; C18:1: 38.297%; C18:2: 0.299%; C18:3: 31.1%. Traditional use: It is not an essential oil but instead is prepared by pressing the carefully dried nuts. This oil from the Polynesian islands has a distinctive herbal odour and a long history of skin repair and care of the skin. Said to be particularly good for removing skin blemishes. Excellent results have been seen on the treatment of badly damaged skin and scar tissue.

Camelia drupifera. Camelia Oil fixed oil. Average carbon number: 17.64. Average molecular weight: 280. C16:0: 9%; C18:0: 1%; C18:1: 80%; C18:2: 9%.

Camelina sativa. Gold of Pleasure Oil fixed oil. Sp.gr.: 0.922. Average carbon number: 18.21. Average molecular weight: 284.1. C16:0: 5%; C18:0: 3.5%; C18:1: 20%; C18:2: 21%; C18:3: 35%; C20:0: 15.5%. Traditional use: Probably introduced into this country by the Romans, this oil has an amazing lubricity which that makes it ideal for those products where spreadability is required.

Camelina sativa. Gold of Pleasure Oil fixed oil. In another source the values were as follows. Saponification value: 191. Iodine value: 154. Average carbon number: 19.539. Average molecular weight: 287. C18:1: 20%; C18:2: 24%; C18:3: 39%; C20:1: 23%. Traditional use: This oil has already found use as a replacement for sperm whale oil. It has potential use in hair and skin care preparations to impart lubricity or in formulations where emollience, moisturising moisturizing and spreading properties are required. It was probably introduced into this country by the Romans for its oil that was used for burning. This oil has an amazing lubricity that makes it ideal for those products where spreadability is required. Folklore: Myagrum, camelina, camline, cheat, flax, oil seed. It is seldom found wild, but is said to grow near Rochester, in the vicinity of the burial place of Hengist and Horsa.

Camellia japonica. Camellia Sasanqua Oil fixed oil. Average carbon number: 17.91. Average molecular weight: 280.04. C16:0: 9%; C18:0: 2%; C18:1: 81%; C18:2: 8%; C18:3: 0.5%. Traditional use: The plant is also known as the Christmas Camellia. This oil (sometimes called sasanqua oil or camellia or white camellia seed oil) has as one of its main components oleic acid (around 80%) with linoleic and palmitic acid present in small amounts. This gives the oil a substantive and very protective effect on the skin and would be considered slightly heavier in skin application when compared to some other oils.

Camellia kissi [Syn. *Camelia drupifera, Thea oleifera*]. Camelia Oil fixed oil. Average carbon number: 17.64. Average molecular weight: 280. C16:0: 9%; C18:0: 1%; C18:1: 80%; C18:2: 9%. Traditional use: An oil is obtained from the seed of the Tea Oil plant that is used for cooking. The oil when applied to the skin leaves a pleasant and substantive feel.

Camellia oleifera. Camellia Oleifera Oil fixed oil. Saponification value: 191. Average carbon number: 18.28. Average molecular weight: 279.69. C16:0: 10.5%; C18:0: 2.60%; C18:1: 79.5%; C18:2: 9.5%; C18:3: 0.598%. Traditional use: Tea oil has been used to manufacture soap, margarine, hair oil, lubricants, paint, synthesis of other high-molecular weight compounds, and also found use as a rustproof oil. Camellia oil has been proven to have its place in all emulsions used in the cosmetology and dermopharmacy fields. Uses include day or night creams, anti-wrinkle compounds, lipstick, hair creams, make-up, anti-sun preparations, rouge, and make-up remover products. Originally produced in China.

Camellia sasanqua. Tea Seed Oil fixed oil. Average carbon number: 17.14. Average molecular weight: 277.69. C16:0: 16%; C18:1: 59%; C18:2: 22%. Traditional use: A non-drying oil is obtained from the seed - used as a hair-dressing and textile oil. Camelia oil is obtained by the maceration of *Thea sinensis* in refined sunflower oil (*Helianthus annuus*). Camelia oil is a light, easily absorbed oil, recommended for facial massage because it does not leave a greasy feel on the skin. Camelia oil is high in oleic acid. It is a yellow green liquid.

Camellia sinensis. Camellia Oil fixed oil. Iodine value: 143. Average carbon number: 17.64. Average molecular weight: 280. C16:0: 9%; C18:0: 2%; C18:1: 80%; C18:2: 8%. Traditional use: A traditional oil used in the Far East and particularly in Japan for the protection and moisturisation of the skin and hair. Traditional use: In the traditional medicine of India, green tea is recorded as a mild excitant, stimulant, diuretic, and astringent, and the leaf-infusion (tea) was formerly used to remedy fungal infections caused by insects. Has antioxidant properties. Folklore: Archaeological evidence indicates that tea leaves steeped in boiling water were consumed by *Homo erectus pekinensis* more than 500,000 years ago. Chinese legend, described in the Cha Ching (tea book) around A.D. 780, attributes tea drinking to one of the earliest Chinese herbalists, King Shen Nong, ca. 2700 B.C. Indian legend claims that tea was brought to China by Siddhartha Guatama Buddha, during his travels in that country .

Canarium album. Chinese Olive Oil fixed oil. Average carbon number: 17.379. Average molecular weight: 277.1. C16:0: 18%; C18:0: 7.83%; C18:1: 30.5%; C18:2: 41.797%; C20:0: 0.390%. Traditional use: Potassium, calcium and magnesium were the predominant mineral elements present in the kernels. Sodium, iron, manganese and zinc were also detected in appreciable amounts. The kernel proteins were rich in arginine, glutamic and aspartic acids (3.19%, 5.02% and 2.47%, respectively) while the limiting amino acids were methionine and lysine.

Canarium indicum. Ngali Nut Oil fixed oil. Average carbon number: 17.34. Average molecular weight: 273.81. C16:0: 33%; C18:0: 15%; C18:1: 35%; C18:2: 17%. Traditional use: A precious oil is extracted from this exotic tree (which also provides a valuable resin) for use as a local remedy for dry skin. The ngali nut, found on the *Canarium indicum* tree and other Canarium species, is indigenous to the Solomon Islands and Papua New Guinea. In both countries it is a valuable food source. Cropping of the nut in the Solomons has been limited and it is collected mainly for use as food or to create an oil that is applied for pain relief. A patent has been taken out in several countries to use ngali nut oil as part of an arthritis relief preparation.

Cannabis sativa. Hemp Oil fixed oil (European source). Sp.gr.: 0.926. Saponification value: 192. Iodine value: 166. Average carbon number: 17.64. Average molecular weight: 263. C18 Epoxy: 6%; C18:1: 12%; C18:2: 55%; C18:3: 25%. Hemp oil is used as an illustration as to how the country of origin, method of extraction, climatic conditions and storage of the oil can result in a difference in the fatty acid composition.

Cannabis sativa. Indian Hemp Oil fixed oil. Sp.gr.: 0. 931. Saponification value: 193. Iodine value: 153. Average carbon number: 17.91. Average molecular weight: 279.5. C16:0: 7%; C18:0: 2.5%; C18:1: 13%; C18:2: 53%; C18:3: 22%; C20:0: 2.5%. Traditional use: A virtually canabinoid-free oil that rivals linseed for its richness and high arachidonic acid content. A perfect choice for skin protection.

Carapa guaianensis. Andiroba Seed Oil fixed oil. Melting Point: 19ºC. Sp.gr.: 0.935. Saponification value: 287. Iodine value: 75. Average carbon number: 17.5. Average molecular weight: 275.69. C16:0: 25%; C16:1: 1%; C18:0: 8%; C18:1: 50%; C18:2: 7%; C18:3: 8%; C20:0: 1%. Traditional use: The anti-inflammatory and insect repellent properties of andiroba oil are attributed to the presence of these limonoids, including a novel one that has been named andirobin. Another limonoid called epoxyazadiradione is found in andiroba oil; it has been documented with in vitro antitumor effects. North American practitioners and consumers are just beginning to learn of andiroba's powerful healing properties. Andiroba oil can be applied topically several times daily to rashes, muscle/joint aches and injuries, wounds, insect bites, boils, and ulcers. It can also be used by itself or combined with other oils as a healing and anti-inflammatory massage oil. The anti-inflammatory property of the oil of the andiroba is because of the fact that the oil contains chemical compounds known as limonoids as a major constituent. This chemical compound present in the herbal oil actively promotes the normal circulation of blood in the skin and aids in bringing relief from the pain and swelling during an injury.

Carica papaya. Papaya Seed Oil fixed oil. Iodine value: 94. Average carbon number: 17.69. Average molecular weight: 278.5. C16:0: 15.4%; C18:0: 6.06%; C18:1: 74.0%; C18:2: 4.40%. Traditional use: On infections, strips of fresh papaya pulp may be used over weeping sores of any kind. The enzyme in the papaya clears up the pus in a short while. Papaya leaves are sometimes used to dress festering wounds. Improper protein breakdown in the system often leads to allergies, papaya is effective in relieving allergies by its ability to denaturise denature proteins. Folklore: *Carica papaya* is known as the Melon Tree, Mamaeire and *Papaya vulgaris*. The juice is used to remove freckles; it is also a strong vermifuge. The leaves are used as a substitute for soap. The juice from the fruit can be dried and is called pawpain. The fresh leaves have been used as dressings for foul wounds. American blacks in various parts of the Deep South have used Adolph's meat tenderiser tenderizer for the relief of itching accompanying mosquito bites, fire ants, hornets and wasps. The meat tenderiser tenderizer contains the enzyme called papain, which makes this a very useful household medicine and home remedy among the black culture.

Carthamus tinctorius. Safflower Oil (Hybrid) fixed oil. Sp.gr.: 0.926. Saponification value: 195. Iodine value: 140. Average carbon number: 17.879. Average molecular weight: 280.6. C16:0: 6%; C18:0: 1%; C18:1: 77%; C18:2: 16%.

Carthamus tinctorius. Safflower Oil fixed oil. Sp.gr.: 0.914. Saponification value: 190. Iodine value: 145. Average carbon number: 17.879. Average molecular weight: 279.5. C16:0: 6%; C18:0: 3%; C18:1: 18%; C18:2: 73%. Traditional use: Safflower oil is mentioned in ancient Egyptian texts and was used to heal old wounds. It has exceptionally high linoleic acid content and is an excellent choice for the replenishment of moisture in skin creams and lotions. Traditional use: The flowers are used for their laxative and diaphoretic properties, also used in children's complaints of measles, fevers and eruptive skin complaints. Safflower oil has excellent skin penetrating properties.

Carum carvii. Caraway Seed oil fixed oil. Sp.gr.: 0.905. Average carbon number: 18.09. Average molecular weight: 281.19. C16:0: 3.60%; C18:1: 60.7%; C18:1 (n-6): 17%;C18:2: 19.6%. The main use is as a flavouring agent and essential oil.

Carya illinoensis. Pecan Nut Oil fixed oil. Sp.gr.: 0.915. Saponification value: 190. Iodine value: 115. Average carbon number: 17.71. Average molecular weight: 280.3. C16:0: 6.40%; C16:1: 0.5%; C18:0: 1.60%; C18:1: 64%; C18:2: 24.8%; C18:3: 1.10%; C20:1: 0.696%. Traditional use: Pecan oil is the highly unsaturated oil extracted from pecan nuts and is a sweet oil, with a flavour between walnut and almond. The nearest comparison in skin feel would be hazelnut or cashew nut oils.

Caryocar brasiliense. Pequi Oil fixed oil. Melting Point: 32°C. Saponification value: 203. Iodine value: 52. Average carbon number: 16.98. Average molecular weight: 269.06. C14:0: 1.39%; C16:0: 48.3%; C18:0: 0.9000%; C18:1: 46%; C18:2: 3.28%. Traditional use: Used in skin care to provide a long lasting, instantly perceivable conditioning benefit. It is used as an effective super-fatting agent in shampoos, conditioners, body washes and soaps.

Ceratonia siliqua. Carob Oil fixed oil. Average carbon number: 16.23. Average molecular weight: 256.8. C16:0: 17.6%; C17:0: 6.40%; C18:0: 40.297%; C18:0 OH: 5.56%; C18:1 OH: 5.56%; C18:2: 22.6%. Traditional use: The gum contained in the seeds (carob gum) is used as a demulcent and lubricant. It is also used in the cosmetic industry to make face packs. Folklore: In ancient Egyptian times it was used for tooth decay, wound healing, drying of a wound, white spots of a burn (with honey), shaking limbs, to stop smells in a man or woman. The Copts used "carob water" in medical treatment, but the nature of the disease is not indicated. Prospero Alpini informs us that, in his day, carob was much used to relieve the stomach. Dioscorides quotes fresh carob pods for this purpose, while the dried pods had the opposite effect.

Chenopodium quinoa. Quinoa Oil. Sp.gr.: 0.891. Saponification value: 190. Iodine value: 129. Average carbon number: 16.3. Average molecular weight: 278.4. C14:0 0.2%; C16:0 9.9%; C16:1 0.1%; C18:0 0.8%; C18:1 24.5%; C18:2 50.2%; C16:3 5.4%; C20:0 0.7%. Quinoa obtained from an Andean grain and is considered a pseudocereal or pseudograin that has been recognized as a complete food due to its protein quality. It contains 15% protein and has an excellent balance of amino acids, polyphenols, phytosterols, and flavonoids. It is also a rich source of omega-6 essential fatty acid and vitamin E that give desirable and efficacious effect on the skin with a feel similar to maize or corn oil.

Citrullus colocynthis. Melon Seed Oil, Apple Bitter Oil fixed oil. Sp.gr.: 0.916. Saponification value: 184. Iodine value: 105. Average carbon number: 17.68. Average molecular weight: 278.3. C16:0: 11.5%; C18:0: 9%; C18:1: 12%; C18:2: 67%.

Citrullus lanatus. Kalahari Melon Oil , Egusi Melon Oil fixed oil. Sp.gr.: 0.930. Saponification value: 192. Iodine value: 110. Average carbon number: 17.71. Average molecular weight: 277.50. C12:0: 0.2%; C14:0: 0.696%; C16:0: 13.5%; C18:0: 13.6%; C18:1: 14.6%; C18:2: 56.8%; C18:3: 0.5%.

Citrullus vulgaris. Watermelon Seed Oil, Ootanga Oil fixed oil. Sp.gr.: 0.95096. Saponification value: 193. Iodine value: 123. Average carbon number: 17.77. Average molecular weight: 278.3. C16:0: 11.3%; C16:1: 0.299%; C17:0: 0.1%; C18:0: 10.1%; C18:1: 18.1%; C18:2: 59.6%; C18:3: 0.4000%.. Traditional Use: Known since the time of the ancient Egyptians, this seed oil has been used for the care of the skin, to maintain its beauty and aid in its repair. The queens of ancient Egypt would refresh themselves by cutting a ripe fruit into slices and, lying down on a bed, would spread them over the face and neck. After an hour they would arise refreshed, with the skin tight and remoisturised. Folklore: The Greeks, the Romans and the Assyrians had no special word for melon, which they simply called ripe cucumber. In ancient Egypt, the Copts used the fruit in soothing applications. "If you take melons, and boil them and annoint inflamed legs with it them, they the legs will heal" [Ebyrs papyrus]. "If you take a roasted melon and grind it with aloe and add wine and annoint the afflicted spot with it, it will heal".

Citrullus vulgaris.Ootanga Oil fixed oil. Sp.gr.: 0.95096. Saponification value: 193. Iodine value: 123. Average carbon number: 17.77. Average molecular weight: 278.3. C16:0: 11.3%; C16:1: 0.299%; C17:0: 0.1%; C18:0: 10.1%; C18:1: 18.1%; C18:2: 59.6%; C18:3: 0.4000%. Traditional use: Known since the time of the ancient Egyptians, this seed oil has been used for the care of the skin, to maintain its beauty and aid in its repair.

Citrus aurantifolia. Lime Seed Oil fixed oil. Sp.gr.: 0.930. Saponification value: 189. Iodine value: 105. Average carbon number: 16.969. Average molecular weight: 273.8. C16:0: 29%; C18:0: 3.60%; C18:1: 23%; C18:2: 38%; C18:3: 3.89%. Traditional use: Rich in essential fatty acids, lime seed oil gives emollient and protective properties to skin care products.

Citrus aurantium dulcis. Orange Seed Oil fixed oil. Sp.gr.: 0.930. Saponification value: 189. Iodine value: 105. Average carbon number: 16.969. Average molecular weight: 273.8. C16:0: 29%; C18:0: 3.60%; C18:1: 23%; C18:2: 38%; C18:3: 3.89%. Traditional use: An oil rich in palmitic, oleic and linoleic fatty acids that is pressed from the seeds of oranges.

Citrus medica limonum. Lemon Seed Oil fixed oil. Sp.gr.: 0.930. Saponification value: 189. Iodine value: 105. Average carbon number: 16.969. Average molecular weight: 273.8. C16:0: 29%; C18:0: 3.60%; C18:1: 23%; C18:2: 38%; C18:3: 3.89%. Traditional use: A moisturising oil with a balanced fatty acid profile produced from the cold pressing of lemon seeds.

Citrus paradisi. Grapefruit Seed Oil fixed oil. Sp.gr.: 0.930. Saponification value: 189. Iodine value: 105. Average carbon number: 16.969. Average molecular weight: 273.8. C16:0: 29%; C18:0: 3.60%; C18:1: 23%; C18:2: 38%; C18:3: 3.89%. Traditional use: Grape seed oil is an excellent massage base oil as it is easily absorbed, has a light texture and imparts a silky feel to the skin.

Citrus sinensis. Orange Seed oil. An oil rich in palmitic, oleic and linoleic fatty acids which is pressed from the seeds of oranges. Average Molecular weight: 272.4. Average carbon number:

17.4. C:0 0.18%; C12:1 0.19%; C14:0 0.8%; C16:0 35.4%; C18:0 4.15%; C18:1 30.4%; C18:2 25.9%; C18:3 2.5%; C20:0 0.4%; C20:1 0.7%.

Cocos nucifera (and) *Gardenia tahitensis*. Monoi fixed oil. Saponification value: 258. Average carbon number: 12.84. Average molecular weight: 211.9. C8: 8%; C10: 7%; C12:0: 48%; C14:0: 19%; C16:0: 8%; C18:0: 3%; C18:1: 5%; C18:2: 2%. Traditional use: The epitome of Tahiti in one gloriously fragranced oil that overpowers the senses with its rich floral bouquet. The pure delight of gardenia flowers infused in skin-loving coconut oil.

Cocos nucifera. Coconut Oil fixed oil. Melting Point: 25°C. Sp.gr.: 0.926. Saponification value: 258. Iodine value: 9. Average carbon number: 12.84. Average molecular weight: 211.9. C8: 8%; C10: 7%; C12:0: 48%; C14:0: 19%; C16:0: 8%; C18:0: 3%; C18:1: 5%; C18:2: 2%. Traditional use: A traditional and trusted moisturising and protective oil from the tropics. One of the most respected oils found in the British Pharmacopoeia. Traditional use: It makes an ideal ointment for the relief of dry, rough and wrinkled skin. Psoriasis and eczema sufferers often see great improvements in these conditions with coconut oil. It makes a useful and simple hair conditioner as it that softens the hair and conditions the scalp. Using the coconut oil as a pre-wash conditioner can help reduce or eliminate dandruff.

Coffea arabica. Coffee Oil fixed oil. Saponification value: 188. Average carbon number: 16.55. Average molecular weight: 272.6. C16:0: 32%; C16:1: 0.9000%; C18:0: 7.5%; C18:1: 8%; C18:2: 46%; C18:3: 1.2%. Traditional use: An unusual and exciting proposition for an emollient. This oil has connotations of being reviving and stimulating. The fatty acids consist chiefly of linoleic, oleic, and palmitic acids, together with smaller amounts of myristic, stearic, and arachidic acids. From the unsaponifiable matter, a phytosterol, sitosterol, cafesterol, caffeol, and tocopherol have been isolated. Among the identified components of the volatile oil present in roasted coffee are: acetaldehyde, furan, furfuraldehyde, furfuryl alcohol, pyridine, hydrogen sulphide, diacetyl, methyl mercaptan, furfuryl mercaptan, dimethyl sulphide, acetylpropionyl, acetic acid, guaiacol, vinyl guaiacol, pyrazine, n-methylpyrrole, and methyl carbinol. All these substances do not pre-exist in the unroasted coffee beans; some are undoubtedly the products of the roasting process and others are produced by the decomposition of the more complex precursors.

Corylus americana. Hazelnut Oil (American) fixed oil. Sp.gr.: 0.917. Saponification value: 194. Iodine value: 86. Average carbon number: 18.449. Average molecular weight: 280.6. C14:0: 0.25%; C16:0: 2.5%; C16:1: 30.5%; C18:0: 0.5%; C18:1: 40%; C18:2: 9.5%; C18:3: 2%; C20:0: 1.5%; C20:1: 8.5%; C22:0: 1%; C22:1: 6.5%; C24:0: 0.5%. Traditional use: This edible pale oil is obtained from the nut. Archeological evidence suggests the cultivation and use of this nut in Mesolithic times. It absorbs quickly and is beneficial for oily or acne prone skin. Because of the absence of data, it is concluded that the available data are insufficient to support the safety of this ingredients in cosmetic products..

Corylus avellana. Hazelnut Oil (European) fixed oil. Saponification value: 193. Iodine value: 96. Average carbon number: 17.879. Average molecular weight: 280.69. C16:0: 6%; C18:0: 3%; C18:1: 76%; C18:2: 15%. Traditional use: A recent paper showed that hazelnut oil has phospholipids that give greater and longer-lasting moisturising potential to cosmetic emulsions. UV absorption, if absorption occurs in the UVA or UVB range, photosensitization may be needed; 28-day dermal toxicity; and two genotoxicity studies, one in a mammalian system, if genotoxicty studies are positive, a two-year dermal carcinogenesis assay using NTP methods.

Because of the absence of data, it is concluded that the available data are insufficient to support the safety of this ingredients in cosmetic products. The oil disperses well on the skin and rubs in easily to leave a pleasant silky emollience.

Crambe abyssinica. Crambe Oil, Abyssinian Mustard Oil fixed oil. Iodine value: 93. Average carbon number: 20.48. Average molecular weight: 316.8. C16:0: 2%; C18:1: 15%; C18:2: 10%; C18:3: 7%; C20:1: 6%; C22:1: 60%. Traditional use: It has a vVery high level of erucic acid. Abyssinian seed oil, moisturises and protects the epidermal layers and the β-sitosterol content maintains the integrity of cellular and tissue structure thereby keeping skin feeling and looking youthful.

Cucumis melo. Melon Seed Oil. Saponification value: 211, Iodine value: 112. Average Molecular weight: 278.7. Average carbon number: 17.4. C16:0 9.5%; C18:0 4.9%; C18:1 19.4%; C18:2 64.1%. Pressed from the seed of the melon, it's extremely high in Omega 6 and 9 fatty acids, it's a light emollient without a greasy feel. The concentrations of individual fatty acids varied from trace quantities to about 64%. Seed proteins were rich in arginine, aspartic and glutamic acids while limiting amino acids were methionine and lysine.

Cucumis sativus. Cucumber Seed Oil. Sp.gr.: 0.922. Saponification value: 185. Iodine value: 125. Average carbon number: 17.9. Average molecular weight: 278.3. C14:0 0.1%; C16:0 11%; C16:1 0.5%;C17:0 0.05%; C18:0 7.5%; C18:1 17.0%; C18:2 64.0%;C18:3 0.5%; C20:1 0.05%; C22:2 0.1%. It also contains tocopherols 600-700ppm and phytosterols 4000-5000ppm. The seeds are rich in oil with a nutty flavour that is said to resemble olive oil and so is used in salad dressings and French cooking. The high levels of phytosterols should make it a useful ingredient in skin care products. It has been shown that phytosterols help the skin strengthen its lipid barrier and restores the moisture balance, smoothing the skin's surface and improving skin elasticity. Phytosterols are also known to stimulate skin cells and encourage the regeneration of healthy skin cells.

Cucurbita foetidissima. Buffalo Gourd Oil, Missouri Gourd Oil fixed oil. Average carbon number: 17.46. Average molecular weight: 278.8. C16:0: 9%; C18:0: 2%; C18:1: 25%; C18:2: 62%. Traditional use: Carotenoids 3mg/kg.

Cucurbita pepo. Pumpkin Oil fixed oil. Sp.gr.: 0.914. Saponification value: 195. Iodine value: 113. Average carbon number: 17.71. Average molecular weight: 277.69. C10: 0.2%; C16:0: 13.4%; C16:1: 0.4000%; C18:0: 10%; C18:1: 20.3%; C18:2: 55.6%. Traditional use: restores skin tissue, supports healing of skin wounds and sores, protects skin from UV damage and absorbs quickly into the skin. There are many cultivars of pumpkin and we have shown two examples to demonstrate the effect that this may have of the composition of the fatty acids.

Cucurbita pepo. Pumpkin Seed Oil fixed oil. Sp.gr.: 0.924. Saponification value: 190. Iodine value: 126. Average carbon number: 17.719. Average molecular weight: 277.8. C16:0: 14.1%; C18:0: 5.28%; C18:1: 30.3%; C18:2: 48.2%; C18:3: 2%. Traditional use: The oil from pumpkin seeds has been used across the world (where it grows) as a treatment for sores, ulcers and other skin problems. Its high sterol and vitamin E content makes it ideal for this purpose. It contain bitter compounds; cucurbitacins B, D, E, G and I, and the glycoside of cucurbitacin E have been recorded. Seeds (as reported from Eritrea) are rich in oil (about 35%) and contain protein 38%, carbohydrate 37% and a-tocopherols.

D

Daucus carota. Carrot Oil fixed oil. Saponification value: 144. Average carbon number: 15.23. Average molecular weight: 279.27. C16:0: 10%; C16:1: 0.640%; C18:0: 2.39%; C18:1: 0.717%; C18:1 (n-6): 59.31%; C18:2: 11.82%; C20:0: 0.810%. Traditional use: A source of beta-carotene and provitamin A. Natural colour and skin nutrient. It is often used in sun care products. It is also said to accelerate the formation of tissue and contributes to an irreproachable skin epithelium. Preparations containing carrot oil are also suited to the care of aging skin with its tendency to cornification (and incipient wrinkling). In the case of dry and scaly skin, carrot oil stimulates the production of sebum, but not to excess. The skin becomes soft and supple as a result. Carrot oil clears the complexion; it gradually dissolves the hardened (cornified) cores of blackheads. The carrot oil contains a-pinene, carotol, daucol, limonene, b-bisabolene, b-elemene, cis-b-bergamotene, geraniol, geranyl acetate, caryophyllene, caryophyllene oxide, asarone, α-terpineol, terpinen 4-ol, γ-decanolactone, coumarin, and β-selinene among others. Other constituents present include palmitic acid, butyric acid and others. Main constituents are carotenes, xanthophylls, sterols, aliphatic and terpene alcohols, tocopherols and hydrocarbons.

Dracocephalum moldavica. Moldavian Dragonhead Seed Oil. Average carbon number: 17.4. Average molecular weight: 278.3. C16:0 5%; C18:0 3%; C18:1 10%; C18:2 18%; C18:3 61%. In addition the oil is rich in phytosterols (β-sitosterol, β-campesterol , Δ5-avenasterol, Δ7-stigmasterol); phospholipids, phytosqualene. These omega fatty acids in addition to the plant sterols give the oil a substantive and beneficial effect on the skin that protects skin hydration and gives a soft but rich after feel. It has good resistance to oxidation. The leaves are used to produce an essential oil.

E

Echium plantagineum. Echium Seed Oil fixed oil. Average carbon number: 17.64. Average molecular weight: 277.8. C16:0: 7.06%; C18:0: 4%; C18:1: 17.3%; C18:2: 15.4%; C18:3: 42.2%; C18:4 (n-3): 12.8%. Traditional use: Echium oil contains the richest source of stearidonic acid (8-14 %) currently found in nature. Uniquely composed of omega-3 and omega-6 fatty acids it has a similar feel on the skin to borage oil. The combination of GLA and SDA makes it ideal for anti-inflammatory and moisturising applications.

Elaeis guineensis. Palm Kernel Oil fixed oil. Melting Point: 24°C. Sp.gr.: 0.865. Saponification value: 202. Iodine value: 37. Average carbon number: 13.32. Average molecular weight: 218.5. C8: 6%; C10: 5%; C12:0: 49%; C14:0: 15%; C16:0: 7%; C18:0: 2%; C18:1: 14%; C18:2: 2%. Traditional use: According to Hartwell (1967—1971), the oil is used as a liniment for indolent tumours. Reported to be anodyne, antidotal, aphrodisiac, diuretic, and vulnerary, oil palm is a folk remedy for cancer, headaches, and rheumatism.

Elaeis guineensis. Palm Oil fixed oil. Melting Point: 35°C. Sp.gr.: 0.924. Saponification value: 202. Iodine value: 54. Average carbon number: 17.3. Average molecular weight: 270.19. C14:0: 1%; C16:1: 42%; C18:0: 4%; C18:1: 44%; C18:2: 10%. Traditional use: The fruit mesocarp oil and palm kernel oil are administered as poison antidote, and used externally with several other herbs as lotion for skin diseases. Traditional use: The fresh sap is used as a laxative, and the partially fermented palm wine is administered to nursing mothers to improve lactation. Soap prepared with ash from the palm fruit-husk is used for the preparation of a soap used for skin infections. A root decoction is used in Nigeria for headache. The leaf extract and juice from young petioles are used as application to fresh wounds. The fruit mesocarp oil and palm kernel oil and are used externally with several other herbs as lotion for skin diseases. Folklore: The oil palm is a feather palm whose trunk is 15-30m tall. The oil palm bears fruit from its third year, produces a full yield from its twelfth12th, and continues to bear fruit until it is 60 years old. An oil palm can live for up to 120 years. The first accounts of the oil palm and its oil came from Portuguese seafarers. "You see oil palms and nothing but oil palms" wrote Captain Gileannes in his logbook when he reached the Guinea Coast in 1434.

Eruca sativa. Rocket Seed Oil fixed oil. Average carbon number: 19.219. Average molecular weight: 308.6. C16:0: 5.06%; C18:0: 1.3%; C18:1: 15.1%; C18:2: 8.3%; C18:3: 14.6%; C20:1: 7.40%; C22:1: 44.7%.

Euterpe oleracea. Acai fixed oil. Average carbon number: 16.6. Average molecular weight: 257.1. Sp.gr. 0.900. Saponification value: 185. Iodine value: 90. C16:0: 18%; C18:0: 3%; C18:1: 51%; C18:2: 10%, C18:3: 2%, C20:0: 1%, C20:1: 1%, C22:0: 1%, C22:1: 1%, C24:1: 1%, Sterols: 7%. Açaí also contains beta-sitosterol (78–91% of total sterols).The oil compartments in açaí fruit contain polyphenols such as procyanidin oligomers and vanillic acid, syringic acid, p-hydroxybenzoic acid, protocatechuic acid, and ferulic acid, which were shown to degrade substantially during storage or exposure to heat. The oil is moisturizing and encourages the regeneration of the epidermis, controls the action of lipids and stimulates cicatrisation.

F

Fragaria ananassa. Strawberry Seed Oil fixed oil. Average carbon number: 17.1. Average molecular weight: 266.03. C18:1: 15%; C18:2: 45%; C18:3: 35%. Traditional use: Natural active ingredient, good oxidative stability, high biological activity, gentle anti-wrinkle active.

G

Garcinia indica. Kokum Butter fixed oil. Melting Point: 41°C. Sp.gr.: 0.8950. Saponification value: 190. Iodine value: 31. Average carbon number: 17.96. Average molecular weight: 283.1. C16:0: 2%; C18:0: 58%; C18:1: 39%; C18:2: 1%. Traditional use: Kokum Butter is obtained from the fruit kernels of the tree, *Garcinia indica* growing in India. It is mainly used as a replacement for Cocoa Butter for which many consider kokum to be an equivalent or better. Traditional medicinal applications of Kokum butter is for the preparation of ointments and for local applications to ulcerations and fissures of hands or lips. It is also used to treat muscle pulls and burns.

Gevuina avellana. Chilean Hazlenut Oil. Sp.gr. 0.922. Iodine value: 87. Saponification value: 190. Average carbon number: 18.0. Average molecular weight: 282.7. C16:0 4.10%; C16:1 22.70%; C18:0 0.80%; C18:1 42.10%; C18:2 9.20%; C20:0 6.30%; C18:3 2.50%; C20:1 1.70%; C22:0 1.90%; C22:1 7.90%. It also contains β-carotene 32.03 +/-0.8 mg/kg, α-carotene >10 mg/kg. The nuts (which compare to Macadamia) yield a valuable edible and cosmetic oil that is has good skin penetrating and UV filtering properties.

Glycine soja. Soybean Oil fixed oil. Melting Point: -16°C. Sp.gr.: 0.925. Saponification value: 191. Iodine value: 130. Average carbon number: 17.859. Average molecular weight: 279.3. C16:0: 7%; C18:0: 4%; C18:1: 29%; C18:2: 54%; C18:3: 6%. Traditional use: This plant has been known and used by the Chinese for more than 4,000 years, though today most of the oil comes from the United States. This oil is a cost-effective base on which to prepare hair and body products where good honest moisturisation is required at a budget price. Soybean is listed as a major starting material for stigmasterol, once known as an anti-stiffness factor. Sitosterol, also a soy by-product, has been used to replace diosgenin in some antihypertensive drugs. This oil is one of the lightest and silkiest when applied to the skin and lacks the greasiness of many vegetable oils.

Gossypium herbaceum. Cotton Seed Oil fixed oil. Melting Point: -1°C. Sp.gr.: 0.917. Saponification value: 195. Iodine value: 105. Average carbon number: 17.6. Average molecular weight: 275.3. C14:0: 1%; C16:0: 22%; C18:0: 2%; C18:1: 32%; C18:2: 44%. Traditional use: The leaves and roots are used for dysentery, emmenagogue; poultice of the leaves and seed kernels used as dressing for sores, bruises and swellings; roots used as abortifacient; seed paste is applied for headache. Gossypol manifested positive healing action against mice infected with herpes sores, working well in healing them externally (usually removed from the oil). In herbal medicine, cottonseed and roots have been used to treat nasal polyps, asthma, diarrhoea, haemorrhoids, dysentery, uterine fibroids and certain cancers, for proud flesh around wounds (swellings).

Guizotia abyssinica. Niger Oil fixed oil. Average carbon number: 17.66. Average molecular weight: 279. C16:0: 8%; C18:0: 8%; C18:1: 8%; C18:2: 75%. Traditional use: The oil from the seeds is used in the treatment of rheumatism. It is also applied to treat burns. A paste of the seeds is applied as a poultice in the treatment of scabies.

Guizotia abyssinica. Noog Abyssinia Oil fixed oil. Sp.gr.: 0.925. Saponification value: 256. Iodine value: 9.13. Average carbon number: 17.789. Average molecular weight: 203.59. C14:0: 2.200%; C16:0: 8%; C18:0: 6.700%; C18:1: 21%; C18:2: 58.3%; C18:3: 2%; C20:0: 1.7%.

H

Harpagophytum procumbens [Syn: *Harpagophytum zeyheri*]. Devil"s Claw fixed oil. It contains: Iridoid glycosides (up to 3 percent%), which are considered pharmacologically active. The fraction of iridoid glycosides consists of harpagoside, procumbide, harpagid, and 8-para-coumaroyl-harpagid. Harpagoside is the primary iridoid glycosides. Iridoids were not considered previously as a particularly important pharmacologically active class of compounds. More recently, extensive investigations into their biological activity in general and their potential pharmacological activity in particular have revealed that iridoids exhibit a wide range of bioactivity. They are now known to be present in a number of folk medicines used as bitter tonics, sedatives, febrifuges, cough medicines, remedies for wounds and skin disorders, and as hypotensives. Additional constituents with probable activity are glycosides of the flavonoids, kaempferol and luteolin, chlorogenic acid and cinnamic acid, the phenylethanoid acteoside, quinone, harpagoquinone, triterpenes like ursolic and oleanic acid and derivatives.Some papers reported on a direct inhibition of cyclooxygenase-2 (COX-2) catalyzed prostaglandin biosynthesis or COX-2 activity by the flavonoid kaempferol as well as ursolic and oleanic acid. We have seen Devil's Claw offered as an oil, but we believe this to be an oily extract where the plant is extracted in sunflower oil (or similar carrier). However, there is the possibility that an oil will be extracted from the roots in the near future and so this entry remains. The main activity for the plant as currently used is in aqueous extracts. It is always wise to check if the offered raw material is an oil or an oily extract.

Helianthus annuus. Sunflower Oil fixed oil. Melting Point: -17°C. Sp.gr.: 0.925. Saponification value: 190. Iodine value: 130. Average carbon number: 17.879. Average molecular weight: 279.69. C16:0: 6%; C18:0: 6%; C18:1: 20%; C18:2: 68%. Traditional use: It is a simple, yet cost-effective emollient oil, well tried and tested for generations in a wide variety of emulsions formulated for face and body products. Tocopherol, or vitamin E, is an important vitamin and natural antioxidant. Traditional Use: The oil has a similar performance to olive oil and almond oil but is significantly cheaper. Folklore: Settlers in Virginia found the Indians using oil from sunflower in bread making as early as 1590. It is used in salves, plasters and liniments for rheumatic pain. It is used externally on cuts and bruises. The dried flower heads have anti-inflammatory properties. Externally it may be used in the same way as tincture of arnica, on bruises and wounds. The oil is a soothing protective emollient. Folklore: Sunflower oil was much revered in Peru by the Aztecs, priestesses crowned with Sunflowers. It was once believed that growing sunflowers near the home gave protection against malaria - the plant does have insecticidal properties.

Hibiscus cannabinus. Ambadi Oil, Jute, Deccan Hemp fixed oil. Saponification value: 193. Iodine value: 100. Average carbon number: 17.68. Average molecular weight: 277.96. C16:0: 15.8%; C18:0: 6.78%; C18:1: 51%; C18:2: 26.3%. Traditional use: The seeds are aphrodisiac[240]. They are added to the diet in order to promote weight increase. Externally, they are used as a poultice on pains and bruises.

Hippophae rhamnoides. Sea Buckthorn Oil (CO_2 extraction) fixed oil. Sp.gr.: 0.915. Saponification value: 165. Iodine value: 82. Average carbon number: 16.76. Average molecular weight: 267. C14:0: 0.4000%; C16:0: 26.8%; C16:1: 27.6%; C18:0: 1%; C18:1: 25.5%; C18:2: 10.6%; C18:3: 6.90%; C20:0: 0.2%; C20:1: 0.1%. Traditional use: Obtained from the maceration and extraction of the fruit into sunflower oil, this is an old and traditional remedy handed down by generations of battling Mongols for the treatment of bruised and battered skin. Ideal for inclusion in "sports" ranges. Traditionally, the oil is used topically for treating haemorrhoids and increased healing after surgical stitching (particularly episiotomy). Significantly improves the skin's appearance and texture by slowing the aging process. Speed up the healing of damaged and infected skin. The berries are also a rich source of vitamin E, carotenoids, flavonoids, sterols including beta sitosterol; stanols, superoxide dismutase (SOD) and polar lipids. In addition to its carotenoid and vitamin E content, the oil from the sea buckthorn berry contains on average 35% of the rare and valuable palmitoleic acid (16:1n-7). This rare fatty acid is a component of skin fat and is known to support cell, tissue and wound healing.

Hippophae rhamnoides. Sea Buckthorn Seed Oil fixed oil. Sp.gr.: 0.915. Saponification value: 165. Average carbon number: 17.03. Average molecular weight: 273.8. C14:0: 0.1%; C16:0: 14.8%; C16:1: 13%; C18:0: 4%; C18:1: 16.1%; C18:2: 49%; C18:3: 0.299%; C20:0: 0.299%; C20:1: 0.1%. Traditional use: It appears in more than 300 ancient prescriptions. Sea buckthorn is used to improve blood circulation, reduce inflammation, regenerate skin and mucous membranes and treat gynaecology disorders. It is an oil rich in EFAs. As a pharmaceutical ingredient it is formulated into creams for radiation burns, sunburn and thermal burns where it is claimed to promote rapid healing. (After the Chernobyl reactor disaster, the authorities impounded all stocks of hippophae in the Soviet Union to treat victims). Formulations exist on the market for skin ulcer treatment.

Humulus lupulus. Hops Oil fixed oil. Traditional use: A light and slightly fragrant oil with connotations of being soothing and relaxing. The female fruiting body contains humulone and lupulone, these are highly bacteriostatic against gram-positive and acid-fast bacteria. Hops are stated to possess sedative, hypnotic, and topical bactericidal properties. Traditionally they have been used for neuralgia, insomnia, excitability, priapism, mucous colitis, topically for crural ulcers, and specifically for restlessness associated with nervous tension headache and/or indigestion.

Hypericum perforatum. St. John''s Wort Oil fixed oil. Traditional use: An orange-red oil that takes its colour from the hypericin it contains. This oil is part of most herbalists'' repertoire for damaged skin, bruises and other skin problems. Hypericum has a reputation as an analgesic, and is used either internally or externally to treat neuralgic pain. The macerated oil can be applied externally for neuralgia and will ease the pain of sciatica. It also soothes burns by lowering the temperature of the skin. The Commission E recognized the anti-inflammatory action of topical oily Hypericum preparations. External use of oily preparations of St. John's wort is approved for treatment and post-therapy of acute and contused injuries, myalgia, and first-degree burns. Wound healing: Hypericum has long been used both orally and topically for healing of wounds and burns, due in part to its antimicrobial effects. In a study of second- and third third-degree burns, burns treated with a topical Hypericum ointment healed at least three times faster and keloid formation was prevented when compared to burns treated with conventional methods. The wound healing effect of oral Hypericum was more pronounced than the effect of topical Calendula for the healing of incision, excision, and dead space wounds. This oil is normally available as an oily extract but for a very short time there was a commercial source of the seed oil. Hypericum Perforatum Oil is the fixed oil obtained from the flowers according to INCI description.

Hyptis suaveolens. Hyptis Oil fixed oil. Average carbon number: 17.87. Average molecular weight: 278.63. C16:0: 8.56%; C18:0: 2.10%; C18:1: 8.13%; C18:2: 81.0%; C18:3: 0.25%. Traditional use: A high concentration of omega-6 lipids makes hyptis oil an ideal choice in products for dry, flaky skin.

I

Irvingia gabonensis. African wild mango butter, African mango, Dika nut, bush mango. Africa, Congo. Family: Irvingiaceae. Traditional use: Methanolic extract of *Irvingia gabonensis* are use in the treatment of bacterial and fungal infections. The material is used as an alternative to shea butter. Moisturizing and softening efficacy on skin. It protects and repairs skin and lips through its film-forming activity, thereby restoring softness and suppleness. A clinical test has shown that it is a better skin moisturizer than Shea butter. It is smoothing on the skin. Chemistry: Melting point: 2.53°C. Saponification value: 190. Iodine value: 4.5. Average carbon number: 13.4. Average molecular weight: 220.64. Average molecular weight: 283.1. C10:0 1.11%; C12:0 36.6%; C14:0 53.71%; C16:0: 5.23%; C18:0: 0.8%; C18:1n-9: 1.82%; C18:2n-6: 0.49%. The minor unsaponifiable fraction contains mainly sterols and tocopherol.

J

Jatropha curcas. Pinhoen Oil, Jatropha Oil, Ratanjyot Oil fixed oil. Melting Point: 9°C. Sp.gr.: 0.919. Saponification value: 195. Iodine value: 102. Average carbon number: 17.69. Average molecular weight: 277.8. C14:0: 0.5%; C16:0: 14.5%; C18:0: 5.5%; C18:1: 50%; C18:2: 29.5%. Traditional use: Curcas oil used as wound disinfectant and as a treatment for rheumatism, skin diseases and other ailments. The oil is also useful externally in cutaneous diseases and rheumatism. Steyn quotes Dragendorff, who states that the seed contains a purgative oil, which is used as a remedy in dropsy, sciatica, paralysis, worms, and skin diseases. Externally it is an esteemed remedy for itches, herpes, and eczema, and it is said to be a cleansing application for wounds and ulcers. Diluted with a bland oil (1 part to 2 or 3), it forms a useful embrocation in chronic rheumatism. It is generally used for adulterating olive oil. Contains α-amyrin, β-sitosterol, stigmasterol, and campesterol, 7-keto-β-sitosterol, stigmast-5-ene-3-β, 7-α-diol, and stigmast-5-ene-3 β, 7 β-diol. The oil of the seeds is very popular throughout West Africa in local medicine as a remedy for sciatica, paralysis, and numerous other skin diseases, e.g. measles, mumps, chicken pox, etc.

Juglans nigra. Black Walnut Oil fixed oil. Saponification value: 136. Iodine value: 191. Average carbon number: 12.42. Average molecular weight: 280. C18:2: 55%; C18:3: 14%. Traditional use: Paracelsus described in his writings the positive characteristics of the nut oil for scalp and hair. The oil is liked by painters for mixing their colours. Walnut oil is known for its lubricity and natural moisturising properties. The *nigra* species is more common to Asia and foreign parts, the variant *regia* is the one most commonly found close to home. The oil from the ripe seeds has been used externally in the treatment of gangrene, leprosy, and wounds.

Juglans regia. Walnut Oil fixed oil. Saponification value: 193. Iodine value: 150. Average carbon number: 17.879. Average molecular weight: 279.3. C16:0: 6%; C18:0: 4%; C18:1: 26%; C18:2: 48%; C18:3: 16%. Traditional use: Probably a native of Persia - but the variety more commonly found in Europe - this nut provides an emollient oil that has been used for its efficacy on dry and damaged skin. In mythology, while man ate acorns, the Gods ate walnuts. The oil from the seed is anthelmintic. It is also used in the treatment of menstrual problems and dry skin conditions.

K

Kigelia africana. Sausage Tree Oil fixed oil. Traditional use: A significant body of scientific literature and patents confirm the validity of many of the traditional uses of kigelia and suggest a number of new applications. Several papers support the use of kigelia fruit extract for treating skin cancer, whilst it has also found a market in Europe and the Far East as the active ingredient in skin skin-tightening and breast breast-firming formulations. Kigelia''s known chemical constituents include: Napthaquinones (including kigelinone), Fatty acids (including vernolic), coumarins (including kigelin), iridoids, caffeic acid, norviburtinal, sterols (including sitosterol and stigmasterol) and flavonoids (including luteolin and 6- hydroxyluteiolin)., The steroids are known to help a range of skin conditions, notably eczema, and the flavonoids have clear hygroscopic and fungicidal properties. Strong anecdotal evidence suggests that it is effective in the treatment of solar keratosis, skin cancer and Kaposi sarcoma, an HIV-related skin ailment. New research by PhytoTrade Africa has supported anti-oxidant and anti-inflammatory properties.

L

Lactuca sativa. Lettuce Seed Oil fixed oil. Traditional use: A decoction of the leaves also makes a good skin wash. Medicinally it is a cooling lotion for sunburn, mild sedative. The juice diluted and applied to the skin, removes blemishes and soreness caused by sunburn or wind, leaving the face soft and smooth. Folklore: Lettuce is under the sign of Venus. In the Assyrian herbal the seeds were used with cumin as an eye poultice.

Lesquerella fendleri. Lesquerella Oil, Bladderpod Oil fixed oil. Saponification value: 168. Iodine value: 106. Average carbon number: 18.87. Average molecular weight: 296.8. C16:0: 1.3%; C16:1: 1.7%; C18:0: 2.10%; C18:1: 18.1%; C18:2: 9.30%; C18:3: 14%; C20:0: 0.2%; C20:1: 1.2%; C20:1 (OH): 51.3%. Traditional use: Derived from the desert of southwest America, this rich oil has properties reminiscent of those possessed by castor oil, for which it can be used as a replacement. The structure of lesquerolic acid, 14-hydroxy, cis-11-eicosenoic acid bears a close relationship to ricinoleic acid found in castor oil.

Leucas aspera. Thumba Oil fixed oil. Melting Point: 23°C. Sp.gr.: 0.926. Saponification value: 173. Iodine value: 120. Average carbon number: 17.82. Average molecular weight: 278.13. C16:0: 13.5%; C18:0: 9.5%; C18:1: 27%; C18:2: 50.5%. Traditional use: The highly unsaturated fatty acid portion resembles linseed oil and contributes significantly to the characteristics of the oil and unique derivative potential.

Licana rigida. Oiticica Oil (Cicoil) fixed oil. Sp.gr.: 0.982. Saponification value: 189. Average carbon number: 17.859. Average molecular weight: 277.38. C16:0: 7%; C18:0: 5%; C18:1: 6%; C18:2: 0%; C18:3: 78%; C18:3 Conj n=5: 4%. Traditional use: The seeds of *L. rigida* Benth. yield Oiticica oil, much like tung oil.

Limnanthes alba. Meadowfoam Seed Oil fixed oil. Saponification value: 167. Iodine value: 91. Average carbon number: 20.5. Average molecular weight: 285.19. C18:2: 0%; C20:1: 63%; C20:2: 12%; C22:1: 15%; C22:2: 10%. Traditional use: An oil that is stable, non-greasy and rapidly absorbed. This oil is ideal for those products where a soft, smooth silky feel is required whether it be on skin or hair. An oil obtained from the seed has similar properties to whale sperm oil and to Jojoba (*Simmondsia chinensis*). The seed contains circa 20% protein, 25 - 30% oil, with 1.56% volatile isothiocyanates. The high concentration of C20 and fatty acids in the seed oil is unique. No other seed oil is known to have as high concentration (>90%) of total fatty acids of chain length greater than C18.

Linum usitatissimum. Linseed Oil fixed oil. Melting Point: -24°C. Sp.gr.: 0.934. Saponification value: 192. Iodine value: 186. Average carbon number: 17.93. Average molecular weight: 278.3. C16:0: 5%; C18:0: 2%; C18:1: 17%; C18:2: 15%; C18:3: 61%; C26: 0.1%. Traditional use: The lignan constituents of flaxseed (not flaxseed oil) possess *in vitro* anti-oxidant and possible oestrogen receptor agonist/antagonist properties. The oil of this plant is used externally to cure affections of the skin. It can be used in cases of burns or hair fall, and also for sebum reduction. Externally it is applied as a hot moist cataplasm, compress, or poultice to reduce inflammation. Flaxseed preparations have had experimental success as anti-carcinogens and in treating lupus nephritis. External use is approved as a cataplasm for local inflammation.

Lithospermum officinale. Millet Oil. See *Panicum miliaceum*

Lunaria annua. Honesty Oil fixed oil. Sp.gr.: 0.8970. Average carbon number: 14.3. Average molecular weight: 348.3. C18:2: 0%; C22:1: 41%; C24:1: 22%. Traditional use: Honesty is also known as the Money Plant, Moonwort and Satin Flower. It has purple or white flowers in the late spring which develop into papery silver disc-shaped fruits. The botanical name for this plant refers to the seed pods, which become luminous as the plant ages and resembles a brightly shining moon. The oil from this plant is known to provide nervonic acid (C24:1) and is present at the highest level compared to all the other oils we surveyed. It helps rebuild the nerve sheaves (the breakdown of which is a key factor in MS), and has many medicinal properties.

Lycopersicon esculentum. Tomato Seed Oil fixed oil. Sp.gr.: 0.918. Saponification value: 160. Iodine value: 126. Average carbon number: 17.44. Average molecular weight: 278. C14:0: 0.2%; C16:0: 13.1%; C16:1: 0.3299%; C17:0: 0.3299%; C17:1: 0.6598%; C18:0: 5.25%; C18:1: 20.5%; C18:2: 55%; C18:3: 2%; C20:0: 0.6598%; C22:0: 0.2%; C22:1: 0.2%. Traditional use: It reduces sebum production by about 13%. It seems that this anti-seborrheic capacity depends on the peculiar lipidic formula that competitively reacts with excess sebum, thus causing its reduction. It has only slight moisturising capability.

M

Macadamia ternifolia. Macadamia Nut Oil fixed oil. Saponification value: 195. Iodine value: 76. Average carbon number: 14.32. Average molecular weight: 276.5. C16:1: 19.5%; C18:1: 60%; C18:2: 0%; C20:1: 2%. Traditional use: An oil from the "King of Nuts", this Hawaiian emollient is reported to have properties akin to those of sebum. It helps to recapture the skin of childhood. It effectively hydrates dry and rough skin and reduces the appearance of the fine lines including those around the eyes. It is ideal for use where penetration and lubrication are essential and provides amazing slip for massaging. It shows excellent absorbancy with protective barrier that does not clog the pores of the skin. It is non-toxic, non-allergenic and non-staining. It is easily removed by soapy water. Other studies have shown that the level of palmitoleic acid in the skin of people suffering from psoriasis is about half that of a healthy individual. This would seem to indicate that the replenishment of the palmitoleic acid may be of benefit.

Macassar schleichera. Kusum Oil or Macassar Oil. M.Pt.: 21C. Sp.gr.: 0.860. Saponification value: 225, Iodine value: 53. Average Molecular weight: 285.1. Average carbon number: 18.2. C14:0 0.01%; C16:0 7.59%; C16:1 1.8%; C18:1 2.83%; C18:2 55.25%; C18:3 0.26%; C20:1 29.54%; C20:2 0.24%; C22:0 1.14%; C24:0 0.03%; C26:1 0.02%. It is from India, and has Ayurverdic properties, particularly being antiseptic and nourishing. Also called Lac tree, Kusum, Gum lac tree , Ceylon oak, Macassar oil tree, Chendola, Kosam, Kosum, Sagdi, Puvam, Kusumb, Rusam, Kussam, Puvam, Pusku. There are hair oils made by fragrancing coconut oil with ylang ylang which is also called macassar oil. This tree is grown for its fruit seeds, which contain edible fat used in illumination and hair-oils. The fruit pulp is edible and unripe fruits are eaten as pickles and young leaves as vegetables.

Madhuca butyracea. Phulwara Butter fixed oil. Meting Point: 45°C. Sp.gr.: 0.859. Saponification value: 196. Iodine value: 96. Average carbon number: 16.87. Average molecular weight: 267.72. C16:0: 56.6%; C18:0: 3.60%; C18:1: 36%; C18:2: 3.78%. Traditional use: Phulwara butter is obtained from healthy seed kernels of *M. butyracea* tree. It is rich in palmitic acid and is an edible food fat that makes an excellent substitute for cocoa butter, but without the typical odour of chocolate.

Madhuca latifolia. See *Bassia latifolia*

Mallotus philippinensis. Kamala Seed Oil fixed oil. Sp.gr.: 0.941. Saponification value: 195. Iodine value: 166. Average carbon number: 17.98. Average molecular weight: 277.62. C16:0: 8%; C18:0: 0.6500%; C18:1: 16.5%; C18:2: 11.6%; C18:3 Conj n=5: 64%. Traditional use: Ferulic and salicylic acids are also found in Dehradun sample that may contribute to the action of *M. philippinensis* in infectious or non-healing wounds. *M. philippinensis* is a plant with very high medicinal value and has been in use since 1000 B.C. It is stated to be effective in non-healing or infectious wounds, dermal problems, intestinal worms etc. Tenifuge (anthelmintic, purgative); tape-worm, sometimes for the round- and seat-rowmsworms; also externally in scabies, skin affections, herpetic ringworm. Phloroglucinol derivatives; rottlerin, isorottlerin, isoallorottlerin (the“ "red”" compound) and methylene-bis-methylphloroacetophenone (the “"yellow" ” compound) (ii) Kamalins I and II of undetermined structure.

Mangifera indica. Mango Butter fixed oil. Saponification value: 190. Iodine value: 60. Average carbon number: 17.5. Average molecular weight: 281.3. C16:0: 7%; C18:0: 43%; C18:1: 46%; C18:2: 2%. Traditional use: Mango kernel oil is known to release a drug like salicylic acid at a much higher rate than standard paraffin base emollient. Mango kernel oil is known to be used in

India for soap making and as an emollient in skin care products. Carotenoids present. An unusual fatty acid, cis-9, cis-15-octadecadienoic acid was isolated from the pulp lipids of mango.

Mangifera indica. Mango Seed Oil fixed oil. Saponification value: 187. Iodine value: 43. Average carbon number: 17.71. Average molecular weight: 281. C14:0: 0.1%; C16:0: 10.1%; C17:0: 0.2%; C18:0: 38.3%; C18:1: 42.7%; C18:2: 5.700%; C18:3: 0.3%; C20:0: 1.5%; C20:1: 0.1%; C22:0: 0.2%. Traditional use: This oil from India is a greatly respected emollient that is often used as a cocoa butter replacement.

Mauritia flexuosa. Buriti Oil fixed oil. Sp.gr.: 0.859. Saponification value: 170. Iodine value: 77.2. Average carbon number: 17.87. Average molecular weight: 277.25. C14:0: 0.1%; C16:0: 18.6%; C18:0: 2%; C18:1: 75.297%; C18:2: 3.10%; C18:3: 2.20%. Traditional use: Invigorates tired skin. It is a natural UV absorber. It contains the free radical scavenger beta-carotene. Suitable for skin care, hair care and sun care applications.

Medicago sativa. Alfalfa Oil fixed oil. Average carbon number: 18.559. Average molecular weight: 281.1. C18:0: 20%; C18:1: 31%; C18:2: 16.8%; C18:3: 35.2%. Traditional use: Reported to reduce the erythema caused by sunburn. Rich in carotenes and lutein.

Melia azadirachta. Neem Oil fixed oil. Sp.gr.: 0.932. Saponification value: 185. Iodine value: 89. Average carbon number: 16.23. Average molecular weight: 275.5. C16:0: 25%; C18:0: 15%; C18:1: 41.5%; C18:2: 11.5%. Traditional use: It contains an aromatic oil, neem (also known as margosa or nimba) is one of India's most respected treatments for problem skin. There are many oils, but few have as many therapeutic properties as neem oil. Neem oil is known for its medicinal and cosmetic properties. As an insecticide, it first served to protect plants. In the form of soap, it can also protect humans against pain, illness and the effects of aging. The oil has moisturizing, regenerating and restructuring properties. As an ointment, neem oil can be used to treat wounds, boils and eczema. It can be used in massage to combat muscle pain, oedema of the joints and fever. It is an antifungal and antiviral agent and also combats lice and other parasites. Applied to the scalp, it has the reputation of retarding baldness and the appearance of white hair.

Mesua ferrea. Nahor Seed Oil fixed oil. Sp.gr.: 0.956. Saponification value: 201. Iodine value: 83. Average carbon number: 17.93. Average molecular weight: 279.19. C14:0: 1.39%; C16:0: 10.8%; C18:0: 12.9%; C18:1: 60%; C18:2: 15%; C20:0: 0.90%. Traditional use: Medicine: In Malaysia and India, a mixture of pounded kernels and seed oil is used for poulticing wounds. The seed-oil is used for treating itch and other skin eruptions, dandruff and against rheumatism. Calophyllolide is effective in reducing the increased capillary permeability.

Momordica charantia. Long Green Monsoon Oil fixed oil. Average carbon number: 17.94. Average molecular weight: 280.24. C16:0: 2%; C18:0: 34%; C18:1: 4%; C18:2: 3%; C18:3 Conj n=5: 56.8%. Traditional use: It is also used for wounds that are refractive to other kinds of treatment, for skin diseases, olive or almond oil infusions of the fruit (minus the seeds) are applied to chapped hands, haemorrhages, burns, etc. and that the mashed fruit is used in the preparation of poultices. Guerero reports that the sap of the leaves is used as a parasiticide, and fruit, when macerated in oil, as a vulnerary. It is also used for wounds that are refractive to other kinds of treatment, for skin diseases.

Moringa oleifera. Moringa Oil, Ben Oil fixed oil. Saponification value: 193. Iodine value: 75. Average carbon number: 18. Average molecular weight: 282.1. C18:0: 1.5%; C18:1: 76.797%; C18:2: 20.1%; C18:3: 1.5%. Traditional use: The oil, applied locally, has also been helpful for arthritic pains, rheumatic and gouty joints. Oil is clear, sweet and odorless, never becoming rancid; consequently it is edible and useful in the manufacture of perfumes and hairdressings.

Moringa pterygosperma. Moringa Oil fixed oil. Saponification value: 187. Iodine value: 67. Average carbon number: 17.969. Average molecular weight: 284.19. C16:0: 9.13%; C16:1: 2.96%; C18:0: 3.28%; C18:1: 72.5%; C18:2: 0.598%; C20:0: 2.39%; C20:1: 2.319%; C22:0: 5.46%; C22:1: 0.5%. Traditional use: Moringa is native to Northern India, where it was first described around 2000 B.C. as a medicinal herb. In ancient Egypt, Moringa oil was used by the more wealthy to anoint the body and to keep skin supple. In Central Africa, it was used to treat skin infections. Moringa seed kernels contain about 40% oil by weight. The oil can be used for soap making and for fine cosmetics and skin care. The fatty acids profile of the oil shows a very high content of oleic acid.

N

Nicotania tabacum. Tobacco Seed Oil fixed oil. Melting Point: 17. Sp.gr.: 0.924. Saponification value: 191. Iodine value: 140. Average carbon number: 16.57. Average molecular weight: 278.69. C16:0: 9.56%; C18:0: 6.28%; C18:1: 21.6%; C18:2: 55.6%. Traditional use: β-sitosterol (433-682g.kg-1) was the main component in sterol fraction, followed by stigmasterol (102-188g.kg-1) and kampesterol (93-149g.kg-1). Unusual high amount of cholesterol (56-112g.kg-1) typical for animal lipids were identified.

Nigella sativa. Black Cumin Oil fixed oil. Iodine value: 117. Average carbon number: 17.059. Average molecular weight: 280.5. C16:0: 10%; C18:1: 35%; C18:2: 45%; C20:1: 1%; C20:2: 4.28%. Traditional use: Contains high level thymoquinone. Nigella sativa (black seed) is an important medicinal herb. In many Arabian, Asian and African countries, black seed oil is used as a natural remedy for a wide range of diseases, including various allergies. The plant''s mechanism of action is still largely unknown. N. sativa seed oil possesses antimicrobial activity against several multidrug multidrug-resistant pathogenic bacteria and may be used topically in susceptible cases.

O

Ocimum basilicum. Basil Seed Oil fixed oil. Average carbon number: 17.559. Average molecular weight: 277.5. C16:0: 8.5%; C18:1: 11%; C18:2: 24.5%; C18:3: 54.5%. Traditional use: The fixed oil obtained from the seeds.

Oenothera biennis. Evening Primrose Oil fixed oil. Saponification value: 186. Iodine value: 152. Average carbon number: 17.879. Average molecular weight: 279.1. C16:0: 6%; C18:0: 2%; C18:1: 8%; C18:2: 75%; C18:3: 9%. Traditional use: A favourite source of GLA, this modern seed oil is a well well-known and much loved moisturisermoisturizer and skin nutrient. Anti-inflammatory for eczema, cyclic mastalgia, rheumatoid arthritis. Novel sterol obtusifoliol. Atopic eczema licensed, but then withdrawn.

Olea europaea. Olive Oil fixed oil. Melting Point: -6ºC. Sp.gr.: 0.916. Saponification value: 190. Iodine value: 85. Average carbon number: 17.84. Average molecular weight: 280.3. C16:0: 8%; C18:0: 2%; C18:1: 84%; C18:2: 6%. Traditional use: An oil that is mentioned in the Bible and was known to the ancient Greeks and Phoenicians, who introduced it into Spain. This oil is legendary for its safe, gentle care and treatment of the skin. Traditional use: It is used in oinments for wounds, burns, dermatosis, stretch marks and breast firming and is anti-inflammatory. The unsaponifiables are anti-inflammatory. The oil is a healing agent on wounds and bums. Martindale: Externally, olive oil is emollient and soothing to inflamed surfaces, and is employed to soften the skin and crusts in eczema and psoriasis, and as a lubricant for massage. It is used to soften ear wax.

Ongokea gore. Ongokea Oil, Boleko Oil, Isano Oil fixed oil. Iodine value: 190. Average carbon number: 17.96. Average molecular weight: 295.8. C16:0: 2%; C18:0: 2%; C18:3: 80%; C18:1: 12%; C18:2: 4%. Traditional use: The seed oil, called 'boleko oil' or 'isano oil', is inedible but can be used as an additive to linseed oil in the manufacture of paints, varnishes and linoleum and to oil for moulding cores in metal foundry. Its major constituent fatty acids, isanic and isanolic acid, are unusual in their reactivity because of a conjugated acetylenic bond system in the middle of an 18-carbon chain. When heated sufficiently, the oil reacts exothermically with violence. This property has useful application in fire-retardant paints.

Opuntia ficus indica. Opuntia Barbary Fig Oil. Sp.gr.: 0.913. Saponification value: 162. Iodine value: 127. Average carbon number: 17.6. Average molecular weight: 276. C14:0: 0.2%; C16:0: 16.2%; C16:1: 1.5%; C18:0: 3.3%; C18:1: 19.9%; C18:2: 57.7%; C18:3: 1.2%. Prickly pear has a long history of traditional Mexican folk medicine use, particularly as a treatment for diabetes. Prickly pear pads have been used as a poultice for rheumatism. The fleshy stems or cladodes have been used to treat wounds, insect bites and cases of skin that is inflamed and helps allay heat. It is also used as a soothing application to the sore breasts of nursing mothers. The isolated oil is emollient and smoothing.

Orbignya cohune. Cohune Oil. Iodine value: 13. Average carbon number: 13.1. Average molecular weight: 215.6. C8: 7.5%; C10: 6.5%; C12:0: 46.5%; C14:0: 16%; C16:0: 9.5%; C18:0: 3%; C18:1: 10%; C18:2: 1%. It is a tropical oil which is obtained from this species of palm indigenous to the rain forests in the Petén region of Guatemala. It has been shown to have a softer feel on the skin than coconut oil or mineral oil. Indigenous people use the oil for cooking, soap-making, skin lubrication and protection as well as for illumination.

Orbignya martiana. Babassu Oil fixed oil. Iodine value: 15. Average carbon number: 13.52. Average molecular weight: 221.3. C8: 6%; C10: 4%; C12:0: 45%; C14:0: 17%; C16:0: 9%; C18:0: 3%; C18:1: 13%; C18:2: 3%. Traditional use: is harvested from the seeds of this oil-rich fruit and provides superior emollience, especially in soaps and hair care products. It produces a non-comedogenic oil, which glides on the skin, with a drier feel than coconut oil, thus creating a non-greasy emollience. It helps to soften the skin and impart a natural lustreluster. It was cited to counteract "dry,", brittle hair that has lost its natural sebum.

Orbignya oleifera. Babassu Oil fixed oil. Saponification value: 250. Iodine value: 15. Average carbon number: 14.23. Average molecular weight: 222.59. C8: 4.5%; C10: 4.5%; C12:0: 47.5%; C14:0: 19%; C16:0: 8%; C18:0: 4.5%; C18:1: 14.5%; C18:2: 2%. Traditional use: Originating from the Brazilian rain forest, this oil is reported to be non-comedogenic. It leaves the skin with a soft, lustrous smoothness. In skin care, babassu butter is used for its soothing, protective and emollient properties. Babassu butter penetrates quickly and does not leave a greasy after feel.

Oryza sativa. Rice Bran Oil fixed oil. Saponification value: 185. Iodine value: 105. Average carbon number: 17.78. Average molecular weight: 276.6. C14:0: 1%; C16:0: 18%; C18:0: 3%; C18:1: 42%; C18:2: 37%. Traditional use: A moisturising oil, rich in gamma oryzanol, tocopherol. Rice bran oil is high in fatty acids and rich in unsaponifiables. Rice bran oil is a natural anti-oxidant and also offers a small degree of sunscreen protection.

P

Panicum miliaceum. Millet Oil fixed oil. Sp.gr.: 0.9383. Saponification value: 191.5. Iodine value: 129. Average carbon number: 17.69. Average molecular weight: 278.3. C16:0: 10.8%; C18:1: 53.797%; C18:2: 34.8%. Traditional use: In the studies on 64 rabbits with the model of a suppurating wound against the background of diabetes mellitus, and as well in treating 29 patients with diabetes mellitus and purulent-inflammatory soft tissue diseases, it was established that the millet oil had the anti-inflammatory and stimulating tissue regeneration effect. The duration of wound treatment in rabbits with the use of millet oil was 16 days shorter than with the use of the sea buckthorn oil, and 21 days shorter that with Wishnevsky's ointment. In patients with *diabetes mellitus*, the duration of wound treatment in use of the millet oil reduced by 16 days when compared to that in using the sea buckthorn oil. Clinical investigations carried out in treatment of purulent wounds in 63 rabbits and in 80 patients have shown high curative efficiency of the local application of millet oil. It has a pronounced anti-inflammatory, antimicrobial effect, stimulates reparative processes in the wounds in combination with a considerable reaction of cells responsible for immune defence.

Papaver orientale. Poppy Seed Oil fixed oil. Sp.gr.: 0.8980. Saponification value: 197. Iodine value: 138. Average carbon number: 17.8. Average molecular weight: 278.3. C16:0: 10%; C18:0: 2%; C18:1: 15%; C18:2: 73%. Traditional use: Containing virtually no opiates, this must be the sister to hemp oil and could be used in any products where an interesting moisturiser is required with an emotive story line in the pack copy.

Papaver somniferum. Poppy Seed Oil fixed oil. Melting Point: -18ºC. Sp.gr.: 0.925. Saponification value: 206. Iodine value: 126. Average carbon number: 17.44. Average molecular weight: 278.1. C16:0: 10%; C18:1: 11%; C18:2: 72%; C18:3: 5%. Traditional use: A high quality edible drying oil is obtained from the seed. It has an almond flavour and makes a good substitute for olive oil. Poppy seed oil appears to be of good quality for human consumption since it is generally rich in polyunsaturated fatty acids. A poultice made of poppy seeds with milk is prescribed by Saran-gadhara in porrigo of the scalp. Sanskrit writers describe poppy seeds as demulcent and nutritive, and useful in cough and asthma.

Parinari curatellifolia. Mobola Plum Oil fixed oil. Sp.gr.: 0.915. Average carbon number: 17.8. Average molecular weight: 278.1. C16:0: 6.9%; C18:0: 6.6%; C18:1: 12.5%; C18:2: 14.8%; C18:3: 1%; C18:3 (n=5): 58. The oil can be extracted from the seeds which contain about 40% oil (c.70% in kernels). The edible oil is used for cooking, paint, varnish, soap and others. The oil from these seeds of the Parinari is an effective moisturizer and helps restructure and protect the skin due to its high content of polyunsaturated fatty acids and particularly eleostearic acid.

Passiflora caerulea. Blue Passion Flower Seed Oil fixed oil. Average carbon number: 16.859. Average molecular weight: 278.18. C16:0: 10.1%; C18:1: 17.6%; C18:2: 63.1%; C18:3: 4%. Traditional use: The fruit is edible, but not especially tasty (slightly blackberry when cooked). The seeds of this particular passion flower has an oil content of around 23% and its oil quality is as good as that of sunflower seeds; therefore, it is a superb edible oil with a beautifully soft smoothing feel on the skin.

Passiflora edulis. Passion Fruit Oil fixed oil. Sp.gr.: 0.921. Saponification value: 195. Iodine value: 135. Average carbon number: 17.17. Average molecular weight: 278.3. C16:0: 10%; C18:1: 16.5%; C18:2: 70%. Traditional use: Passion flower oil is widely used for dermal application of conditions such as psoriasis and eczema, and it is said to have healing properties.

It is used as a balm in cases of skin cancer where it is apparently effective in healing skin lesions. An edible oil is obtained from the seed. It is a light non-sticky oil traditionally used to soothe, protect and moisturise moisturize the skin. Containing approximately 70% linoleic acid (unsaturated omega-6 fatty acid), passion fruit oil contains anti-oxidants and can be used to improve skin elasticity. It can also be used in hair softening products and treating dry flaky scalp.

Passiflora incarnata. Passion Flower Oil fixed oil. Saponification value: 183. Iodine value: 132. Average carbon number: 17.84. Average molecular weight: 278.83. C16:0: 8%; C18:0: 2%; C18:1: 12%; C18:2: 77%; C18:3: 1%. Traditional use: A light, gentle oil with connotations of being soothing and relaxing. It leaves a natural soft feel to the skin without being over-occlusive.

Pentaclethra macroloba. Pracaxi Oil. Saponification value: 172. Iodine value: 85. Average carbon number: 19.1. Average molecular weight: 298.2. C12:0: 0.12%; C14:0: 0.13%; C16:0: 5.0%; C16:1: 5.0%; C18:0: 5.0%; C18:1 44%; C18:2: 12%; C18:3: 0.5%; C20:0: 1.0%; C20:1: 1.5%; C22:0: 12.25%; C22:1: 1.0%; C24:0: 12.25%. Moisturising oil, reduces wrinkles and fine lines. It is used in Latin American hospitals for scarring and pigmentation disorders. In the Amazon region Pracaxi Oil has been used for generations for treatment of skin spots, skin depigmentation, severe acne and acne scars. Pracaxi helps to hydrate and promote cellular renewal. It is a powerful antiseptic and is often used in hospitals to cleanse skin after surgery and treat skin infections. It can be used during pregnancy to help in the prevention of stretch marks. It is a moisturizing oil that reduces wrinkles and fine lines. Pracaxi helps to protect skin hydration and promote cellular renewal. The people of the Amazon river use the oil to treat bites from snakes and the healing of ulcers. The seeds produce an oil that is used for cooking. It is an excellent skin and hair protectant that brings brightness and softness.

Pentadesma butyracea. Butter tree, kanya, kanya butter, tallow tree, krinda, kpangnan, tama. Family: Clusiaceae. West Africa, from Guinea, Sierra-Leone and Cote d''Ivoire, Togo to the Democratic Republic of Congo, extending eastwards into Tanzania and Uganda. Traditional use: It is a high oil producing species and the odourless oil extracted from the seeds is used as a vegetable butter, and to make candles and soaps. Chemistry: The seed fat is used as an insecticide for lice. Butter Tree fat extracted from the seeds has been analysed analyzed for its chemical and physical constants and fatty acid composition and compared with those of the better known cocoa butter and shea butter. Butter Tree kernels, shea butter kernels and cocoa beans contained 50.0%, 52.1% and 53.4% fat, respectively. Butter Tree fat, Cocoa butter and Shea butter are similar in several of their characteristics, particularly slip point, saponification number, solidification point and fatty acid composition; but butter tree fat has a much lower unsaponifiable matter content (1.5-1.8%) than shea butter (7.3-9.0%). This profile aesthetically gives the butter a distinct texture. Both are markedly different from cocoa butter and cocoa butter replacement fats in respect of their melting points and fatty acid composition. The unsaponifiable fraction shows, for the sterolic composition, a predominance of stigmasterol (nearly 68% of the total sterols and has anti-inflammatory activity) whilst while the β-tocopherol is the main tocopherol.

Perilla frutescens. Perilla Oil. Shiso Oil fixed oil. Sp.gr.: 0.933. Saponification value: 192. Iodine value: 197. Average carbon number: 17.46. Average molecular weight: 279.3. C18:1: 15%; C18:2: 18%; C18:3: 64%. Traditional use: The seeds are the source of a drying oil resembling linseed oil and comprising glycerides of linoleic, oleic and palmitic acids. *Perilla frutescens* (Shiso oil) that contains perillaldehyde (74%) and limonene (12.8%) has antimicrobial activity, mainly due to the perillaldehyde. It was especially effective against *Propionibacterium acnes*, *Staphylococcus aureus* (both of which can cause acne) and novel antioxidants isolated. Its essential oils provide for a strong taste whose intensity might be compared to that of mint or fennel. [beta]-caryophyllene (24.2%), thujopsene (13.0—20.8%), perillaldehyde (14.2—15.1%) and (Z)-[beta]-farnesene (3.3—10.9%) were major.

Persea gratissima. Avocado Butter fixed oil. Saponification value: 187. Iodine value: 64. Average carbon number: 17.96. Average molecular weight: 277.1. C16:0: 20%; C18:1: 67%; C18:2: 15%. Traditional use: It increases skin elasticity and encourages healthy skin. Avocado oil is cold cold-pressed and refined for stable shelf life. It has been used in African skin treatments for centuries. This highly therapeutic oil is rich in vitamins A, B1, B2, B5 (panthothenic acid), vitamins D and E, minerals, protein, lecithin and fatty acids is a useful, penetrating nutrient for dry skin and eczema. Avocado oil is said to have healing and regenerating qualities.

Persea gratissima. Avocado Oil fixed oil. Saponification value: 187. Iodine value: 86. Average carbon number: 17.559. Average molecular weight: 276.5. C16:0: 22%; C18:0: 3%; C18:1: 62%; C18:2: 13%. Traditional use: A light, fast penetrating oil that was reported to be absorbed faster by the skin than corn, soybean, almond and olive oils. ß-sitosterol, campesterol, stigmsterol, brassicasterol, delta5-Avenastenol, tocopherols and other unidentified sterols.

Pinus pinea. Pine Nut Oil fixed oil. Average carbon number: 16.85. Average molecular weight: 279.63. C16:0: 4%; C16:1: 0.2%; C17:0: 0.1%; C18:0: 2.10%; C18:1: 24.5%; C18:2: 47.1%; C18:3: 16.1%. Traditional use: A Mediterranean delicacy, the edible seeds are known as "pignons" or "pinocchi" and they yield a novel moisturising oil. *P. pinea* oil unsaponifiable matter contained very high levels of phytosterols (4298 mg kg-1 of total extracted lipids), of which beta-sitosterol was the most abundant (74%). Aliphatic alcohol contents were 1365 mg kg-1 of total extracted lipids, of which octacosanol was the most abundant (41%). Two alcohols (hexacosanol and octacosanol), which are usually absent in common vegetable oils, were described for *P. pinea* oils.

Pistacia vera. Pistachio Nut Oil fixed oil. Sp.gr.: 0.916. Saponification value: 191. Iodine value: 85. Average carbon number: 17.62. Average molecular weight: 279.6. C16:0: 10%; C18:0: 3%; C18:1: 69%; C18:2: 17%. Traditional use: the two major volatile constituents, α-pinene and terpinolene, are compounds with interesting antibacterial and antifungal properties. Additionally, terpinolene has been identified as an antioxidant agent that can prevent LDL oxidation and also as an insecticide agent useful in food storage.

Pistacia vera. Pistachio Nut Oil fixed oil. Saponification value: 192. Iodine value: 99. Average carbon number: 17.649. Average molecular weight: 278. C16:0: 13.5%; C16:1: 2.25%; C18:0: 2.5%; C18:1: 65.5%; C18:2: 15.5%; C18:3: 0.550%. Traditional use: An oil that is substantive and protective to the harshest of external conditions. It compares favourably with peanut oil. An edible oil is obtained from the seed but is not produced commercially due to the high price of the seed.

Plukenetia volubilis. Sacha Oil fixed oil. Sp.gr.: 0.929. Saponification value: 230. Iodine value: 189. Average carbon number: 17.94. Average molecular weight: 278.81. C16:0: 3.85%; C18:0: 2.549%; C18:1: 8.27%; C18:2: 36.797%; C18:3: 48.609%. Traditional use: Protect skin. Preserve lipidic balance, soften skin, improve epidermis texture, anti-ageing skin care, regenerative skin care, healing skin care, anti-stretch marks skin care. The linoleic and linolenic acids are essential for the epidermis lipidic barrier formation. In addition, the linoleic acid allows reduces comedones and the irritation associated with them.

Pongamia glabra. Karanja Oil or Pongamia seed oil fixed oil. Melting Point: 2°C. Sp.gr.: 0.936. Saponification value: 193. Iodine value: 102. Average carbon number: 16.52. Average molecular weight: 278. C16:0: 18.57%; C18:0: 29.64%; C18:1: 44.24%; C18:2: 0%; C28:1: 0.88%. Traditional use: Its isolation and application in cosmetics is patented worldwide. Pongamia Extract exhibits effective UV absorbing properties. It enhances the UV protection of cosmetic products and broadens the UV spectrum, especially into the UVA region. The oil is known to have value in folk medicine for the treatment of rheumatism, as well as human and animal skin diseases. It is effective in enhancing the pigmentation of skin affected by leucoderma or scabies.

Portulaca oleracea. Purslane Oil fixed oil. Average carbon number: 16.98. Average molecular weight: 276.69. C16:0: 15%; C18:0: 4%; C18:1: 18%; C18:2: 33%; C18:3: 26%. Traditional use: Due to its high content of nutrients, especially antioxidants (vitamins A and C, α-tocopherol, β-carotene, glutathione) and omega-3 fatty acids, and its wound wound-healing and antimicrobial effects as well as its traditional use in the topical treatment of inflammatory conditions, purslane is a highly likely candidate as a useful cosmetic ingredient. The seed contains (per 100g ZMB) 21g protein, 18.9g fat 3.4g ash[218]. Fatty acids of the seeds are 10.9% palmitic, 3.7% stearic, 1.3% behenic, 28.7% oleic, 38.9% linoleic and 9.9% linolenic.

Pouteria sapota. Sapote Oil or Mammy Apple Oil. Average Molecular weight: 279.6. Average carbon number: 15.8. C16:0 12.0%; C18:0 27.2%; C18:1 50.0%. Tree typical of Mexico, Nicaragua, Guatemala, has many medicinal uses in South America. It is a semi solid oil with a sweet scent (reminiscent of marzipan), sapote oil is extracted from the edible sapote fruit. The kernel yields about 60% oil. Clinical tests prove that sapote oil is an effective oil to stop hair loss, promote hair growth and ideal for dermatitis. An excellent oil for all parts of the body.

Prunus amygdalus dulcis. Sweet Almond Oil fixed oil. Sp.gr.: 0.917. Saponification value: 98. Iodine value: 102. Average carbon number: 17.87. Average molecular weight: 280.3. C6: 0.1%; C16:0: 7%; C18:0: 1%; C18:1: 73%; C18:2: 19%. Traditional use: Much loved for generations, listed in the British Pharmacopoeia and an excellent choice for even the most simple of moisturisers or massage oils. Almond oil should be in every formulator's palette. almond oil has been widely employed in a number of muscle oils, anti-wrinkle oils, so-called skin foods, as well as in brilliantines. Folklore: The rod of Aaron was an almond twig, and the fruit of the almond was one of the subjects selected for the decoration of the golden candlestick employed in the tabernacle. The Jews still carry rods of almond blossom to the synagogues on great festivals.

Prunus armeniaca. Apricot Kernel Oil fixed oil. Sp.gr.: 0.921. Saponification value: 192. Iodine value: 102. Average carbon number: 17.859. Average molecular weight: 280.1. C16:0: 7%; C18:1: 66%; C18:2: 27%. Traditional use: A skin conditioning agent that is emollient, non-greasy and ideal for dry, tired and mature skins. Apricot Kernel Oil has an excellent texture that is great for all skin types, but preferred by many for the benefits that it renders for prematurely aged skin and skin that is dry and irritated. It may also be used as a cosmetic ingredient for softening and moisturizing.

Prunus avium. Cherry Pit Oil fixed oil. Saponification value: 192. Iodine value: 132. Average carbon number: 20.66. Average molecular weight: 278.69. C16:0: 10%; C16:1: 1%; C18:0: 3%; C18:1: 39%; C18:2: 50%; C18:3: 13%. Traditional use: An oil with an interesting profile of fatty acids. It moisturises moisturizes and protects the skin to leave it soft and smooth. An oil with an interesting profile of fatty acids. It moisturizes and protects the skin to leave it soft and smooth.

Prunus cerasus. Bitter Cherry Kernel Oil fixed oil. Saponification value: 198. Iodine value: 118. Average carbon number: 16.02. Average molecular weight: 281.1. C18:1: 35%; C18:2: 45%; C18:3 Conj n=5: 9%. Traditional use: Said to date back as far as 300 B.C., cherries were named after the Turkish town of Cerasus. Also known as the acid cherry, sour cherry, pie cherry or tart cherry. The oil from the stones or pits is intermediate in application weight and skin feel. It is also known as the Montmorency cherry. Discovered in the Pontus region, on the southern shores of the Black Sea, the cherry was introduced to Rome by Lucullus in around 60 B.C. There are records of cherries being established in Britain by A.D. 46.

Prunus domestica. Plum Kernel Oil fixed oil. Saponification value: 181. Iodine value: 90. Average carbon number: 15.66. Average molecular weight: 282.1. C18:1: 71%; C18:2: 16%. Traditional use: A precious oil that is rarely seen but similar to almond oil in texture and lubricity. This oil protects the most delicate of skin from moisture loss. A plum oil is used to help a damson in distress from turning into a prune.

Prunus persica. Peach Kernel Oil fixed oil. Sp.gr.: 0.921. Saponification value: 192. Iodine value: 113. Average carbon number: 17.28. Average molecular weight: 279.6. C16:0: 9%; C18:1: 67%; C18:2: 21%. Traditional use: A skin conditioning agent that is emollient, non-greasy and ideal for dry, tired and mature skins. It can be used as an equivalent to apricot kernel oil.

Prunus persica. Peach Kernel Oil fixed oil. Melting Point: -15ºC. Saponification value: 190. Iodine value: 105. Average carbon number: 17.89. Average molecular weight: 280.69. C14:0: 0.1%; C16:0: 5.5%; C16:1: 1%; C17:0: 0.1%; C17:1: 0.1%; C18:0: 2.25%; C18:1: 64%; C18:2: 24%; C18:3: 1.5%; C20:0: 0.5%; C20:1: 0.5%; C22:0: 0.299%; C22:1: 0.1%. Traditional use: A skin conditioning agent that is emollient, non-greasy and ideal for dry, tired and mature skins.

Punica granatum. Pomegranate Oil fixed oil. Saponification value: 191. Iodine value: 235. Average carbon number: 17.649. Average molecular weight: 276.8. C16:0: 4.28%; C18:0: 2.5%; C18:1: 11.1%; C18:2: 13.5%; C18:3: 4.78%; C18:3 Conj n=5: 61.8%; C20:1: 0.40%.

Punica granatum. Pomegranate Oil fixed oil. Saponification value: 191. Iodine value: 235. Average carbon number: 18.649. Average molecular weight: 276.8. C16:0: 8.5%; C16:1: 2.89%; C18:0: 5.06%; C18:1: 11.5%; C18:2: 11%; C18:3: 65.797%; C18:3 Conj n=5: 0.1%. Traditional use: Antioxidant and eicosanoid enzyme inhibition properties of pomegranate seed oil. The seeds of pomegranate, that ancient symbol of fertility, were found to contain an oestrone identical with the genuine hormone. *Punica granatum* seeds are the best source of plant oestrone to date. The oil from the seeds of the pomegranate provides a powerful anti-oxidant benefit, fighting free radicals that damage and age the skin. But pomegranate oil is also a potent source of punicic acid, an Omega omega-5 conjugated fatty acid, beneficial phytoestrogen and a rare plant plant-based source of CLA. In the first study, laboratory-grown breast cancer cells were treated for three days with pomegranate seed oil. The researchers observed apoptosis in 37- 56% percent of the cancer cells, depending upon the dose of oil applied.

R

Rhus succedanea. Japan Wax fixed oil. Melting Point: 49ºC. Saponification value: 220. Iodine value: 8. Average carbon number: 16.32. Average molecular weight: 259. C16:1: 84%; C18:0:

7%; C18:1: 9%; C18:2: 0%. Traditional use: Japan wax is a by-product of lacquer manufacture. It is not a true wax but a fat that contains 10-15% palmitin, stearin, and olein with about 1% japanic acid (1,21-heneicosanedioic acid). That product is not a true wax but is more like a vegetable tallow found in the kernel and outer skin of the berries of Rhus and Toxicodendron species, including those yielding Japanese lacquer. Other figures quoted say it contains a high amount of palmitic acid triglycerides (93-97%), long chain dicarboxylic acids including C22 and C23 chains (4-5.5%) and free alcohols (12-1.6%). Its melting point is 45—53° C. That wax is much used in Japan in cosmetics, ointments and to make candles but becomes rancid with age. Japan wax has been used for the world famous Sumo wrestlers and Geishas for their hair styling for centuries.

Ribes nigrum. Blackcurrant Seed Oil fixed oil. Saponification value: 190. Iodine value: 141. Average carbon number: 18. Average molecular weight: 281.1. C18:0: 3%; C18:1: 40%; C18:2: 45%; C18:3: 12%. Traditional use: A rich source of GLA and a superb moisturiser, which can be used in place of evening primrose or borage seed oils. A valuable nutritional oil containing a balanced profile of essential fatty acids that when used topically can promote healthier healthier-looking skin.

Ribes rubrum. Red Currant Seed Oil fixed oil. Average carbon number: 18.359. C18:1: 24%; C18:2: 42%; C18:3: 32%; C18:4 (n-3): 4%. Traditional use: One of the rare sources of stearidonic acid (SDA) and gamma-linolenic acid (GLA), important omega-3 and -6 fatty acids respectively. SDA and GLA are anti-inflammatory and anti-proliferative substances acting as precursors for anti-inflammatory eicosanoids and longer chain fatty acids. The synthesis of SDA and GLA in the body is often inefficient and slow.

Ricinodendron rautanenii. Manketti Nut Oil— – see *Schinziophyton rautenenii*

Ricinus communis. Castor Oil fixed oil. Melting Point: -18°C. Sp.gr.: 0.96297. Saponification value: 179. Iodine value: 85. Average carbon number: 18.16. Average molecular weight: 296.3. C16:0: 1%; C18:0: 1%; C18:0 OH: 89%; C18:1 OH: 89%; C18:1: 3%; C18:2: 6%. Traditional use: It is a very glossy oil on the skin. Used in lipsticks, lip balms and lip salves. Also used in transparent soaps and hair grooming products. It also cleans and softens the hair. The oil is an effective rub for inflamed skin, bruises, to prevent falling hair and to grow new hair (where hair follicles are not totally withered). Used externally for ringworm, itch, piles, sores, abscesses; hair wash for dandruff.

Rosa canina. Rose Hips Oil fixed oil. Saponification value: 180. Iodine value: 187. Average carbon number: 17.559. Average molecular weight: 279.19. C16:0: 4%; C18:0: 2%; C18:1: 15%; C18:2: 45%; C18:3: 32%. Traditional use: Tissue regeneration with good results in maintaining the tissue texture and the skin freshness in ageing prevention and attenuation of scars and stains. It has an effect in the skin cell membranes, the defence mechanisms, the growth and other physiologic and biochemical processes related to tissue regeneration, which explain its tissue tissue-regenerating properties (scars, skin burns, early aging etc.) It is rich in vitamin F. It has been tested on cheloids, acne scars, multiple erythema, surgical scars etc.

Rosa rubiginosa. Rosehip Seed Oil fixed oil. Iodine value: 179. Average carbon number: 17.91. Average molecular weight: 279.13. C14:0: 0.260%; C16:0: 3.5%; C16:1: 0.179%; C18:0: 1.82%; C18:1: 15.8%; C18:2: 44.6%; C18:3: 33.797%. Traditional use: The rose of Mosqueta oil, which was once described as the "Fountain of Youth". This oil is remarkable for its benefits to damaged, scarred and distressed skin. Palmitic, palmitoleic, stearic, oleic, linoleic, a-linolenic, g-linolenic, arachidic and eicosanic acids, epicatechin, flavonoids, gallocatechin, isoquercitrin, kaempferol-3-glucoside, leucoanthocyanins, lycopene, magnesium, rubidium, rubixanthin, alpha-tocopherol, xanthophylls, tretinoin and zeaxanthin.

Rubus chamaemorus. Cloudberry Seed Oil fixed oil.

Rubus fruticosus. Blackberry Seed Oil fixed oil. Traditional use: Rich in antioxidants, essential fatty acids and vitamin C, this oil can help improve blemished, damaged and maturing skin.

Rubus idaeus. Raspberry Seed Oil fixed oil. Saponification value: 187. Iodine value: 161. Average carbon number: 17.98. Average molecular weight: 279.6. C16:0: 1%; C18:0: 2%; C18:1: 15%; C18:2: 35%; C18:3: 47%. Traditional use: The seeds are first dried and the pulp of the fruit is removed. When the seeds are fully dried, they are crushed to produce a powder that is then pressed to produce the oil. This is a sweet, slightly aromatic oil that has a beautiful skin feel and freshness somewhat reminiscent of grape seed oil. Raspberry seed oil has very good soothing and softening properties. Its high levels of essential fatty acids, linoleic and linolenic, not synthesized in the body, give it hydrating, regenerating and restructuring properties. They contribute to the integrity of cell walls and contribute to the suppleness and beauty of the skin. Raspberry seed oil thus has anti-aging properties, imparting renewed elasticity to the epidermis. Raspberry seed oil showed absorbance in the UV-B and UV-C ranges with potential for use as a broad broad-spectrum UV protectant. The seed oil was rich in tocopherols with the following composition (mg/100 g): α-tocopherol 71; γ-tocopherol 272; δ-tocopherol 17.4; and total vitamin E equivalent of 97. The oil had good oxidation resistance and storage stability.

S

Salvadora oleoides. Khakan-Pilu Nut Oil fixed oil. Melting Point: 41°C. Sp.gr.: 0.866. Saponification value: 245. Average carbon number: 14.09. Average molecular weight: 229.46. C10: 1.2%; C12:0: 31%; C14:0: 39.2%; C16:0: 19.1%; C18:1: 8.63%; C18:2: 0.696%. Traditional use: The sweet fruits are eaten and seeds yield a green oil, which is said to be medicinal. From the seeds an oil is expressed, which is used as a stimulating application in painful rheumatic affections and after childbirth.

Salvia hispanica. Chia Oil fixed oil. Saponification value: 196. Iodine value: 193. Average carbon number: 17.09. Average molecular weight: 277.8. C14:0: 0.1%; C16:0: 6.700%; C16:1: 0.1%; C17:0: 0.2%; C17:1: 0.1%; C18:0: 3%; C18:1: 6.90%; C18:2: 18.8%; C18:3: 58.7%; C18:3 Conj n=5: 0.1%; C20:0: 0.299%; C20:1: 0.1%; C20:2: 0.2%; C22:0: 0.1%; C24:0: 0.2%. Traditional use: A rich luxuriant oil with a wonderful ancient Aztec storyline. It contains an abundance of linolenic acid that helps to explain its substantive feel on the skin. Total phytosterols in the oil ranged from 7 to 17 g/kg. β-sitosterol accounted for up to 74% of the total unsaponified fraction. The seeds contained less than 0.5 g/kg of squalene, . In cosmetics, the oil combines real skin advantages, anti-inflammatory and anti-allergic properties, favouring the dermal tissue's flexibility with a possible application on mucous membranes. Moreover, the marketing of any product containing chia seed oil could easily be supported by the high levels of omega-3.

Schleichera oleosa. Kosum Oil, Macassa Oil fixed oil. Melting Point: 46°C. Sp.gr.: 0.910. Saponification value: 227. Iodine value: 55. Average carbon number: 16.87. Average molecular weight: 289.05. C14:0: 1%; C16:0: 6.5%; C18:0: 4%; C18:1: 53.5%; C18:2: 3.78%; C20:0: 20.3%; C24:0: 2.5%. Traditional use: Oil extracted from the seed, called "kusum oil", is a valuable component of true Macassar oil used in hairdressing; it is also used for culinary and lighting purpose and in traditional medicine it is applied to cure itching, acne and other skin afflictions.

Schinziophyton rautenenii, Mongongo Oil fixed oil. Sp.gr.: 0.925. Saponification value: 210. Iodine value: 110. Average carbon number: 18. Average molecular weight: 280.1. C18:0: 5.5%; C18:1: 15.1%; C18:2: 38%; C18:3: 12%; C18:3 Conj n=5: 29.3%. Traditional use: Also known as Mongongo oil, Manketti seed oil is obtained from the tree of the same name. The manketti tree is native to Southern Africa, where the oil is used as a nourishing hair treatment. It is cold-pressed from dried manketti seeds and is high in linoleic, a-eleostearic and oleic acids. The resultant oil is emollient and regenerative with a natural UV protection for hair, nails and skin. around 560mg per 100 grams of kernel) of vitamin E (almost entirely as g-tocopherol). The large proportion of this essential linoleic acid (EFA), as well as the conjugated trienoic acid (α-eleostearic) mean that the oil is of huge importance for skin protection, both for hydration but also for the restructuring and regeneration of the epidermis. In particular the α Eleostearic acid reacts rapidly with UV light producing polymerisation polymerization and providing a protective layer.

Sclerocarya birrea. Marula Oil fixed oil. Saponification value: 192. Iodine value: 75. Average carbon number: 17.78. Average molecular weight: 279.6. C16:0: 11.1%; C18:0: 7.200%; C18:1: 75.9%; C18:2: 5.78%. Traditional use: An oil from the fruit of a tree much revered by the indigenous people of Southern Africa, who extract the oil themselves for cracked, dry or damaged skin. It is a prized cosmetic oil for both skin and hair, being similar to olive oil in composition. The seeds are so full of oil that a squeeze with the hand can release a rich yield. This healing oil is used as a cosmetic, by Southern African women, and is massaged onto the skin of their face, feet and hands. Across the generations it has proven to protect against dry, cracking skin and its moisturising properties are so effective that it is also used to treat leather and in preserving meat. Marula oil has also shown to improve skin hydration, skin smoothness and reduce redness.

Sesamum indicum. Sesame Oil fixed oil. Sp.gr.: 0.923. Saponification value: 193. Iodine value: 112. Average carbon number: 17.84. Average molecular weight: 279.69. C16:0: 8%; C18:0: 5%; C18:1: 47%; C18:2: 40%. Traditional use: Also known as gingilli oil, this oil has been known since earliest antiquity, from ancient Egypt to the Indian continent, as a soothing, gentle emollient. It is an extremely good substitute for olive oil and has excellent longevity in massage preparations. Mixed with lime water, the oil is used externally to treat burns, boils and ulcers. A new antioxidant sesamol and tocopherol were obtained from seeds. The seeds and seed oil is are used for medicinal purposes. As the seed oil (sesame oil) has various properties, it is the most widely used oil, by ayurvedic practitioners and pharmacy. Externally, the massage with tila oil reduces the dryness of the skin and alleviates the vata dosha. The oil can also be used in barrier creams to protect the skin from harmful UV light radiation. Because of the presence of sesamin and sesamolin, sesame oil has a strong antioxidant power; it increases the elasticity of the skin and prevents aging. It prevents thickening of the epidermis and dehydration.

Shorea robusta. Shorea Robusta Oil, Sal Seed Oil fixed oil. Melting Point: 43°C. Sp.gr.: 0.868. Saponification value: 190. Iodine value: 37. Average carbon number: 17.35. Average molecular weight: 284.6. C16:0: 5%; C18:0: 41.5%; C18:1: 38.5%; C18:2: 2.5%; C20:0: 8.5%. Traditional use: From the sal tree and sometimes called chus oil or sal butter, this oil from East India is a widely used emollient and skin protectant. The seed oil is used as good remedy for skin diseases and scabies. used Used for ointments of skin diseases. Exceptionally good oxidative stability due to very low content of polyunsaturated fatty acids. Prevents drying of the skin and development of wrinkles. It helps reduce the degeneration of skin cells and restore skin flexibility.

Shorea stenoptera. Illipe Butter fixed oil. Melting Point: 36ºC. Saponification value: 194. Iodine value: 35. Average carbon number: 17.46. Average molecular weight: 278.6. C16:0: 18%; C18:0: 46%; C18:1: 35%; C18:2: 0%. Traditional use: Prevents drying of the skin that leads to the development of wrinkles. Reduces degeneration of skin cells and restores skin flexibility and elasticity. Of minor components Illipe Butter contains sterols (90% being β-sitosterol) and triterpene alcohols. It also gives emolliency to treated areas of the skin. The triglyceride composition is very close to that of cocoa butter. Also called Borneo Tallow. Folklore: Illipe butter is obtained from the nuts of the Illipe tree, a wildly growing tree in Africa, Asia and South America. The first fruits are obtained when the tree is 15-20 years old. In the middle of the 18th century, the Dutch salesmen introduced the illipe fruits on the market. But before this trading started the natives of Borneo, the Dayakers, used the fat for medical care and cooking.

Silybum marianum. Milk Thistle Oil. Sp.gr. 0.915. Iodine value: 70. Average carbon number: 18.2. Average molecular weight: 284.4. C14:0 0.1% ; C16:0 12%; C18:0 5%; C18:1 23.6%; C18:2 64.6; C18:3 0.2%. It contains a rich blend of phytosterols: cholesterol 3.81%; campesterol 6.92%; Δ7-campestanol 0.07%; stigmasterol 7.83%; '7-campesterol 4.24%; clerosterol 0.37%; β-sitosterol 35.74%; Δ5-avenasterol 3.74%; Δ5,24-stigmastadienol 1.06%; Δ7-stigmasterol 22.55%; Δ7-avenasterol 3.29%; fucosterol 1.32%. The oil has a soothing and calming effect on the skin and has been used traditionally on problem skin conditions. The seeds contain silymarin that is a mixture of flavonolignans, such as silybin, isosilybin, silydianin and silychristin. Studies have shown that silymarin has a similar effect to that of retinoids keratinocyte proliferation and the production of extracellular matrix (ECM) proteins, such as type I collagen, elastin and laminin. Topical treatments of silymarin improved wrinkles in human skin.

Simarouba glauca. Aceituno Oil fixed oil. Melting Point: 28ºC. Saponification value: 191. Iodine value: 52. Average carbon number: 17.3. Average molecular weight: 279.8. C16:0: 12%; C18:0: 28%; C18:1: 58%; C18:2: 0%. Traditional use: No skin data available.

Simmondsia chinensis. Jojoba Oil fixed oil. Melting Point: 11ºC. Saponification value: 94. Iodine value: 84. Average carbon number: 19.85. Average molecular weight: 307.6. C12:0: 0.2%; C14:0: 0.299%; C16:0: 4%; C16:1: 1%; C17:0: 0.1%; C17:1: 0.1%; C18:0: 1%; C18:1: 14%; C18:2: 5%; C18:3: 1%; C20:0: 0.5%; C20:1: 56%; C22:0: 0.5%; C22:1: 16%; C24:0: 1%. Traditional use: Indians and Mexicans have for a long time used jojoba oil as a hair conditioner and restorer. There is a substantial body of anecdotal evidence that suggests that the wax is beneficial in alleviating minor skin irritations. Jojoba oil has been shown to significantly soften the skin as measured by viscoelastic dynamometry. Jojoba oil may be effective in alleviating the symptoms of psoriasis and in controlling acne outbreaks. It has been used medicinally for the treatment of burns and sores etc. Topical irritations such as sunburn and chapped skin appear to respond to topical jojoba therapy. A substantial body of anecdotal evidence that suggests that the wax is beneficial in alleviating minor skin irritations. It is a liquid wax that acts as a skin protectant and emollient. Folklore: Indians and Mexicans have for a long time used jojoba oil as a hair conditioner and restorer, and in medicine, cooking and rituals.

Sisymbrium irio. Sisymbrium Irio Oil fixed oil. Average carbon number: 18.23. Average molecular weight: 284.43. C16:0: 15%; C18:1: 18%; C18:2: 15%; C18:3: 35%; C20:0: 7.5%;

C22:1: 9.5%. Traditional use: This seed oil was much loved by the Romans and is still used today by the beautiful women of Asia for improvement of the complexion. It can be used with great benefit in skin care preparations. Romans used this oil to massage the body, whilst while in India it has been used for the treatment of many skin disorders and also to improve the complexion.

T

Taraktogenos kurzii. Chaulmoogra Oil, Maroti Oil fixed oil. Melting Point: 22ºC. Sp.gr.: 0.94195. Saponification value: 213. Iodine value: 98. Average carbon number: 16.01. Average molecular weight: 261.3. C14:0: 12.1%; C16:0: 3%; C16:1: 48.7%; C18:0: 27%; C18:1: 6.5%; C18:2: 0%. Traditional use: An oil native to Burma and China, it is an Indian remedy for problem skins, particularly for dry, desquamative skin conditions and sores. Chaulmoogra contains strongly antibacterial chemicals, two of which, hydnocarpic and chaulmoogric acids, are responsible for destroying the bacterium and so is useful for the treatment of acne and its healing properties. It is used in compositions for infections and regulating sebaceous activity. (French patent 2518402). It is also used in lipsticks for healing and repairing chapped lips. (Erno Laslo). It is formulated in to after-shave lotion to reduce burning and itch. (Erno Laslo). It has also been used for the treatment of hypopigmentation. Commercial products have used chaulmoogra in dandruff treatment lotions to slow down desquamation, and calm burning and itching. (Phytosqualame).

Telfairia pedata. Oyster Nut Oil fixed oil. Saponification value: 197. Iodine value: 101. Average carbon number: 15.98. Average molecular weight: 274.99. C16:0: 24.3%; C18:0: 18.1%; C18:1: 11.4%; C18:2: 32.6%; C18:3: 5%. Traditional use: A novel oil that offers skin conditioning and protection against the loss of precious skin hydration.

Theobroma cacao. Cocoa Butter fixed oil. Melting Point: 34ºC. Sp.gr.: 0.96897. Saponification value: 194. Iodine value: 37. Average carbon number: 17.46. Average molecular weight: 276.1. C16:0: 27%; C18:0: 35%; C18:1: 35%; C18:2: 3%. Traditional use: A traditional African remedy for dry skin, suitable for the most delicate of skin types. It has excellent emollient properties and is used to soften and protect chapped hands and lips. It has also been employed in the formation of suppositories and pessaries, for rectal, vaginal, and other difficulties. It likewise enters into preparations for rough or chafed skin, chapped lips, sore nipples, various cosmetics, pomatums, and fancy soaps; and has also been used for coating pills.

Theobroma grandiflorum. Cupuacu Butter fixed oil. Iodine value: 40. Average carbon number: 17.1. Average molecular weight: 271.19. C14:0: 7%; C16:0: 31%; C18:0: 44%; C18:1: 5%; C18:2: 11%; C18:3: 2%. Traditional use: Cupuacu is a small to medium tree in the rain forest canopy that belongs to the chocolate family. It has an excellent occlusivity that is not too heavy on application. It is a very good replacement material for cocoa butter to which it has a similar feel. Imparts gloss to the hair.

Trichilia emetica. Mafura Butter, Mafura butter, Natal Mahogany, Cape Mahogany, Christmas bells. Family: Meliaceae. Sp.gr.: 0.905. Saponification value: 200. Iodine value: 70. Average carbon number: 17.3 Average molecular weight: 273.1. C16:0: 15.5%; C18:0: 2.75%; C18:1: 50.0%; C18:2: 10.5%; C18:3: 1.5%. It is found across Southern Africa in low altitude, frost-free areas where rich alluvial soils are present, mainly along rivers and the coast. The oil is easily extracted by immersing the seeds in hot water, soaking them for several hours and then crushing the seeds. This pressing process releases a solid, yellow fatty butter (Mafura butter) with a high melting point. Traditional use: The tree's leaves, bark and seeds have a wide variety of traditional medicinal uses such as treatment of stomach and intestinal ailments, as an emetic, purgative and the oil is used for rheumatism treatment. It has healing properties: the seed oil forms the basis for a leprosy remedy, is used as a cure for rheumatism and to heal wounds. The conditioning and colouring properties of Mafura butter have made it a popular hair care product, as well as being used on the skin to nourish and revitalise. The oil has also been shown to have some antimicrobial and anti-inflammatory activity due to the presence of a number of limonoids such as trichilin A.

Trichodesma zeylanicum. Wild Borage Oil fixed oil. Traditional use: This oil is not totally unrelated to our borage (*Borago officinalis*) since it is from the same family. It originates from Tanzania, where it is used for its emollient and soothing properties. It may also be found in Pakistan, where it is used for similar purposes. The seed oil is used for its emollient properties in Tanzania, India and Pakistan.

Triticum vulgare. Wheatgerm Oil fixed oil. Sp.gr.: 0.929. Saponification value: 184. Iodine value: 130. Average carbon number: 17.6. Average molecular weight: 277.19. C16:0: 15%; C18:0: 3%; C18:1: 20%; C18:2: 55%; C18:3: 7%. Traditional use: Contains one of the highest levels of natural vitamin E and is a valuable additive to any skin care product where care and protection of the skin is important. It contains around 3000 ppm tocopherols. Traditional Use: Wheat germ oil is rich in vitamin E. The germ (the heart of the wheat) is now proved a vital food and medicine for the human body. Wheat germ oil is one of the best natural skin conditioning ingredients available. It is used in creams, facial scrubs and cleansers, hair care products and body lotions to help promote a healthy-looking skin. Folklore: More than 1000 kilos of wheat grain are needed to product just one 1 kilo of this deep golden orange oil. The germ makes up only 2% of a single grain but holds one of nature's richest stores of both essential fatty acids and vitamin E.

V

Vaccinium angustifolium. Blueberry Seed Oil fixed oil. Traditional use: An light oil rich in essential fatty acids and natural antioxidants that absorbs quickly into skin.

Vaccinium macrocarpon. Cranberry Seed Oil fixed oil. Saponification value: 190. Iodine value: 164. Average carbon number: 17.89. Average molecular weight: 279.1. C16:0: 5.5%; C18:0: 1.5%; C18:1: 24.8%; C18:2: 37%; C18:3: 31.1%. Traditional use: near Near one-to-one balance of omega-3 and omega-6 fatty acids and a healthy mixture of antioxidants including very high levels of tocotrienols and tocopherols. At total of 70% essential fatty acids makes cranberry seed oil an excellent emollient, ideal for use in lip, skin, and baby care applications. It supplies the skin with critical essential nutrients and it contains a blend of omega omega-3, -6 and -9 essential fatty acids that are crucial to skin, heart, brain and total cellular health. Uniquely balanced 3, 6, and 9, phytosterols, phospholipids and other powerful antioxidants reside in this highly stable oil.

Vaccinium myrtillus. Bilberry Seed Oil fixed oil. Traditional use: Blue Tocol offers a full spectrum of all natural isomers of Tocopherols tocopherols and Tocotrienolstocotrienols. It is very rich in essential fatty acids (over more than 70%) and phytosterols. Ultrapure Blue Tocol is produced from seeds of Bilberry (European Blueberry), and is ideal for nourishing and protecting skin.

Vaccinium vitis-idaea. Lingonberry Oil fixed oil. Traditional use: Lingonberry seed oil is among the richest source of alpha-linoleic acid, an omega-3 fatty acid with clear anti-ageing and skin skin-lightening effects. In addition, the seed oil combines high levels of plant sterols and tocotrienols.

Vanilla planifolia. Orchid Oil fixed oil. Traditional use: A light, delicate oil that is perfect for providing a light moisturisation to those products where the after-skin feel should be a whisper.

Vateria indica. Dhupa Butter, Malabar Tallow. India. Melting Point: 35ºC. Sp.gr.: 0.912. Saponification value: 189. Iodine value: 39. Average carbon number: 18.039. Average molecular weight: 281.02. C16:0: 11.5%; C18:0: 41%; C18:1: 45%; C18:2: 1.2%; C20:0: 2.5%. Dhupa butter is cold-pressed from the dried seeds of the Valeria indica tree, which grows in India. It is high in stearic, palmitic, oleic (mainly as distearo-oleic glycerol), linoleic, linolenic and arachidic acids and has a long shelf life. It is used as cocoa butter substitute. Dhupa butter is emollient, softening, and aids spreadability when added to creams, balms and lotions. It has been applied as a remedy against rheumatism and epidermal pains. The physical properties are comparable to cocoa butter. It has a good softening effect and spreadability on the skin. It is protective against sunlight (UV-protective) and the butter has traditionally been used for soap-making in India.

Vernonia galamensis. Veronia Oil fixed oil. Iodine value: 106. Average carbon number: 17.96. Average molecular weight: 296.5. C16:0: 2%; C18:0: 3%; C18 Epoxy: 79%; C18:1: 5%; C18:2: 11%. Traditional use: Derivatives of vernonia oil have shown great promise in a number of areas. It has been used as a topical wound healer for both humans and animalsin topical treatment for many skin diseases including psoriasis, eczema, scalp dermatitis and certain types of body ulcers. It is used as a carrier for existing drugs to prolong their 'half-life'. For example, increasing the half-life of a drug in the human body from minutes to hours. Acting as a slow slow-release agent for drugs in the body. It has a low viscosity that allows quick drying.

Virola surinamensis. Ucuhuba Oil fixed oil. Average carbon number: 14.14. Average molecular weight: 232.09. C12:0: 13%; C14:0: 69%; C16:0: 7%; C18:1: 5%; C18:2: 5%. Traditional use: The tree contains hallucinogenic materials. The tree is an endangered species. No skin data on oil.

Vitis vinifera. Grape Seed Oil fixed oil. Saponification value: 192. Iodine value: 131. Average carbon number: 17.8. Average molecular weight: 276.69. C16:0: 10%; C18:1: 3%; C18:2: 17%; C18:3: 70%. Traditional use: A slightly green, low-odour oil that is ideal as a carrier for essential oils in massage oils and other delicate colour/fragrance products. It is has antioxidant properties.

WX

Ximenia americana. Ximenia Seed Oil, Seaside Plum Oil fixed oil. Saponification value: 180. Iodine value: 143. Average carbon number: 20.71. Average molecular weight: 321.3. C16:0: 2%; C18:0: 1%; C18:1: 56.5%; C18:2: 2%; C18:3: 1%; C20:1: 2.200%; C22:1: 2%; C24:0: 2%; C24:1: 10%; C26: 2.5%; C26:1: 7.200%; C28: 1.5%; C28:1: 10%. Traditional use: The seed oil, extracted in various ways, is edible and used in cooking. However, it's principal use is as an emollient. Ximenia oil is used by bushmen on their bows and bow strings, whilst while the women and girls use it to anoint their bodies and hair. It is a very effective hair oil, rich conditioner and skin softener. Chapped skin is often soothed by massaging Ximenia oil into the affected area. Interestingly, the presence of the active ximenyic acid has shown to increase vasomotion of cutaneous small arteries, in addition to having anti-inflammatory properties, improving the functioning of sebaceous tissues and being an effective treatment for dry skin prone to senescence.

Z

Zea mays. Corn Oil fixed oil. Saponification value: 187. Iodine value: 121. Average carbon number: 16.84. Average molecular weight: 278.1. C16:0: 13%; C18:0: 13%; C18:1: 18%; C18:2: 51%. Traditional use: Corn oil is also known as maize oil; maydol; mazola. The crude oil may contain up to 2% phospholipids (vegetable lecithin, inositol esters). Used as a hair dressing.

Chapter 2

A Review of Butters

Introduction

The information available for butters is often very confused confusing and makes identification difficult. The presence of numerous synonyms does not make the situation any easier. It is hoped that this review will make the identification and understanding of the butters named in the PCPC International Cosmetic Ingredient Dictionary and the PCPC Buyers' Guide (Cosing) more simple.

Aesandra butyracea [Syn. *Diploknema butyracea, Bassia butyracea*] Chiuri Tree, Chiuri Butter, Phulwara butter. Nepal. Family: Sapotaceae. The main product of the tree is ""ghee"" or butter, extracted from the seeds, and named Chiuri ghee or Phulwara butter. The seed produces an oil rich in fatty acids that is mainly used as vegetable butter. The vegetable butter is also used in lighting lamps, confectionery, pharmaceutical, vegetable ghee production, candle manufacturing and soap soap-making. Folklore: Chiuri has a very important cultural value in Nepal where the people of the Chepang community give Chiuri seedlings as dowries to daughters indicating its significance in the livelihood of the Chepang community. The juice is also consumed to quench thirst. It has been found to be effective for rheumatism.

Astrocaryum murumuru. Murumuru Butter fixed oil. Family: Arecaceae. Traditional use: A very soft butter from the heart of the Brazilian rain forest. This butter is a substantive and protective material that makes a perfect, if not better, replacement for cocoa butter, illipe butter and others. Chemistry: Saponification value: 245. Iodine value: 25. Average carbon number: 13.89. Average molecular weight: 226.47. C8: 1%; C10: 0.9000%; C12:0: 44%; C14:0: 28%; C16:0: 9%; C18:0: 3%; C18:1: 10%; C18:2: 4%; C20:0: 0.1%.

Astrocaryum tucuma. Tucuma Butter. Upper Amazon and Rio Negro. Family: Arecaceae. Traditional use: The fleshy part of the fruit is esteemed for food by the Indians. The yellowish, fibrous pulp is eaten by the natives. The large fruits are edible and valued for their flavour in South America. A light-coloured butter is obtained from the seeds which has a unique, characteristic odour and behaviour similar to Murumuru butter. The butter's natural gloss brings a desirable shine to dry, damaged hair and can act as a wonderful skin moisturizer, exhibiting great lubricity. Tucuma butter is solid at room temperature, but melts immediately on contact with the skin. Saponification value: 231.4. Iodine value: 12.5. Average carbon number: 13.49. Average molecular weight: 221.47. C8: 1.92%; C10: 1.95%; C12:0: 50.157%; C14:0: 24.44%; C16:0: 6.21%; C18:0: 2.33%; C18:1: 8.35%; C18:2: 4.16%; C20:0: 0.1%; C24:0 0.06%.

Bassia butyracea. [Syn: *Diploknema butyracea*] Family: Sapotaceae. Indian-Butter, Phoolwa-Oil Plant. is a plant from the East Indies. The pulp of the fruit is eatable. The juice is extracted from the flowers and made into sugar by the natives. It is sold in the Calcutta bazaar and has all the appearance of date sugar, to which it is equal if not superior in quality. An oil is extracted from the seeds, and the oil cake is eaten as also is the pure vegetable butter which is called chooris and is sold inexpensively.

Bassia latifolia. Illipe Butter. Melting Point: 36ºC. Saponification value: 185. Iodine value: 32. Average carbon number: 16.81. Average molecular weight: 276.19. C16:0: 23.5%; C18:0: 22%; C18:1: 35.5%; C18:2: 15%. Traditional use: Another of the African butters, used by natives for the treatment of dry skin and especially for the care and protection of babies' "delicate skin."

Mahuwa oil has emollient properties and is used in skin disease, rheumatism and headache. It is also a laxative and considered useful in habitual constipation, piles and haemorrhoids and as an emetic. Tribal people also used it as an illuminant and hair fixer.

Bassia latifolia. Mowrah Butter fixed oil. Melting Point: 39ºC. Saponification value: 291. Iodine value: 62. Average carbon number: 17.35. Average molecular weight: 276.3. C16:0: 23.5%; C18:0: 21%; C18:1: 38.5%; C18:2: 16%. Traditional use: [Syn. *Bassia longifolia*] The butter is extracted from the seed kernels and further processed and refined to obtain a yellowish butter which that has a very mild odour. The butter is a solid at room temperature, but melts readily on contact with the skin. It will help prevent drying of the skin and so help reduce the formation of wrinkles. Helps soften and smooth the skin. Medicinally, Bassia oil is used as an emollient application to the skin.

Butyrospermum parkii. Shea Butter. Saponification value: 180. Iodine value: 3. Average carbon number: 17.6. Average molecular weight: 278.1. C16:0: 20%; C18:0: 45%; C18:1: 33%; C18:2: 2%. Traditional use: This rich buttery oil from Central Africa is used for the protection and care of skin cracked and dehydrated by the elements. Beurre de karite is an elegant addition to products crafted for the smoothing and replenishment of dry skins. Shea butter is a suitable base for topical medicines. Its application relieves rheumatic and joint pains and heals wounds, swellings, dermatitis, bruises and other skin problems. It is used traditionally to relieve inflammation of the nostrils. Shea butter has natural antioxidant properties and is said to contain a small quantity of allantoin which that is renowned for its healing qualities. It is also said to protect the skin against external aggressions, and sun rays. It is an effective skin emollient and skin smoother. Folklore: In Africa, the fat (shea butter) is used as an ointment for rheumatic pains and boils. A decoction from the bark is used to facilitate child delivery and ease labour pains. The leaf extract is dispensed for headaches and as an eye bath.

Butyrospermum parkii. Karite Butter (example from West Africa). Melting Point: 38ºC. Saponification value: 180. Iodine value: 65. Average carbon number: 18.42. Average molecular weight: 281.6. C16:0: 6%; C18:0: 40.5%; C18:1: 50.5%; C18:2: 6%. Traditional use: The high proportion of unsaponifiable matter, consisting of 60–70% triterpene alcohols, gives shea butter creams good penetrative properties that are particularly useful in cosmetics. Allantoin, another unsaponifiable compound, is responsible for the anti-inflammatory and healing effect on the skin. Shea Butter and Karite butters are from trees indigenous to Africa, occurring in Mali, Cameroon, Congo, Côte d'Ivoire, Ghana, Guinea, Togo, Nigeria, Senegal, Sudan, Burkina Faso and Uganda. The geographical location of the source has a significant impact on the composition of the butter as illustrated above.

Carpotroche brasiliensis. Sapucainha Butter, Fruta-de-lepra. Family: Flacourtiaceae. Traditional uses: It is used to treat skin diseases and leprosy. Chemistry: Gynocardin and tetraphyllin B were isolated from the seeds and pericarp. The cyanogenic glycosides occur in an approximately equimolar mixture in the seeds and in the pericarp. The oil extracted from its seeds also contains, as major constituents, the same cyclopentenyl fatty acids hydnocarpic (40.5%), chaulmoogric (14.0%) and gorlic (16.1%) acids found in the better known chaulmoogra oil prepared from the seeds of various species of *Hydnocarpus* spp. These acids are known to be related to the pharmacological activities of these plants and to their use as anti-leprotic agents.

Cocos nucifera, Coconut butter. This is not available in nature, although the oil is usually solid at room temperature. The commercial offer for this material contains coconut oil blended with octyldodecanol and beeswax.

Elaeis guineensis. Palm Butter. Flow point: 33-37°C. It is a natural (non-hydrogenated) stearin fraction of palm fat that is a solid butter at room temperature that has great antioxidant power. The combination of the fatty acids and triglycerides present make palm butter particularly stable. The stability is further improved by the presence of natural antioxidants β-carotene, tocopherol and tocotrienols that also give palm butter its anti-aging properties and free radical scavenger action. Tocotrienols are more potent therapeutically than tocopherols and tend to be 40 to 50 times stronger in oxidative stability. In addition, skin benefit is enhanced by the presence of sterols (campesterol, stigmasterol 300mg/kg),

Garcinia Indica. Kokum Butter. Melting Point: 41°C. Saponification value: 190. Iodine value: 31. Average carbon number: 17.96. Average molecular weight: 283.1. C16:0: 2%; C18:0: 58%; C18:1: 39%; C18:2: 1%. Traditional use: Kokum Butter is obtained from the fruit kernels of the tree, Garcinia indica growing in India. It is mainly used as a replacement for cocoa butter, for which many consider kokum to be an equivalent or better. Traditional medicinal applications of Kokum butter is for the preparation of ointments and for local applications to ulcerations and fissures of the hands or lips. It is also used to treat muscle pulls and burns.

Helianthus annuus. Sunflower butter. It might have been expected that a butter could be isolated from this oil, but commercial material is the sunflower seed oil blended with materials like glycine soja (soybean) lipids and beeswax. Increasingly worrying is suppliers are hydrogenating the fixed oil and then re-introducing this hydrogenated material back in to the oil to form a butter. Even more worrying is that a number of organic certifications have been issued for this synthetic butter.

Irvingia gabonensis. African wild mango butter, African mango, Dika nut, bush mango. Africa, Congo. Family: Irvingiaceae. Traditional use: Methanolic extract of *Irvingia gabonensis* are is used in the treatment of bacterial and fungal infections. The material is used as an alternative to shea butter. Moisturizing and softening efficacy on skin. It protects and repairs skin and lips through its film-forming activity, thereby restoring softness and suppleness. A clinical test has shown that it is a better skin moisturizer than shea butter and is smoothing on the skin. Chemistry: Melting point: 4.5°C. Saponification value: 190. Iodine value: 4.5. Average carbon number: 13.4. Average molecular weight: 220.64. Average molecular weight: 283.1. C10:0 1.11%; C12:0 36.6%; C14:0 53.71%; C16:0: 5.23%; C18:0: 0.8%; C18:1n-9: 1.82%; C18:2n-6: 0.49%. The minor unsaponifiable fraction contains mainly sterols and tocopherol.

Madhuca butyracea. Phulwara Butter. Melting Point: 45°C. Saponification value: 196. Iodine value: 96. Average carbon number: 16.87. Average molecular weight: 267.72. C16:0: 56.6%; C18:0: 3.60%; C18:1: 36%; C18:2: 3.78%. Traditional use: Phulwara butter is obtained from healthy seed kernels of the *Madhuca butyracea* tree. It is rich in palmitic acid and is an edible food fat that makes an excellent substitute for cocoa butter, but without the typical odour of chocolate.

Mangifera indica. Mango Butter. Saponification value: 190. Iodine value: 60. Average carbon number: 17.5. Average molecular weight: 281.3. C16:0: 7%; C18:0: 43%; C18:1: 46%; C18:2: 2%. Traditional use: Mango kernel oil is known to release a drug like salicylic acid at a much higher rate than standard paraffin base emollient. Mango kernel oil is known to be used in India for soap soap-making and as an emollient in skin care products. Carotenoids present. An unusual fatty acid, cis-9, cis-15-octadecadienoic acid was isolated from the pulp lipids of mango. Mango butter has been used in the rainforests and tropics for its skin softening, soothing, moisturizing and protective properties and to restore flexibility. It has a protective effect against UV radiation and has been found to provide relief from the dryness of eczema and psoriasis.

Olea europaea. Olive butter. This does not occur naturally and products claiming to be olive butter are usually blends, either olive oil mixed with beeswax or olive oil mixed with hydrogenated olive oil or another higher melting point material, e.g. beeswax, glyceryl monostearate, etc. It is a practice that is confusing those not trained in the natural chemistry of plant materials.

Pentadesma butyracea. Butter tree, kanya, kanya butter, tallow tree, krinda, kpangnan, tama. Family: Clusiaceae. West Africa, from Guinea, Sierra-Leone and Cote d'Ivoire, Togo to the Democratic Republic of Congo, extending eastwards into Tanzania and Uganda. Traditional use: It is a high oil producing species and the odourless oil extracted from the seeds is used as a vegetable butter, and to make candles and soaps. The seed fat is used as an insecticide for lice. Butter tree fat extracted from the seeds has been analysed for its chemical and physical constants and fatty acid composition and compared with those of the better better-known cocoa butter and shea butter. Butter tree kernels, shea butter kernels and cocoa beans contained 50.0%, 52.1% and 53.4% fat, respectively. Butter tree fat, cocoa butter and shea butter are similar in several of their characteristics, particularly slip point, saponification number, solidification point and fatty acid composition; but butter tree fat has a much lower unsaponifiable matter content (1.5—1.8%) than shea butter (7.3—9.0%). This profile aesthetically gives the butter a distinct texture. Both are markedly different from cocoa butter and cocoa butter replacement fats in respect of their melting points and fatty acid compositions. The unsaponifiable fraction shows, for the sterolic composition, a predominance of stigmasterol (nearly 68% of the total sterols and has anti-inflammatory activity) whilst the β-tocopherol is the main tocopherol.

Persea gratissima. Avocado Butter. Saponification value: 187. Iodine value: 64. Average carbon number: 17.96. Average molecular weight: 277.1. C16:0: 20%; C18:1: 67%; C18:2: 15%. Traditional use: A complex blend of vitamins A and E and other active materials is reported, which increases skin elasticity and encourages healthy skin. avocado Avocado oil is cold cold-pressed and refined for stable shelf life. Avocado oil has been used in African skin treatments for centuries. This highly therapeutic oil is rich in vitamins A, B1, B2, B5 (panthothenic acid), vitamins D and E, minerals, protein, lecithin and fatty acids. It is a useful, penetrating nutrient for dry skin and eczema. Avocado oil is said to have healing and regenerating qualities. There are some avocado butters which are not pure isolated fatty acids, but contain hydrogenated avocado oil or other higher melting point materials. The real avocado butter may be expected to be more costly than the blend.

Sclerocarya Birrea. Marula Butter. The stone within the Marula fruit contains several kernels, which are rich in oil. Marula Oil is extracted using simple pressing and filtration techniques. The process does not use solvents and gives a high purity oil for cosmetic applications. The Marula oil is rich in oleic acid (see previous chapter for the composition). Marula butter is a higher melting point version of Marula oil and provides improved skin hydration and smoothness.

Sesamum indicum. Sesame seed butter. This is not available naturally and is usually hydrogenated sesame oil or its blend with sesame oil.

Shorea robusta. Shorea Robusta Oil, Sal Seed Oil. Melting Point: 43°C. Saponification value: 190. Iodine value: 37. Average carbon number: 17.35. Average molecular weight: 284.6. C16:0: 5%; C18:0: 41.5%; C18:1: 38.5%; C18:2: 2.5%; C20:0: 8.5%. Traditional use: From the sal tree and sometimes called chus oil or sal butter, this oil from East India is a widely used emollient and skin protectant. The seed oil is used as good remedy for skin diseases and scabies. It has exceptionally good oxidative stability due to very low content of polyunsaturated fatty acids. It prevents drying of the skin and development of wrinkles, reduces degeneration of skin cells and helps restore skin flexibility.

Shorea stenoptera. Illipe Butter also called Borneo Tallow.. Melting Point: 35°C. Saponification value: 189. Iodine value: 33. Average carbon number: 17.66. Average molecular weight: 278.8. C16:0: 17%; C18:0: 45%; C18:1: 35%; C18:2: 2%; C18:3: 1%. Traditional use: Prevents drying of the skin which that leads to the development of wrinkles. It reduces the degeneration of skin cells and restores skin flexibility and elasticity. It also gives emolliency to treated areas of the skin. The triglyceride composition is very close to that of cocoa butter. Folklore: Illipe butter is obtained from the nuts of the Illipe tree, a wildly growing tree in Africa, Asia and South America. The first fruits are obtained when the tree is 15-20 years old. In the middle of the 18th century, the Dutch salesmen introduced the illipe fruits on the market. But before this trading started, the Dayakers, natives of Borneo, the Dayakers, used the fat for medical care and cooking.

Simmondsia chinensis. Jojoba Butter. There appears to be the genuine material and also a blended material in commercial supply. In one case, jojoba butter is obtained by specific cold pressing select seed kernels of *S. chinensis*. The resultant oil is then blended with hydrogenated vegetable oil to modify the rheology to create a "butter" which is soft and pliable and suitable for a variety of cosmetic applications. In another example it was actually hydrogentated jojoba oil, and in another it was a blend of the oil with glycine soja (soybean) lipids. However, in amongst the confusion of the supplier market it was possible to find a product that was jojoba butter [INCI: Simmondsia Chinensis (Jojoba) Butter] with no additives or adulteration.

Theobroma cacao. Cocoa Butter. Melting Point: 34°C. Saponification value: 194. Iodine value: 37. Average carbon number: 17.46. Average molecular weight: 276.1. C16:0: 27%; C18:0: 35%; C18:1: 35%; C18:2: 3%. Traditional use: A traditional African remedy for dry skin, suitable for the most delicate of skin types. It has excellent emollient properties and is used to soften and protect chapped hands and lips. It has also been employed in the formation of suppositories and pessaries, for rectal, vaginal, and other difficulties. It likewise enters into preparations for rough or chafed skin, chapped lips, sore nipples, various cosmetics, pomatums, and fancy soaps; and has also been used for coating pills. Said to soothe the skin after sunburn or windburn, cocoa butter is used widely as a suppository and ointment base, as an emollient and as an ingredient in various topical cosmetic preparations. Cocoa butter has been reported to be a source of natural antioxidants. It has excellent emollient properties and is used to soften and protect chapped hands and lips. Folklore: Cocoa Butter is made from the expressed oil of the chocolate nut. It came originally from Central and South America and was cultivated by the Mayas. In pre-Columbian civilisations civilizations it was a very important food and drink, given its high energy content. The regard in which it was held is reflected in the generic name which means "food of the Gods,", while the specific name is derived from the Aztec word 'Kakawa'.

Theobroma grandiflorum. Cupuacu Butter. Flow point: 27-33°C. Iodine value: 40. Average carbon number: 17.1. Average molecular weight: 271.19. C14:0: 7%; C16:0: 31%; C18:0: 44%; C18:1: 5%; C18:2: 11%; C18:3: 2%. Traditional use: Cupuacu is a small to medium tree in the

Rainforest rain forest canopy which that belongs to the Chocolate family. It has an excellent occlusivity that is not too heavy on application. It is a very good replacement material for cocoa butter to which it has a similar feel. It imparts gloss to the hair.

Trichilia emetica. Mafura Butter, Mafura butter, Natal Mahogany, Cape Mahogany, Christmas bells. Family: Meliaceae. found Found across Southern Africa in low altitude, frost-free areas where rich alluvial soils are present, mainly along rivers and the coast. The oil is easily extracted by immersing the seeds in hot water, soaking them for several hours and then crushing the seeds. This pressing process releases a solid, yellow, fatty butter with a high melting point. Traditional use: The tree's leaves, bark and seeds have a wide variety of traditional medicinal uses such as treatment of the stomach and intestinal ailments, as an emetic, purgative and the oil is used for rheumatism treatment. It has healing properties. The seed oil forms the basis for a leprosy remedy, is used as a cure for rheumatism and to heal wounds. The conditioning and colouring properties of Mafura butter have made it a popular hair care product, as well as being used on the skin to nourish and revitalise. The oil has also been shown to have some antimicrobial and anti-inflammatory activity due to the presence of a number of limonoids such as trichilin A. Chemistry: Specific gravity: 0.905. Saponification value: 200. Iodine value: 70. Average carbon number: 17.3 Average molecular weight: 273.1. C16:0: 15.5%; C18:0: 2.75%; C18:1: 50.0%; C18:2: 10.5%; C18:3: 1.5%.

Trichilia roka. Trichilia Roka Seed Butter. Mafura butter, Natal mahogany, Ethiopian mahogany, Christmas bells. The Latin name is a synonym for *T. emetica* discussed above but has been allocated its own INCI name.

Vateria indica. Dhupa Butter, Malabar Tallow. India. Dhupa butter is cold-pressed from the dried seeds of the Valeria indica tree, which that grows in India. It is used as a cocoa butter substitute. Dhupa butter is emollient, softening, and aids spreadability when added to creams, balms and lotions. It has been used as a remedy for rheumatism and epidermal pains. The physical properties are comparable to cocoa butter. It has a good softening effect and spreadability on the skin. It is protective against sunlight (UV-protective) and the butter has traditionally been used for soap-making in India. Chemistry: It is high in stearic, palmitic, oleic (mainly as distearo-oleic glycerol), linoleic, linolenic and arachidic acids and has a long shelf life. Melting point: 35°C. Specific gravity: 0.912. Saponification value: 189. Iodine value: 42. Average carbon number: 18.0 Average molecular weight: 281. C16:0: 11.5%; C18:0: 41.0%; C18:1: 45.0%; C18:2: 1.2%; C20:0: 2.5%.

Virola sebifera. Ucuuba Butter. [INCI: Virola Sebifera Seed Oil]. Ucuuba Butter is native to Central and South America and is cold pressed from the seeds of the Ucuuba tree. It is said to have anti-inflammatory and antiseptic properties and is ideal for treating acne, eczema and dry or irritated skin. Exceptionally rich in essential fatty acids, Ucuuba Butter is considered to have anti-aging properties and can be used to replenish tone and moisture to dry and mature skin. Virola fat has been used traditionally in candle manufacture. Native tribes also use the seeds, impaled on sticks, as candle lights. The fat and pulverised kernels find use in traditional medicines. C12:0 15.8%, C14:0 73.2%, C16:0 4.7%, C18:1 5.7%.

Vitis vinifera. Grapeseed butter. This is not available in nature and commercial supplies of the butter are either hydrogenated grapeseed oil or a blend of the oil and the hydrogenated derivative.

Conclusion

The growth in natural butters continues and we it is hoped that the popularity of these materials will spur the extraction of natural waxes which that would be equally acceptable to the natural chemist who is limited by the current choices on offer.

Personally, I find the description of adulterated blends as the natural butter to be offensive if not totally misleading and possibly illegal.

Chapter 3

Fats and Waxes

Introduction

The previous chapter dealt specifically with butters, but those sources can become almost wax-like depending on the extraction, source and composition. Temperature and concentration of use are also a key factor in formulation and a material like coconut oil could be solid at ambient temperatures in some cooler climates. Where a material has been mentioned before we have added new additional information.

The natural world is abundant in fatty acids that are beneficial to both body and skin, internally and externally. The table below in this chapter gives some of the examples according to carbon numbers and double bonds present.

Fatty acids consist of the elements carbon (C), hydrogen (H) and oxygen (O) arranged as a carbon chain skeleton with a carboxyl group (-COOH) at one end. Saturated fatty acids (SFAs) have all the hydrogen that the carbon atoms can hold, and therefore, have no double bonds between the carbons. Mono-unsaturated fatty acids (MUFAs) have only one double bond. Poly-unsaturated fatty acids (PUFAs) have more than one double bond.

Omega-3 (ω3) and omega-6 (ω6) fatty acids are unsaturated "Essential Fatty Acids" (EFAs) that need to be included in the diet because the human metabolism cannot create them from other fatty acids. They are extremely important in skin care and most commonly found in evening primrose oil (*Oenothera biennis*), blackcurrant seed oil (*Ribes nigrum*), borage oil (*Borago officinalis*) and more recently discovered in high level in Inca Inchi oil (*Plukenetia volubilis*). Since these fatty acids are polyunsaturated, the terms n-3 PUFAs and n-6 PUFAs are applied to omega-3 and omega-6 fatty acids, respectively. These fatty acids use the Greek alphabet (α,β,γ,...,ω) to identify the location of the double bonds. The "alpha" carbon is the carbon closest to the carboxyl group (carbon number 2), and the "omega" is the last carbon of the chain because omega is the last letter of the Greek alphabet. Linoleic acid is an omega-6 fatty acid because it has a double bond six carbons away from the "omega" carbon.

Linoleic acid is an another important fatty acid in the care of the skin and the body because it is Linoleic acid which in the body plays an important role in lowering cholesterol levels. Alpha-linolenic acid is an omega-3 fatty acid because it has a double bond three carbons away from the "omega" carbon. By subtracting the highest double-bond locant in the scientific name from the number of carbons in the fatty acid, we can obtain its classification can be obtained. For arachidonic acid, we subtract 14 from 20 to obtain 6; therefore, it is an omega-6 fatty acid. This type of terminology is sometimes applied to oleic acid, which is an omega-9 fatty acid.

Triglyceride, more properly known as triacylglycerol , TAG or triacylglyceride is glyceride in which the glycerol is esterified with three fatty acids. It is the main constituent of vegetable oil and animal fats. Triglycerides have lower densities than water (i.e they float on water), and at normal room temperatures may be solid or liquid. When solid, they are called "fats" or "butters," and when liquid they are called "oils".

Glycerol is a trihydric alcohol (containing three -OH hydroxyl groups) that can combine with up to three fatty acids to form monoglycerides, diglycerides, and triglycerides. Fatty acids may combine with any of the three hydroxyl groups to create a wide diversity of compounds. Monoglycerides, diglycerides, and triglycerides are classified as esters, which are compounds created by the reaction between acids and alcohols that release water (H_2O) as a by-product.

Common fatty acids

Chemical names and descriptions of some common fatty acids

Common Name	Carbon Atoms	Double Bonds	Scientific Name
Butyric acid	4	0	butanoic acid
Caproic Acid	6	0	hexanoic acid
Caprylic Acid	8	0	octanoic acid
Capric Acid	10	0	decanoic acid
Lauric Acid	12	0	dodecanoic acid
Myristic Acid	14	0	tetradecanoic acid
Palmitic Acid	16	0	hexadecanoic acid
Palmitoleic Acid	16	1	9-hexadecenoic acids
Stearic Acid	18	0	octadecanoic acid
Oleic Acid	18	1	9-octadecenoic acids
Ricinoleic acid	18	1	12-hydroxy-9-octadecenoic acids
Vaccenic Acid	18	1	11-octadecenoic acids
Linoleic Acid	18	2	9,12-octadecadienoic acid
Alpha-Linolenic Acid (ALA)	18	3	9,12,15-octadecatrienoic acid
Gamma-Linolenic Acid (GLA)	18	3	6,9,12-octadecatrienoic acid
Arachidic Acid	20	0	eicosanoic acid
Gadoleic Acid	20	1	9-eicosenoic acids
Arachidonic Acid (AA)	20	4	5,8,11,14-eicosatetraenoic acid
EPA	20	5	5,8,11,14,17-eicosapentaenoic acid
Behenic acid	22	0	docosanoic acid
Erucic acid	22	1	13-docosenoic acids
DHA	22	6	4,7,10,13,16,19-docosahexaenoic acid
Lignoceric acid	24	0	tetracosanoic acid

The chemical formula is RCOO-CH2CH(-OOCR')CH2-OOCR", where R, R', and R" are longer alkyl chains. The three fatty acids RCOOH, R'COOH and R"COOH can be all different, all the same, or only two the same.

Chain lengths of the fatty acids in naturally occurring triglycerides can be of varying lengths but 16, 18 and 20 carbons are the most common. Natural fatty acids found in plants and animals are typically composed only of an even numbers of carbon atoms due to the way they are bio-synthesised from acetyl CoA. In chemical structure, acetyl-CoA is the thioester between coenzyme A (a thiol) and acetic acid (an acyl group carrier).

Most natural fats contain a complex mixture of individual triglycerides; because of this, they melt over a broad range of temperatures. Cocoa butter is unusual in that it is composed of only a few triglycerides, one of which contains palmitic, oleic and stearic acids in that order. This gives rise to a fairly sharp melting point, causing chocolate to melt in the mouth without feeling greasy.

Traditional Waxes

Any plant that can yield an oil will likely yield a wax, though it has to be accepted that the yield may not always be commercially attractive. Similarly those plants that have the ability to produce resins, will also be likely to have an associated resinous wax.

A wax can be described in the simplest of terms as a blend, of fatty acids and their derivatives, or of glycerides and their derivatives, that forms a solid at normal ambient temperatures. The individual components of higher molecular weight may alone be considered waxes, e.g. palmitic or stearic acids. It is also considered fair to include fats in the category of waxes.

In the 18th and 19th centuries the number of waxes available was probably greater than today. In this chapter we have described a number of waxes and fats that have virtually vanished but which could be of great commercial interest. There is often great confusion on the identity of the source of these plants and we have attempted to clarify the correct botanical classification.

Acacia decurrens. Mimosa Flower Wax

A floral wax that is highly perfumed and obtained from the sludge left behind after distillation of the essential oils from the flowers of the plant.

Astrocarya murumuru. Murumuru Butter

It is a natural fatty product obtained by cold-pressing the palm tree kernels of different species of Astrocaryum genus. It is originally from the Amazon region and presents an interesting fatty composition rich in oleic, lauric and myristic fatty acids. It can improve the emollience and hydration to the skin and the hair.

Butyrospermum parkii. Shea Butter or Beurre de Karite

Shea Butter has natural antioxidant properties and contains a small quantity of allantoin, which is renowned for its healing qualities. Shea butter is a natural fat obtained from fruits of the Shea tree, one of the Sapotaceae. Shea trees grow naturally on the lateral slopes of the savanna zones of the hinterland of West Africa and throughout that continent's equatorial region (where rainfall is not too high) and also in parts of southern Sudan.

It contains relatively high level of unsaponifiables, comprising between 65–75% terpenic alcohols and between 3.5–8% phytosterols. The fatty acids of shea butter have been found to consist of between 45–60% of oleic acid and 30–40% of stearic acid as well as smaller amounts of myristic, palmitic and linoleic acids. Free fatty acids constitute between 3–5% of the shea butter composition. The pure material has an SPF value of 3, and the literature describes the 15% use as having slight light-protective effects that can be recognized. It has a good active molecule delivery performance and salicylic acid and benzoic acid were released from an ointment base prepared from shea butter 70%, arachis oil 20% and hard paraffin 10% at a faster rate than from white soft paraffin BP or simple ointment BP.

It has good skin absorption properties, is skin moisturizing and improves its suppleness. For this reason it is used in the treatment of dry and cracked skin that might occur in sunburn, eczema and dermatitis. It leaves the skin feeling soft.

Cananga odorata. Ylang Ylang Wax

A floral wax that is highly perfumed and obtained from the sludge left behind after distillation of the essential oils from the flowers of the plant.

Cera alba. Beeswax

This is another favorite ingredient of lipsticks and emulsion formulae. There is white beeswax, yellow beeswax, bleached beeswax, white wax and yellow wax. Beeswax is the wax obtained from the honeycomb of the honeybee. Beeswax absolute is a pale yellow solid with a mild, sweet, and oily odor reminiscent of good linseed oil with a trace of honey notes, depending on sources. Beeswax (white and yellow) is the refined wax from honeycombs. The wax is a secretion from bees of the genus Apis e.g., Apis dorsata, A. indica, A. florea and the domesticated A. mellifera. Beeswax is the wax produced by worker bees from nectar and pollen. These substances are converted in the body of the bee to wax, which then exudes through wax pores on the ventral surface into eight small molds and there hardens into small scales. When required it is plucked off, made plastic with saliva and used for building the honeycomb and capping the cells. Beeswax forms about one=tenth of the total weight of the honeycomb.

The oldest emulsifier was probably the beeswax/borax system used to formulate cold creams and brilliantine hair dressings. The source of the wax has great influence on its color and attributes, in much the same way that the location of the hive is reflected in the condition and taste of the honey. It is not accepted as an ingredient by vegans and many vegetarians.

India wax differs from official wax chiefly in its lower acid value. A wax is also produced in India by the so-called Kota bees belonging to the genus Melipona. They are minute stingless insects that furnish a sticky, dark-colored wax resembling, in physical and chemical characteristics, the propolis of the honey bee rather than the true beeswax.

Ceroplastes ceriferus. Indian Wax, Arjun Wax, Indian Wax Scale; Indian White Wax Scale; Japanese Wax Scale.

It is an insect wax obtained from the insect *Ceroplastes ceriferus*.

Ceroxylon andicola.

It is from an Ecuadoran palm from the slopes of the Andes, which exudes a yellow wax from the rings of its trunk which can be used as a beeswax substitute and though quite a resinous wax is mixed with tallow to make candles. It seems to be a natural wax that is no longer available to the cosmetic industry. The South Americans call the tree 'Palma de cera' (literally wax palm) and this could be a promising lead for a new natural wax. We believe this is also the wax known as 'Cire des Andaquies', however, other sources refer to Andiquies Wax as produced by a wild bee found in the Orinoco and Amazon valleys in South America, used in making candles. It may well be that two waxes share the same name.

Cinnamomum pedunculatum. Koya Wax, Koga Wax or Japanese Cinnamon Wax.

This is available commercially (mainly from the Far East). The tree is widely distributed in the southern islands of Japan, where it is known as Yabunikkei. It is softer than Japan wax but further information is not available.

Citrus aurantium amara. Bitter Orange Wax

An aromatic wax that is highly perfumed and obtained from the sludge left behind after distillation of the essential oils from the flowers of the plant. Both the flower and peel waxes are available.

Citrus aurantium dulcis. Orange Wax

An aromatic wax that is highly perfumed and obtained from the sludge left behind after distillation of the essential oils from the flowers of the plant. Both the flower and peel waxes are available.

Citrus medica limonum. Lemon Peel Wax

A wax that is highly perfumed and obtained from the sludge left behind after distillation of the essential oils from the flowers of the plant.

Coccus ceriferus. Chinese Wax. Chinese insect wax or cochineal wax is produced from the secretion of the *Coccus ceriferus*, deposited on ash trees (*Fraxinus chinensis*). The wax is relatively hard, melting between 79 and 83°C, and is white and odourless. It is used for making candles and polishes and as a protective layer on paper and textiles. See also *Fraxinus chinensis*.

Copernicia cerifera. Carnauba

There is a clue from the Latin name *cerifera*, since this literally means wax-maker (examples would include *Benincasa cerifera* (Winter Melon) and *Myrica cerifera* (Bayberry).

Carnauba wax comes from the Brazilian Wax Palm or Carnauba Palm Tree, and is obtained by shaking the large palmate leaves that are coated with the wax. Wax is also present on the leaf buds, which are so abundantly coated in the wax that it falls to the ground in the height of summer and can be collected. It has been used for generations as a component of lipstick preparations, tablet coating, furniture polishes, mascaras, scented candles, brilliantine hair dressings, etc.

Cocos nucifera. Coconut Wax

It may be cheating to call coconut oil a wax, but at lower temperatures it certainly forms a solid mass that melts at skin temperature. There have been a number of samples seen where the oil has been fractionated to give the higher carbon number fatty acids, and these are solid at higher temperatures. The emollient and refatting effects are too well known to require detailed description.

It is rich in glycerides. It contains oil, comprising the glycerides, trimyristin, trilaurin, triolein, tristearin and tripalmitin; also the glycerides of caprylic, capric and caproic acids. The oil, kernels, seed, leaves, sap are nutritive and anthelmintic.

Coconut oil-based soaps are useful for marine purposes because they are not easily precipitated by salt water or salty solutions. In the British Pharmaceutical Codex (1973) it says that coconut oil is used as an ointment basis, particularly in preparations intended for application to the scalp.

All parts of the plant are useful in Africa. The oil is used for cosmetics and food. The coconut milk is a refreshing drink and is mainly used as a poison antidote. It is not usually recommended for those with cardiac problems. The oil from the nuts is valued as an emollient and used as an ingredient in remedies for skin infections.

Elaeis guineensis. Oil Palm

It is a tall palm with a straight trunk and black rough bark. The leaves are up to 5 cm long and feather-shaped with spines at the lower leaf bases. Flowers are in separate male and female spikes clustered around the top of the stem. The fruit is a dark red stone fruit. Palm wine is obtained from the base of the terminal bud and palm oil from the fruit of the pulp.

The oil palm is a tropical palm tree. There are two species of oil palm, the better-known is the one originating from Guinea, Africa, and was first illustrated by Nicholaas Jacquin in 1763, hence its name, *Elaeis guineensis* Jacq.

The reddish fruit is about the size of a large plum and grows in large bunches. A bunch of fruits can weigh between 10 to 40 kilograms each. Each fruit contains a single seed (the palm kernel) surrounded by a soft oily pulp. Oil is extracted from both the pulp of the fruit (palm oil, which is edible) and the kernel (palm kernel oil, used mainly for soap manufacture).

For every 100 kilograms of fruit bunches, typically 22 kilograms of palm oil and 1.6 kilograms of palm kernel oil can be extracted. Palm oil contains saturated palmitic acid, oleic acid and linoleic acid, giving it a higher unsaturated acid content than palm kernel or coconut oils. Palm oil is used for manufacture of soaps and candles, and more recently, in manufacture of margarine and cooking fats. Palm kernel oil is extracted from the kernel of endosperm and contains about 50% oil. Similar to coconut oil, with a high content of saturated acids, mainly lauric, it is solid at normal temperatures in temperate areas, and is nearly colorless, varying from white to slightly yellow. This non-drying oil is used in edible fats, in making ice cream and mayonnaise, in baked goods and confectioneries, and in the manufacture of soaps and detergents.

The oil is used as a liniment for indolent tumors. Reported to be anodyne, antidotal, aphrodisiac, diuretic, and vulnerary, oil palm is a folk remedy for cancer, headaches and rheumatism.

Ficus ceriflua [Syn: F. Cerifera]. Fig Wax.

It is a vegetable wax, obtained from the sap of the wax fig-tree, *Ficus ceriflua* or *F. cerifera*, found in Java and Sumatra. It is used in making candles and also known as Java wax or getah wax. In Java, wax is extracted from *F. ceriflua* and another related species *F. toxícaria* that is employed in batik work and once again used for the making of candles.

Fraxinus chinensis. Sela Wax, Pela Wax or Chinese Wax.

The wax is abundantly distributed in the vegetable world and in some cases its production is stimulated by the attacks of insects. In this example Chinese wax is produced an insect called *Coccus pela* puncturing the young branches of *Ligustrum lucidum* and *Fraxinus chinensis* on which it lives. Chinese wax melts at 82° C. (180° F.) and consists almost entirely of cerotyl (C26) cerotate.

Galactodendron utile [Syn. *Brosimum galactodendron*]. Cow Tree Wax or Milk-Tree Wax.

It is a resinous, yellowish, waxy solid contained in the sap (30–35%) of the South American cow-tree, milk-tree, or palo de vaca. It is used by the natives for making candles which are reported to burn very brightly. Cow-tree wax is obtained by boiling and evaporating down the milk of the cow tree; it softens at 104° F., melts at 140°F, and is insoluble in cold alcohol, but dissolves completely in boiling alcohol and is saponified by alkalies. It resembles beeswax more than any other kind of wax.

Jasminium officinale. Jasmine Wax

A floral wax that is highly perfumed and obtained from the sludge left behind after distillation of the essential oils from the flowers of the plant.

Langsdorffia hypogæa. Balanophore Wax.

It is a wax which has the consistency of natural beeswax and is obtained from the parenchymatous tissue of Langsdorffia hypogæa, a native of Brazil. Candles prepared from the rhizomes, which are rich in wax, are sold in the markets of Colombia under the name siejos.

Lanolin Cera. Lanolin Wax

This was is obtained by removing the excess oils and sebum from wool and fractionating out the waxes by molecular distillation. It is a first class material, an excellent skin emollient and at the same time provides a protective barrier. It can be used universally across emulsions and lipsticks. The debate continues on the allergy of lanolin, with most of the evidence pointing to lanolin not being a problem.

Lavandula angustifolia. Lavender Flower Wax

A floral wax that is highly perfumed and obtained from the sludge left behind after distillation of the essential oils from the flowers of the plant. They are usually too highly fragranced to be used as normal waxes and can be expensive. However, they are a delicious indulgence if the product can bear the cost.

Ligustrum obtusifolium [Syn. *Lingustrum ibota*]. Ibota Wax.

A commercial insect wax is produced on the branches as a result of eggs being laid by insects according to one report. Another says that the wax is produced by the plant due to the stimulation of the feeding insects, however, there is another theory that the wax is produced by the insects themselves. It is used for candles and as a polish for earthenware pots, book edges etc. It is white, very solid (melting point 42° C or 108° F) and is hard to find commercially.

Mangifera indica. Mango Butter

Mythologically, the mango tree derives its existence from the ashes of the Daughter of the Sun God. For Indian poets, the tree has a peculiar significance. Its flowers are of great beauty and the fruit is much esteemed. The giant mango tree stands for "power and strength."; earlier, the Sanskrit name amra was often used as a prefix to give a person or thing the attributes of honor or admiration.

The mango tree, *Mangifera indica* of the Anacardiaceae or cashew family, is one of the most productive plants of the tropical region. The fruit has a yellow to pale green skin with a red blush, and a yellow to orange flesh with a large flat stone. The stone, which is the seed, consists of a hard fibrous shell and an inner kernel from which the fat is obtained. The oil content of the kernel is about 10% (dry material). During the harvest season, the stones are then further sun-dried to avoid hydrolytic splitting of the fat molecules and fatty acids. The soft fat is extracted from the stone, neutralized and bleached, and then distilled to produce a soft pale-colored fat. The Mango kernel fat is fractionated to obtain a unique product with a lower melting point than traditional mango kernel oil. The crude mango kernel oil contains mainly triglycerides, but also impurities that have to be removed. Through the refining process free fatty acids, peroxides, metal and phosphatides are removed. The result is a pure, pale yellow, oxidation-stable product free from odor.

It has a very pleasant feeling when applied as the pure oil to the skin. The main part of the oil consists of triglycerides and gives high emolliency to the skin. A very slight sweet scent gives the final touch of natural feeling. As a food ingredient, the material is used as a replacement for cocoa butter.

Myrica cerifera . Bayberry Wax

It is also known as American Bayberries, Candle Berry, Wax Berry, Wax Myrtle, Tallow Shrub and American Vegetable Wax. It is an ornamental shrub that has green berries covered with pale blue, lavender or grayish-white aromatic wax. The berries, or berry wax, which possesses mild astringent properties, can be obtained by boiling the berries. The wax that is collected from the surface can be used to make fragrant candles.

The bark, leaves and wax from the fruit of the wax myrtle have been made into astringent teas, gargles and douches for diarrhoea, hemorrhages, sore throat, and poultices for wounds, cuts and bruises. The wax was said to be highly effective against dysentery, one tribe of American Indians took the decoction of the stems and leaves for fever.

The status of commercial availability is not known at the current time although a number of sources are shown in the CTFA Cosmetic Ingredient Dictionary 2008.

Myristica becuhyba. Becuiba Tallow, Becuiba Wax.

This fat is obtained from the seeds of a Brazilian species *Myristica becuhyba* by expression. It resembles expressed oil of nutmeg, except in its taste, which is sharp and acidulous. Fusing point: 47° C. (116.6° F.). The nut yield a medicinal balsam used for rheumatism.

Myristica otoba. Otoba Butter.

Obtained from the fruit of *Myristica otoba*. A nearly colorless or yellowish fat, the odor resembling that of nutmegs when fresh, but becoming brownish in color and disagreeable in odor vith age. It fuses at 38° C. (100.4° F.). It contains myristin, olein, and otobit. The latter forms colorless, odorless, tasteless, prismatic crystals, which fuse at 133° C. (271.4° F.). Cold alcohol sparingly dissolves them.

Myristica surinamensis. Ucuhuba Fat

A yellow solid fat obtained from Ucuhuba nuts, the fruit of *Myristica surinamensis*.

Narcisus poeticus. Narcisus Flower Wax

A floral wax that is highly perfumed and obtained from the sludge left behind after distillation of the essential oils from the flowers of the plant.

Orbygnia martiana . Babassu Palm Fat

The oil originates from the Maya Biosphere Reserve in the Petén region of northern Guatemala, an area where the Cohune Palm grows wild. Here the oil is extracted by local inhabitants, who first harvest the nuts (similar in texture to the coconut) from the forest floor and then remove the woody outer shell that encloses the oval seed containing the oil. These seeds resemble the meat of the coconut in appearance, taste and smell but contain more oil - typically, 65-72%. The oil is extracted using a press and then filtered.

10Kg of fat can be obtained from 15Kg of the flat seeds. The oil is white and, like coconut oil, has a pleasant, nutty smell. Also like coconut oil, it can be used in the manufacture of margarine. The inaccessible nature of the source of babassu has restricted its exploitation.

This exotic South American kernel is an oil with a melting point close to human body temperature. It was first introduced onto the market by Croda. The material is non-comedogenic and super refined babassu oil glides on the skin creating a rich, non-greasy emollience with a drier feel than coconut oil. It helps to soften the skin and impart a natural lustre.

Oryza sativa. Rice Bran Wax

This is prepared from rice bran oil, and contains mainly fatty acids and esters of higher alcohols. It also contains vitamins E and F, as well as gamma-oryzanol, which is showing exceptional skin benefits with properties similar to vitamin E.

Papaver somniferum. Opium wax.

It is a white crystalline substance, a mixture of ceryl cerotate and palmitate, from the seed-vessels of the poppy. This is a by-product of the production of the opium drug and so unlikely to be available commercially.

Pedilanthus pavonis. Candelilla

This wax has similar uses to carnauba and the wax is obtained from a wax coating that surrounds the whole of the plant, which comes from northern Mexico or southern Texas.

There are many chemists who prefer this wax to carnauba for its structural effect in lipstick formulation.

Persea Americana. Avocado Wax.

This material is rich in a phytosterol similar to the cholesterol found in lanolin. The material again provides excellent skin emolliency, but with additional skin soothing and healing properties.

Polianthes tuberose. Tuberose wax

A floral wax that is highly perfumed and obtained from the sludge left behind after distillation of the essential oils from the flowers of the plant.

Propolis Cera. Propolis Wax

Propolis is found in the beehive, and is a tough waxy cement that is used by the bee to hold the hive together and repair any cracks. Analysis of this material shows it to be almost identical to the resin found on the leaf buds of the Birch (Betula verrucosa). It is already used in alternative medicine and has been examined for its antibiotic properties, but it is likely that even greater information on this material will appear in the future.

Rhus succedanea. Japan wax

Rhus Succedanea Fruit Wax (Japan wax) is a wax obtained from the mesocarp of the fruit of *Rhus succedanea* or sumac tree. It is widely called "shoro" (raw fat) in Japan, but more precisely named "haze fat" (sumac fat) from the chemical viewpoint. This wax imparts good feel to the skin, and there is substantial herbal evidence to show that this material has traditional herbal benefits when applied topically. Japan wax is a byproduct of lacquer manufacture. It is not a true wax but a fat that contains 10-15% palmitin, stearin, and olein with about 1% japanic acid (1,21-heneicosanedioic acid). Japan wax was originally from China and Japan. It contains 10% glycerin and is rather soft, melting between 50 - 52°C. It is used for polishes, cosmetics, in the production of pastels and crayon and for treating leather.

There are numerous species. For example, there is the Sweet Sumach (*Rhus aromatica*), from Canada and the United States, also the Smooth Sumach (*R. glabra*), which in various parts of the world is also known as the Dwarf Sumac, Mountain Sumac and Scarlet Sumac. It is also known as Indian Salt due to the powder found on the berries. There is also a Russian species (*R. cariaria*), and it is the leaves of this species that can be irritating to the skin when touched.

Rhus verniciflua [Syn. *Toxicodendron verniciflua*]. Berry Wax

Rhus Verniciflua Peel Wax is a wax obtained from the peel of the fruit. It is a small Asiatic tree (also called the varnish tree) that yields a toxic exudate from which lacquer is obtained. It is sometimes called "Berry Wax" in commerce. The fruits have a high fat content, and have been used to produce a wax for candle making, cooking and for skin and hair care. Many parts of the plant are poisonous and (as in the related *Rhus radicans*, or Poison ivy), severe skin rashes can occur with contact with the raw sap. The wax softens (that is perfectly free of toxins) is used to protect the skin and lips.

Rosa multiflora. Rose Flower wax

A floral wax that is highly perfumed and obtained from the sludge left behind after distillation of the essential oils from the flowers of the plant.

Saccharum officinarum. Sugar Cane Wax

A wax can be obtained by boiling up the stems of sugar cane. Obviously with such an abundance of the raw material, together with the acceptable aesthetics of such an idea, this would be very welcome to any consumer. Sugar-cane wax a white crystalline substance found on the surface of the stem of the sugarcane, particularly the violet cane, which can easily be scraped off, and is sometimes seen in the melted state floating on the surface of evaporating cane juice. It appears to consist chiefly of ceryl cerotate and palmitate. The commercial sample seen to date was almost black in color and came from China. If this could be considerably cleaned up, then it would be most interesting as a new cosmetic ingredient.

Sapium sebiferum [Syn: *Stillingia sebifera, Croton sebiferus*]. Chinese Vegetable Tallow.

It is used in making candles and is an easily disintegrated substance that melts at 37°C. The wax from the seed is used as a lard substitute or in cacao butter. The seed contains 8.1-9.2% protein and 40.5-50.7% fat.

The seed is coated with a wax which makes up about 24% of the seed and is used locally to make candles and soap. It has an excellent burning quality, and gives a clear bright flame which does not smell. Pure tallow fat is known in commerce as Pi-yu.

The wax is separated from the seed by heating them in hot water and skimming off the wax as it floats to the surface. The wax is solid at temperatures below 40°C. It is said to blacken grey hair. The seed contains about 20% of a drying oil that is used to make candles and soap and is also used to make varnishes and native paints because of its quick-drying properties. It is also used in machine oils and as a crude lamp oil. The pure oil expressed from the inner part of the seeds is known in commerce as Ting-yu.

Shellac Cera. Shellac Wax

Shellac is not obtained directly from a plant, but a tiny insect (*Kerria lacca*) that depends on some very specialized plants on which to secrete its extraordinary resin that is a sort of protective greenhouse for its developing young. These include *Caesalpina nuga*, *C. sepiaria*, *Cajanus cajan* (Congo Pea), *Schleichera oleosa* (Indian Kusum Tree), *Lizyphus mauritiana* (Indian Ber-Tree) and *Samanea saman* (Thailand Rain Tree). There are also other species, and it is the strain of the insect and the species of the host tree that determine the color and properties of the final shellac. Shellac is an excellent film-former and has been used extensively in mascara formulations. There is also good evidence for its use as a hair conditioner. In the old days, when phonographs revolved at 78 rpm, shellac was a fundamental constituent in the composition of records.

Beeswax (Cera alba)

This is another favourite ingredient of lipsticks and emulsion formulae. There is white beeswax, yellow beeswax, bleached beeswax, white wax and yellow wax. Beeswax is the wax obtained from the honeycomb of the honeybee. Beeswax absolute is a pale -yellow solid with a mild, sweet, and oily odour reminiscent of good linseed oil with a trace of honey notes, depending on sources. Beeswax (white and yellow) is the refined wax from honeycombs. The wax is a secretion from bees of the genus Apis e.g., Apis dorsata, A. indica, A. florea and the domesticated A. mellifera. Beeswax is the wax produced by worker bees from nectar and pollen. These substances are converted in the body of the bee to wax, which then exudes through wax pores on the ventral surface into 8 eight small moulds and there hardens into small scales. When required it is plucked off, made plastic with saliva and used for building the honey comb and capping the cells. Beeswax forms about one tenth of the total weight of the honeycomb.

The oldest emulsifier was probably the beeswax/borax system used to formulate cold creams and brilliantine hair dressings. The source of the wax has great influence on its colour and attributes, in much the same way that the location of the hive is reflected in the condition and taste of the honey. It is not accepted as an ingredient by Vegans vegans and many vegetarians.

India wax differs from official wax chiefly in its lower acid value. A wax is also produced in India by the so-called Kota bees belonging to the genus Melipona, . they They are minute stingless insects , whichthat furnish a sticky, dark-colored wax, resembling, in physical and chemical characteristics, the propolis of the honey bee rather than the true beeswax.

Shorea stenoptera. Borneo Tallow, Illipe Butter (Shorea stenoptera)

Illipe butter is obtained from the nuts of the Illipe tree, a wildly growing tree that grows wild in Africa, Asia and South America. The first fruits are obtained when the tree is 15-–20 years old. The nut contains three seeds, from which the yellow oil is expressed. The oil content of the seeds is 45-–50%. Illipe oil is also known as Borneo tallow (*Shorea stenoptera*). In Borneo there are about 30 known varieties of the Illipe tree, but only about 13 of them are suitable for cosmetic application, because of their suitable triglyceride composition and fatty acid content. It is closer to cocoa butter than any other fat. Its triglyceride composition is uniform and consists mainly of stearic-oleic-stearic triglycerides. It has high oxidation stability.

The seeds contain palmitic acid (17%), stearic acid (45%), oleic acid (35%), linoleic acid (0.5%), linolenic acid (nil), arachidic acid (2%). The content of sterols, triterpenic alcohols and tocopherol makes up about 1% (i.e. the unsaponifiable content). In mg/Kg, the following sterol materials are contained: stigmasterol (944), avenasterol (281). The triterpenic alcohols are ß-amyrin (3060), butyrospermol (9945), α-amyrin (17595), lupeol (6120), others (1148). The content of the tocopherol varies according to the source.

Like all vegetable fats, illipe butter presents soothing, anti-drying and protective virtues. Used externally, illipe butter is recommended to heal sores and mouth ulcers. Illipe butter has emollient virtues and it reinforces the skin lipidic barrier and keeps the skin moisturized.

Simmondsia chinensis. Jojoba

Jojoba is a desert shrub indigenous to Arizona, California and Northern Mexico. It grows in a number of deserts, including Israel's Negev desert. The mature plant produces five to ten pounds of seeds. The oil is obtained from the seeds.

Jojoba seeds produce 50% by weight a colorless, odorless oil comprising (97%) straight chain monoesters of C20 and C22 alcohols and acids with two double bonds. The acids have been identified as a mixture of cis-11-eicosenoic (C-20 and cis-13-docosenoic (C-22, erucic) acids. The alcohols have been identified as mixtures of cis-11-eicosenol, cis-13-docosenol and cis-15-tetracosenol (C-24). These alcohols are potentially valuable in the production of detergents, wetting agents, dibasic acids, etc. Also included are small quantities of sterols (less than 0.5% of a total mixture of campesterol, stigmasterol and sitosterol). Jojoba oil is essentially triglyceride. It is the only "liquid wax" of its type, so far as can be ascertained.

The Indians and Mexicans have, for a long time, used jojoba oil as a hair conditioner and restorer. The wax may be of value in the management of acne and psoriasis. It is a unique liquid mixture of long chain linear monoesters, which is also an excellent emollient.

Additionally it has been shown to be non-comedogenic and non-antigenic. Jojoba is readily absorbed by the skin and is miscible with sebum. As a result, it leaves no greasy after feel.

Jojoba Oil appears to be effective in controlling transpirational water loss, having the water vapor porosity necessary to permit the skin to function normally while still providing emolliency. Jojoba effectively controls skin sloughing without the greasiness associated with occlusion such as petrolatum. As an added benefit, jojoba oil has been shown to significantly soften the skin as measured by viscoelastic dynamometry.

Theobroma cacao. Cocoa Butter

It came originally from Central and South America and was cultivated by the Mayas. In pre-Columbian civilisations it was a very important food and drink, given its high energy content. The regard in which it was held is reflected in the generic name which means "food of the Gods,", while the specific name is derived from the Aztec word 'Kakawa'. Another source explains the words 'theo broma' are Greek for 'food "food of the Gods' Gods" and that Cocoa cocoa was named Theobroma by Linnaeus. Cortez described the use of a beverage called chocalatl, based on the seeds of *T. cacao*, among the members of the Aztec court.

The major producers are the Ivory Coast, Brazil, Ghana and Malaysia. The three main commercial products obtained from cacao seeds are cocoa powder, cocoa butter and cocoa extracts. Following curing and fermentation, the beans are dried and roasted to yield the desired flavour, colour and aroma. Oil of Theobroma, or cocoa butter, is a yellowish white solid, with an odour resembling that of cocoa.

Cocoa contains the alkaloids theobromine (0.5– to 2.7%) caffeine (about 0.25% in cocoa), trigonelline and others. The characteristic bitter tatsetaste is due to the reaction of diketopiperazines with theobromine during roasting. Theobromine is produced commercially from cocoa husks. Cocoa butter contains triglycerides consisting mainly of oleic, stearic and palmitic acids. About three quarters of the fats are present as monounsaturates.

It has excellent emollient properties and is used to soften and protect chapped hands and lips.

Virola sebifera. Ocuba Wax, Virola Tallow.

It is a subcrystalline, yellowish fat, melting at 45 - 50° C. (113 - 122° F.), dissolving wholly in alcohol, obtained from the fruit of a shrub, the *Virola sebifera*, [Syn. *Myristica sebifera*]. Other references describe Ocuba as a waxy fat obtained from the fruit of the myrtle *Myristica ocuba*; melting point, 40°C; used in candles. There is a great deal of confusion in the exact identity of the actual species from which these fats are obtained. In the New World two or perhaps three genera of the Myristicaceae are involved in the production of vegetable tallows: Virola, Dialyanthera, and possibly Osteophloem. Vegetable tallows originating at Belém do Pará should be mostly from *Virola surinamensis*; those originating from Colombia should be from *Dialyanthera otoba*; and tallow from southern Brazil should be from *Virola sebifera*, or from *V. oleifera*.

CONCLUSIONS

The choices of natural fats, butters and waxes are increasing as suppliers fill the demand for new and exotic materials. In this section it is hoped that new potential crops can be identified for future natural sources of raw material.

CHAPTER 4

Triterpenoidal Saponins

Natural surfactant and emulsifier

Introduction

The saponins are found in many plants, but first got their name from the Soapwort plant or *Saponaria officinalis*, the roots of which were traditionally used as a natural soap. It received its name from the Latin sapo meaning "soap." These saponins are the plant's 'immune system'.

Saponins are widely distributed in the plant kingdom and many of these saponin-producing plants have soap as a part of their name:

Soapwort - *Saponaria officinalis*
Soaproot - *Chloragalum pomeridianum*
Soapbark - *Saponaria officinalis*
Soapnut- *Sapindus mukerossi*

The presence of the Latin name *saponaria* might also be a clue to the saponin content of the plant.

Jaboncillo - *Sapindus saponaria*
Soapwort - *Saponaria officinalis*
Soapbark - *Saponaria officinalis*

Commercial saponin has been obtained from Quillaja bark or from soap-root by extraction with water and subsequently with alcohol. The resultant saponin powder is not a pure chemical compound but a mixture. It is powerfully sternutatory (causes sneezing).
Saponin has been employed in foam baths and in hair shampoo. It may also be used as an emulsifier. They form oil-in-water emulsions and act as protective colloids.

Safety and efficacy

Saponins cause haemolysis of blood and if added to defibrinated blood may cause destruction of the red blood corpuscles. Dermatologically, when applied to normal, healthy, unbroken skin, saponin is not a primary irritant under the normal conditions of use and there is no evidence that it possesses sensitising sensitizing properties.

When taken by mouth, saponins are comparatively harmless. Sarsaparilla, for example, is rich in saponins but is widely used in the preparation of non-alcoholic beverages.

Several in -vivo tests show an anti-inflammatory activity of crude saponin and hederagenin isolated from *Sapindus mukorossi*. Crude saponin and hederagenin inhibited the development of carrageenan-induced oedema in the rat hind paw as well as on granuloma and exudate formations induced by croton oil in rats. Additional crude saponin caused a significant inhibition of the development of hind paw oedema associated with adjuvant arthritis in rats and inhibited the increase in vascular permeability and the number of writhings induced by acetic acid in mice.

Chemistry

Saponins are glycoside compounds that are often referred to as natural detergents because of their ability to form foaming solutions in water. The majority of naturally occurring saponins are of the triterpenoidal type, with the steroidal based saponins forming a much smaller class. The steroidal saponins are based on a backbone of a (C30) triterpenoid saponin nucleus attached via C3 and an ether bond to a sugar side chain, whereas the steroidal are based on a choline (C27) steroid backbone.

The aglycone of the triterpenoidal derivative is known as a sapogenin, whereas the steroidal aglycone derivatives are known as saraponins. The non-saccharide portion (aglycone) of the saponin molecule is called the "genin" or "sapogenin". Saponins are divided into three main classes depending on the type of sapogenin present:

- Triterpene glycosides - there are over more than 350 sapogenins and more than 750 triterpene glycosides in the triterpene glycoside class.
- Steroid glycosides
- Steroid alkaloid glycosides

The ability of a saponin to foam is caused by the combination of the non-polar sapogenin and the water-soluble side chain present on the molecule. The foams tend to be stable and have been used in fire extinguishers as the foaming agent. They are also used to produce foam in beer and are responsible for the natural foam in root beer. They have been used as the foaming agent in toothpaste and are employed by local people where the plants occur as a shampoo and laundry detergent.

Typical soap plants include Yucca (*Yucca schidigera*), Soapwort (*Saponaria officinalis*), Soapbark (*Quillaia saponaria*), Soaproot (*Chlorogalum pomeridianum*) and Soapnut (*Sapindus* spp).

Triterpenoidal Saponins

These are represented by lupeol, α- and β-amyrin. These saponins are found in a whole variety of foodstuffs, e.g. beans, lentils, soya beans, spinach and oats contain significant amounts.

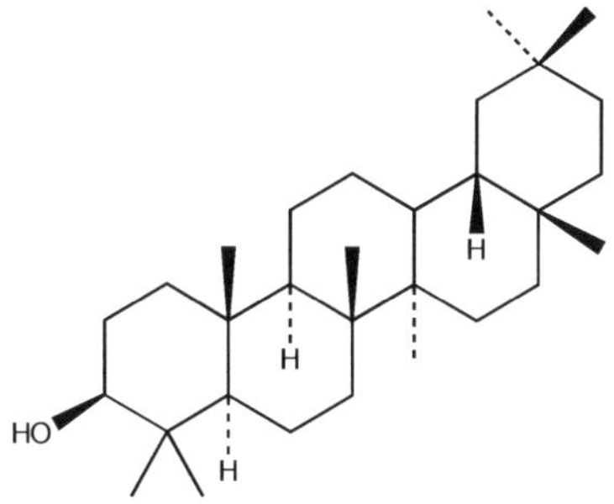

HO H H H

Figure 1. β-amyrin

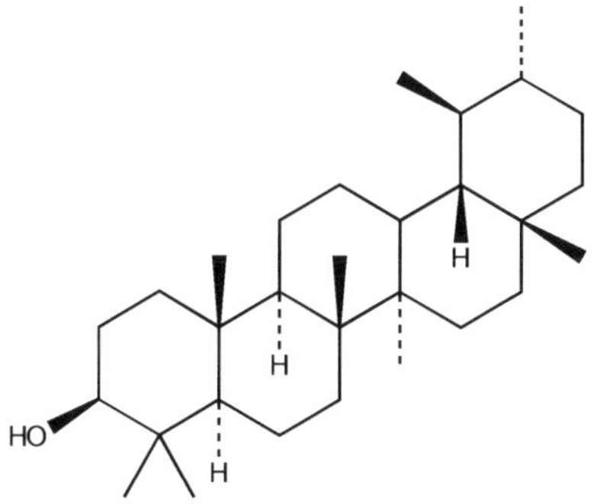

HO H H H

Figure 2. α-amyrin

HO H H H H

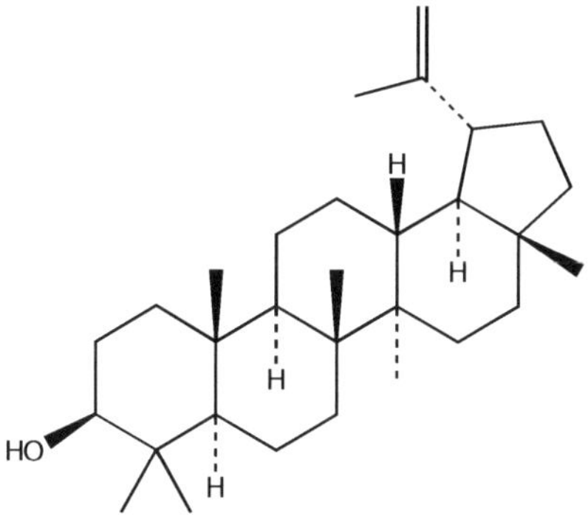

Figure 3g. Lupeol

Sapindus mukurossi
Indian Soapnut

The soapnut tree (*Sapindus* species, ; Family: Sapindaceae) is known as Arishta, Aristaka or Phenila in Sanskrit, and Ritha or Reetha in Hindi. The Sapinaceae family are is a family of about 150 general and 2,000 species, ; the genera include Paullinia (180 spp), Sapindus (13 spp), Cardiospermum, Eriocoelum, Blighia and Radlkofera. [Evans].

The soapberry Soapberry family comprises nearly 2,000 species, which that are primarily tropical.

The Soapnut tree, also known as Soapberry is known by the Indian names Ritha, Doadni, Doda or Dodan. In China and Japan, it has been used as a remedy for centuries. In Japan the pericarp (shell) is called 'enmei-hi', which means "life life-prolonging pericarp,", and in China 'Wu-Huan-Zi', the "non-illness fruit" .

It is found over throughout most of the hilly regions of India. There are two major species. in In the north, it is *Sapindus mukurossi*, ; and in the south, the species it is *S. emarginatus* or *S. trifoliatus*. The common name for these species is Indian filbert or soap nut.

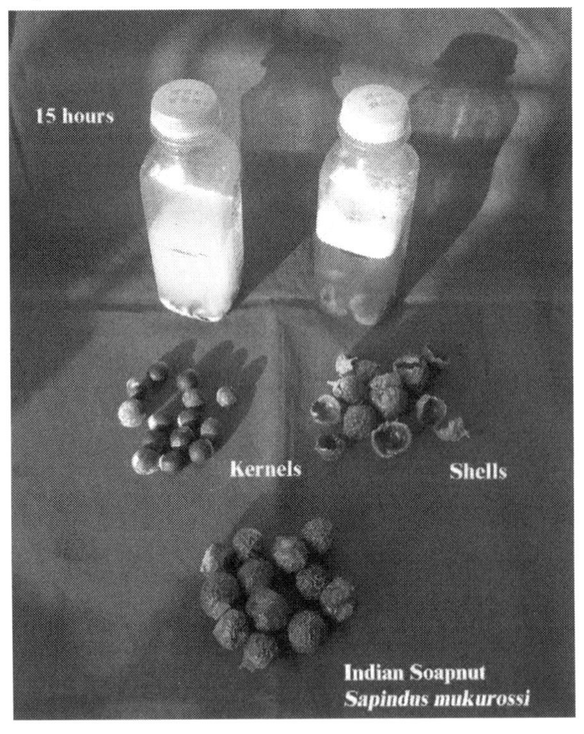

Sapindus mukorossi is one of the most important sources for saponins. The pericarp contains 10-11.5 % saponins, and is used locally for protection against pests and microorganisms.

The powdered fruits (nuts) of these deciduous trees are traditionally used as a mild detergent for washing hair, as a hair tonic and for the management of dental caries. The leaves are used in a bath to relieve joint pain and the roots are used in the management of gout and rheumatism.

The pericarp (shell) is a rich source of saponins that function as a mild detergent, cleanser, surfactant, antimicrobial and anti-inflammatory agent. Since ancient times, *S. mukurossi* has been used in East Asia and the Himalayan regions as a detergent for shawls and silks.

The two triterpenoidal saponins containing hederagenin aglycone moiety have been isolated from the butanol soluble fraction of the fruits (without shells) of *S. mukorossi* Gaertn.

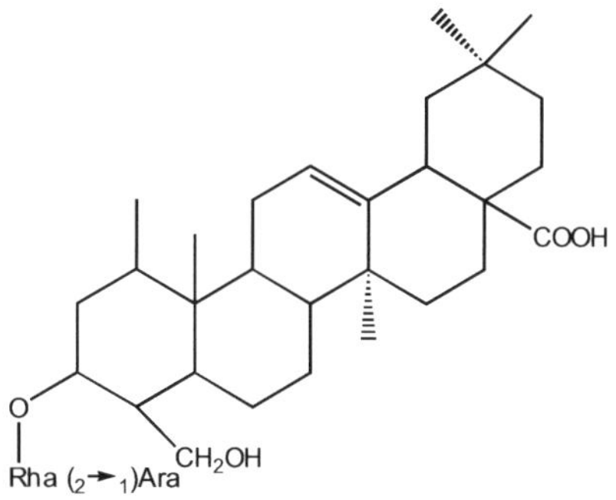

COOH

O

CH_2OH

Rha (2→1)Ara

Figure 4. Sapindoside A

Sapindus spp - Soapnuts, a review of the species

Sapindus delavayi (Chinese name: Pyi-shiau-tzu, ; Japanese name: Hishoushi)

Sapindus detergens is the Chinese Soapberry.

Sapindus drummondii. There is also Western Soapberry (*S. drummondii*) found in Texas. The common name is derived from the fact that the fruits, when crushed in water, create great quantities of suds and were used by West Indian/Mexican natives as a laundry soap, floor wax and varnish.

Sapindus indica is another Soapberry.

Sapindus laurifolia Vahl. [*S. trifoliatus* Linn, *S. marginatus* Willd.] has the Kannada name Antawalamara. It is said that the fruits are crushed and used as a soap.

Sapindus marginatus or Soapberry is also found in the southern United States and Mexico; which is also used as a substitute for soap.

Sapindus oahuensis Hillebr is Lonomea or āulu and is a species of flowering tree in the soapberry family, Sapindaceae, that is endemic to the islands of Kaua΄i and O΄ahu in Hawai΄i. Sadly this species is threatened by habitat loss.

Sapindus saponaria var. *drummondii* (Hook. & Arn.) L. Benson or Western Soapberry. This was previously classified as *S. drumondii* Hook. & Arn. (see above). The other common names include wild China-tree, soapberry, Indian soap-plant, cherrioni, jaboncillo. It is similar to the wingleafed soapberry.

Sapindus saponaria L. var. *saponaria* or Wingleafed Soapberry is found from Arizona to Florida.

Sapindus saponaria L. (Sapindaceae), popularly known as 'sabão-de-soldado' and 'saboeiro', is a medium-sized deciduous tree occurring in the tropics, e.g. America and India, where the fruit is used as a soap and as a remedy against ulcers, scabies, joint pain, inflammations and skin lesions caused by fungi [Tsuzuki et al.]. Extracts from the dried pericarp of *S. saponaria* L. (Sapindaceae) fruits were investigated for their antifungal activity against clinical isolates of yeasts Candida albicans and C. non-albicans from vaginal secretions of women with vulvovaginal candidiasis. Four clinical isolates of *C. albicans*, ; a single clinical isolated of each of the species *C. parapsilosis*, *C. glabrata*, *C. tropicalis*, ; and the strain of *C. albicans* were used. The hydroalcoholic extract was bioactivity-directed against a clinical isolate of *C. parapsilosis*, and showed strong activity. The n-BuOH extract and one fraction showed strong activity against all isolates tested. The saponins isolated showed strong activity against *C. parapsilosis*.

Sapindus trifolatus also contains saponins but has smaller fruits that contain less saponin.

The complete nut weighs around 4 g (the inner kernel around 1.7 g and the outer shell or pericarp 2.3 g). The inner kernel has a tough black outer skin that houses a woody pulp, which that has little or no surfactant qualities. The pericarp, on the other hand, produces a tight creamy foam that has outstanding stability even after 24 hours.

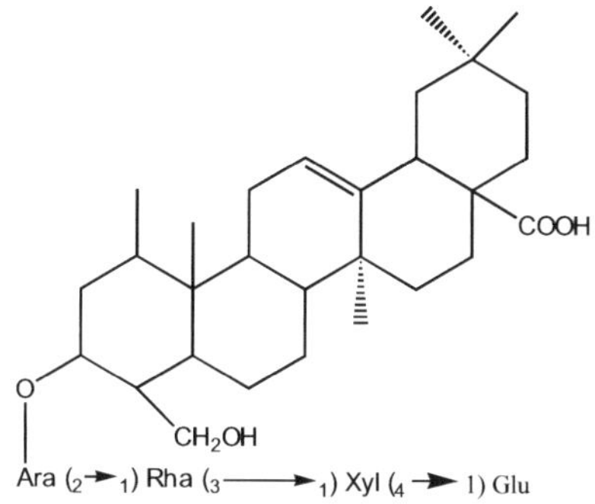

COOH
O
CH_2OH
Ara (2→1) Rha (3→1) Xyl (4→1) Glu

Figure 5. Sapindoside B

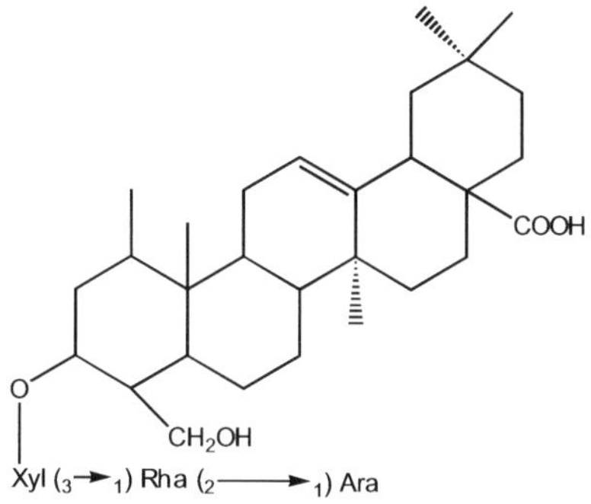

COOH
O
CH_2OH
Xyl (3→1) Rha (2→1) Ara

Figure 6. Sapindoside C

Commercial suppliers

There are a number of good sources for this material with high activity.

Bio-Botanica Inc.
Bio-Saponins

INCI Name: Saponins and is composed of Smilax Aristolochiaefolia (Sarsparilla) Root Extract, Dioscorea Villosa (Wild Yam) Root Extract, Quillaja Saponaria (Quillaja) Bark Extract, Yucca Schidigera (Yucca) Extract and Glycerin.

CAS No: 11006-75-0, 8047-15-2, EINECS No: 232-462-6

Het-Cam Evaluation
The blend was evaluated for Ocular Irritancy Potential with the Hen's Egg Test, utilizing the ChorioAllantoic Membrane (HET-CAM). Under the conditions of the test, the results indicated that the product at 10% showed practically no ocular irritation potential.

Surface Tension
It was also compared to Tween-80 and Sodium sodium Lauryl lauryl Sulfate sulfate (SLS) on an equivalent dry weight basis. The Surface Tension was measured using a Laboratory Surface Tensiometer. The surface tension reduction capability of the blend was similar to Tween-80 and only slightly less than SLS.

Foaming Index
The foaming index is the highest degree of dilution of an aqueous decoction of a material, which that produces persistent foam under the conditions specified. The foaming index was determined from an aqueous solution of the blend, Tween-80 and SLS. The solutions were prepared on an equivalent solids basis. Results showed that the blend was better than the other two at low concentrations (0.1%).

Foam Stability
Foam stability was determined on 1% solutions of the three test materials. The solutions were vigorously shaken to establish "good foam;"; then the foam height was determined as a function of time. The new blend had better foam stability than Tween-80 or SLS.

Kaden Biochemicals through Symrise
Neo Actipone Soap Nutshell

INCI: Sapindus Mukurossi Peel Extract

It is approved by Ecocert SAS F 32600 according to the Ecocert standards for Natural and Organic Cosmetics. The extract is readily biodegradable (OECD 301B).

Neo Actipone® Soap Nutshell is obtained from the soap nut tree (Sapindus mukorossi), which grows in warm temperate to tropical regions. The wild harvested fruits are small orange coloured drupes with 1-2 cm diameter, known as soap nuts 100 % natural detergent. In India the soap nuts are commonly used for washing hair, clothes and for polishing jewelry.

Exhibiting an excellent cleaning performance Neo Actipone® Soap Nutshell is a natural and mild alternative to synthetic surfactants.

Desert King International
Andean QD Ultra

Andean QD ultra is a spray-dried purified (available as powdered and aqueous solution) Chilean Soap Bark Tree (Quillaja Saponaria Molina), with an average composition as follows:

Moisture content	2.0– – 7.0%
Fibre Fiber	0.01– – 0.5%
Proteins	3.5– – 7.0%
Ash	6.5– – 12.0%
Fats	0.01– – 0.5%
Carbohydrates	73.0– – 87.94%

Triterpenic saponins solutions are acidic, but neutralization with alkali (e.g. triethanolamine, sodium hydroxide, arginine etc.) does not modify their stability although a slight browning of the colour is observed that can easily be returned by restoring the pH with lactic acid.

A 10% saponins solution will maintain transparency by successive ethanol addition up to a maximum of 20%, but the opacity increases at higher levels.

Purified quillaja triterpenic saponins, when used alone, have an average foaming power which that is about 40% lower than for most high-foaming surfactants. It was noted that a blend prepared with coco-glucoside, in the ratio 80:20, has foaming power better than when it was prepared with sodium lauryl ether sulfate (SLES). Foam quality remains in all cases, optimal, i.e. small, homogeneous bubbles.

There is positive synergic interaction between natural polymers and Quillaja triterpenic saponins in aqueous solution. Saponins dramatically increased the thickening power of carrageenan-type and xanthan gum polymers. The presence of polymer also increased the foaming power of the saponins solution. The addition of both polymers noticeably incrcased the foam stability, without decreasing its height and quality.

Dr H. Schmittmann
Saponin 374

This is sold as a powder, but we are unable to ascertain the source or activity is unable to be ascertained at the time of writing.

Sabinsa Corporation (through Unifect in UKthe United Kingdom)

Sapindin

A standardized, powdered extract prepared from the fruit (nuts) of Sapindus trifoliatus. Sapindus extract has been used as a detergent for centuries and is listed in the Japanese Cosmetic Ingredient Codex (JCIC) as an ingredient authorized for use in cosmetics by the Ministry of Health and Welfare in Japan. Dermal toxicity studies on the saponin fraction indicated that the material did not show any dermal irritation, sensitization, phototoxicity or photosensitization effects [Tanaka].

INCI Name: Sapindus Trifoliatus Fruit Extract

CAS No: 223748-41-2

Chemistry: Saponins - minimum 60% saponins

Cosmeceutical Applications: Natural surfactant (hair care), antimicrobial. [Sabinsa]

Conclusions

The use of saponins as detergents and emulsifiers has been on the wish list for nearly twenty 20 years and it is only recently that the commercial availability and quality of the material has been developed. It is believed that the potential benefits would justify further investigation in a wide range of personal care products.

Chapter 5

Emulsifiers

Introduction

The subject of emulsifiers when discussing naturals is where many of the difficulties begin. The first realisation of potential problems is when the use of all ethoxylated materials have to be discarded and so the beloved Polysorbates are the first casualty along with all the other non-ionic emulsifiers. The time has come to be selective with your ambitions or a martyr to the cause.

Anionic Emulsions

The first thought is to go back in time and look at some of the emulsions that were around in the 60s and 70s. These were often based on TEA stearate/palmitate soaps and it has to be said that the skin feel was always creamy and luxurious, the application was always smooth and substantial compared to the crisper non-ionic systems.

There was a downside (but maybe that is not true?) which was that the emulsions were always alkaline so pH 7 and higher. It was around about this time that Johnson & Johnson had started the campaign that the pH of the skin was 5.5 and that anything greater than this would not be good for the skin. This discussion is not for these pages, but one should always examine and question old beliefs!

It was said that the anionic emulsions were irritant. In many decades of use, it has to be said that in practice this was not true. In a very few cases there were complaints but in the most part for general toiletries and basic skin care these emulsions were well-liked and well tolerated.

Neutralised fatty acids

The task facing the chemist is to find fatty acids that are of pure vegetable extraction and this is not a problem at all, since there are many vegetable sources of stearic and palmitic acid available on the market. Lotions might consider the saponification of oleic acid fractionated from sunflower oil (although some dermatologists may be concerned that oleic acid may compromise the integrity of the lamellar layers of the skin).

Clearly one cannot use triethanolamine to saponify these fatty acids and so one has to look at sodium or potassium hydroxide to achieve this process. At once there are two problems. Caustic soda and caustic potash are both synthetic, unless you a source can be found that is made from wood ash or from a method using hydroelectricity. It is a real challenge to be pure in these cases and maybe one should accept the term "nature identical" and be content.

Alternatively one can purchase a pre-made organic liquid soap (e.g. Castille Soap) from Stephensons and be rid of the ethics of the problem as to where the caustic was obtained. After all there is an organic version available so that has to be perfect, does it not?

Saponified fixed oil

The use of fixed oil as an emulsifier has a certain appeal. Some formulators have looked at palmates and palm kernelates as a source of their emulsifier. However, in theory any oil that contains a reasonable level of fatty acids of suitable chain length or carbon number will be effective.

The secret is to use just enough to form the emulsion and no more. Use thickeners and humectants to reduce coalescence of the phases and perhaps look at secondary surface active materials to further strengthen the emulsion.

In Chapter 4 we reviewed the triterpenoidal saponins and it might well be that this is the perfect material to act as a secondary emulsifier. There may be sufficient surface activity for it to be the primary emulsifier. These materials are now available as 100% active powders, so will be highly effective (unlike the old days when one used a herbal extract that barely had sufficient material to raise a froth when shaken!)

Lecithin

Lecithin does not only come from eggs, there are sources that come from soy bean, so those interested in animal-free or vegan suitable can use these materials. However, be patient and be warned, for these emulsifiers are not easy to use and may come with added materials that do not sit comfortably with your 'all natural' concept. Lecithin is the popular and commercial name for a naturally occurring mixture of phosphatides (also called phospholipids or, more recently by biochemists, phosphoglycerides). Lecithin does not only come from eggs, there are sources that come from soy bean, so those interested in animal-free or vegan suitable can use these materials.

However, be patient and be warned, for these emulsifiers are not easy to use and may come with added materials that do not sit comfortably with your 'all natural' concept.

Lucas Meyer should be the first port of call as they seem to have the widest catalogue, were the first on the lecithin scene and appear to know more than many other suppliers.

AMISOL™ SOFT is an optimized combination of phospholipids, phytosterols and vegetable lipids for the formulation of O/W emulsions. This lamellar emulsifier mimics the lipid structure of the stratum corneum for a perfect biocompatibility. Very soft, it can be used for the most delicate applications. AMISOL™ SOFT brings a velvety and light feel to the skin.

INCI Name: Behenyl Alcohol (and) Glyceryl Stearate (and) Lecithin (and) Glycine Soja (Soybean) Sterols.

Hydrogenated Lecithin is available as a co-emulsifier, but this is not a material that will ever be found in nature.

The Emulmetiks are used for sprayable milk and the like.

Casein

Casein is a phosphoprotein obtained from milk. It is used to enhance the physical properties of foods, such as whipping, foaming, water binding, thickening, emulsification, texture, and to improve their nutrition. Caseine and its salts are available from a number of suppliers in the industry and should be worthy of exploration.

Beeswax/Borax

These are water in oil emulsions and probably the oldest way of forming an emulsion known to the cosmetic chemist. It was the cold cream used for the removal of make up that first commercially used this technology. The use of beeswax in emulsions was known from the times of the early Greeks, Romans and Egyptians, but it was the addition of the borax that held the secret to emulsion stability.

The borax saponified the free fatty acids in the beeswax to form a soap that then worked as an emulsifying agent. It was the borax that reduced the amount of energy that needed to be put into the emulsion and so reduced manufacturing time and also gave vastly improved emulsion stability. These creams were somewhat stiff and hard and required softening, so they were made cosmetically more acceptable by the addition of petrolatum or vegetable oils. In the example given by Sagarin the formula has spermaceti wax (no longer used) that was obtained from the sperm whale.

Formulation for a typical beeswax-borax cold cream (taken from Sagarin, 1957, p. 83)

Spermaceti	12.5%
White beeswax	12.0%
Almond oil	56.0%
Borax	0.5%
Rose water	5.0%
Distilled water	14.0%
Rose oil	q.s.

The enemy of these water-in-oil creams was oxidation and rancidity of the natural fixed oils. The system was also not self-preserving and so had to be kept sealed and preferably cool to reduce the risk of spoilage organisms from growing in the product. Some have described these systems as smooth spreading and soft, but there are many of us who consider them heavy, occlusive and extremely greasy.

It is at this point that the formulator has come to the end of the 'all natural' line and must compromise if they wish to continue. These suggestions are composed of the soaps, sucrose esters and alternatives to ethoxylated materials that have been chosen as examples of suitable compromises for natural emulsifiers.

C12-16 Alcohols.

There is a natural emulsifier that is composed of a blend of lecithin (and) C12-16 Alcohols (and) Palmitic Acid. Supplier: Lucas Meyer – Biophilic S.

Cetearyl Glucoside.

This material is obtained by combining Glucose with Cetearyl Alcohol (described above). The result is a gentle emulsifier that holds together oil and water in this delicate cream at the same timing leaving the skin very soft and smooth.

Cetearyl Olivate.

It is most often provided as a mixture. Cetearyl Olivate, Sorbitan Olivate is a new non ethoxylated, mild emulsifier. It gives emulsions a pleasant silky touch, moisturising properties and good spreadability, because of its origin from olive oil. The gel network stabilizes the emulsion, absorbs on the skin and reduces the transepidermal water-loss without solubilizing the skin proteins and lipids. CO/SO is free from soap and ethoxylated ingredients. While the Cetearyl Ester derivative stabilizes the liquid crystals, Sorbitan Olivate enhances the emolliency properties and provides an easier dispersion of powders (UV filters and pigments are easily dispersed at high percentages). Furthermore CO/SO combines the liquid crystal structure with the oleic component derived from olive oil, which is responsible for its easy skin penetration, with a soft, silky and smooth after-feel. The substantivity of its composition is very similar to human sebum and provides retention of the skin moisture and increases the active ingredients' resistance to water and/or sweat. CO/SO works by forming liquid crystals in emulsions which increase the hydration capacity on our skin. In this structure the amount of intralamellar water can reach 70% percent, and all this water is immediately available for the skin as soon as the cream is spread and the network is broken. This structure is also responsible for the cooling effect one feels while spreading.

Coco-Glucoside.

Works as a surfactant, foaming agent, conditioner and emulsifier. It helps increase the foaming capacity of a solution, and is particularly useful in hair care products, in which it has the ability to smooth out the hair structure and increase manageability. As an anionic surfactant, it mildly cleanses the skin/ hair by helping water to mix with oil and dirt so that they can be rinsed off. As an emulsifier, it keeps the oil and water parts of an emulsion from separating, and it also enhances the properties of primary cleansing and moisturizing agents contained in a product. This ingredient is compatible with all skin types and gentle enough to be used in baby products. The Duhring Chamber Test lists it as having the lowest irritation score of all common surfactants. You may find it in a variety of cosmetic products such as body wash, shampoo, cleanser, conditioner, hair dye, liquid hand soap, exfoliant/ scrub, acne treatment, facial moisturizer and baby soap.

Corn Starch Modified.

Modified starch, starch derivatives, are prepared by physically, enzymatically, or chemically treating native starch, thereby changing the properties of the starch. Modified starches are used in practically all starch applications, such as in food products as a thickening agent, stabilizer or emulsifier; in pharmaceuticals as a disintegrant; or in paper as a binder.

Decyl Glucoside.

Surfactant; foaming agent, cosurfactant for industrial prods., cosmetics, and dermopharmaceuticals; degreaser; emulsifier; dispersant; wetting agent.

Dilauryl Citrate.

Usually a part of a mixture. Dilauryl Citrate is a highly efficient emulsifier mix developed for emulsion systems (can be cold mix). Function: Emollient/ skin conditioning.

Glyceryl Oleate.

Glycerol monooleate is a clear amber or pale yellow liquid. Glycerol monooleate is also oil soluble, giving it particularly desirable properties as a food emulsion. Glycerol monooleate can form a micro-emulsion in water. The hydrophilic-lipophilic balance (HLB) of glycerol monooleate is 3.8. Glycerol monooleate is prepared by esterifying glycerin with food-grade oleic acid in the presence of a suitable catalyst to form a monoglyceride ester. It is a synthetic surface-active chemical widely used as a nonionic surfactant and emulsant. It is produced by the reaction of glycerine and oleic acid over a catalyst to form a monoglyceride ester. It is used as an antifoam in juice processing and as a lipophilic emulsifier for water-in-oil applications. It also serves as a moisturizer, emulsifier, and flavoring agent. Various forms of glycerol oleate are widely used in cosmetics. It is also widely used as an excipient in antibiotics and other drugs.

Glyceryl Stearate.

Glyceryl stearate is one of the most commonly used skin-conditioning agents and emollients used in skin care products. It is listed in most pharmacopoeias. Mixed with other materials, it is also used as an emulsifier to help prevent oil and water from separating in creams and lotions. The Food and Drug Administration (FDA) includes Glyceryl Stearate (also called glyceryl monostearate) in its list of direct food additives affirmed as Generally Recognized As Safe (GRAS). The safety of Glyceryl Stearate and Glyceryl Stearate SE has been assessed by the Cosmetic Ingredient Review (CIR) Expert Panel. The CIR Expert Panel evaluated the scientific data and concluded that Glyceryl Stearate and Glyceryl Stearate SE were safe for use in cosmetics and personal care products. Function: Emollient/ emulsifying.

Hydrogenated Lecithin.

Within the lecithin derivatives, there is hydrogenated soy lecithin, which is hydrogenated at the site of the unsaturated bonds to the alkyl groups. It can be used as emulsifier and moisturizer for creams and lotions. It is considered a natural emulsifier, dispersing agent for pigments, preparation of liposomes, is EO free and from a vegetable source. The lecithin source promotes subcutaneous absorption, has long-lasting moisturisation, helps reduce skin irritation caused by surfactant. Supplier: Barnet Products Corporation - Lecinol S-10. Supplier: Lucas Meyer: Hydrogentated Lecithin (and) C12-16 Alcohols (and) Palmitic Acid – Biophilic-H

Hydroxystearyl Alcohol and **Hydroxystearyl Glucoside**.

The emulsifier is produced with vegetable oil extracted from Ricinus seeds. It is free of preservatives and does not require a solvent or any other synthetic products in manufacturing. The emulsifier reportedly absorbs quickly into the superficial layers of the skin without causing a "soaping effect" (white traces) in natural emulsions rich in vegetable oils and natural gum. It reportedly emulsifies all oil types including esters, mineral oil, vegetable oil and silicone. The emulsifier promotes the lamellar phase in the continuous aqueous phase, contributing to stratum corneum hydration and skin comfort. The emulsifier is said to spread easy, imparting a light and soft skin feel. The company found it to have a long-lasting, eight hour moisturizing effect in vivo. An in vivo test conducted by the company showed the emulsifier to improve skin state and appearance perceived by 91% of volunteers after 14 days of use. Simulgreen 18-2 from Seppic

Inulin Lauryl Carbamate.

Inulin lauryl carbamate is a natural stabilizer and emulsifier derived from chicory, that keeps oil in water or water in oil lotions and creams consistent and smooth. It also provides a moisturising effect. There are no adverse effects expected or reported from the use of this naturally derived material. Function: Emulsion stabilising/ surfactant. The actual or estimated LD_{50} value: 2,000 mg/kg body weight.

Methyl Gluceth-10.

Poly(oxy-1,2-ethanediyl), .alpha.-hydro-.omega.-hydroxy-, ether with methyl .beta.-d-glucopyranoside (4:1). It is a versatile, gentle, effective humectant, emollient and foaming agent. It imparts a smooth and gentle feel to the skin. It is an extremely effective humectant recommended for use in both rinse off and leave on skin care systems such as lotions, creams and body cleansing formulations. It is derived from corn. It is an easy-to-use liquid form that can be readily formulated into a wide range of products. It can be formulated with other humectancy agents such as glycerin. It often demonstrates synergistic performance in these systems. In addition it improves the sensory properties of these formulations. No adverse effects have been observed or are expected from the topical application of this material topically at the levels employed. Function: Emulsifier/ humectant/ moisturiser.

Myristyl Myristate.

Myristyl Myristate is a wax ester. It is originally formed as a white or yellowish waxy solid, and used as a skin conditioning agent, emulsifier and opacifier in skin care products and cosmetics. It provides a pleasant, soft feel to formulations and provides a dry, powdery after-feel to lotions and creams. Myristyl Myristate is an excellent emulsion enhancer that imparts a white opacifying effect or glossy appearance, and is an effective thickening agent as well. Myristyl Myristate is considered safe and is approved by the CIR for use in cosmetics. Some studies have linked it to mild irritation when applied directly to the skin but this isunlikely in normal use in creams and emulsions. Function: Emollient/ opacifying/ skin conditioning. Myristyl Myristate. CIR: Maximum "as used" concentration for safe as used conclusion: up to 20%.

Polyglyceryl-4 Laurate/Succinate (and) Aqua.

The supplier says "NatraGem E145 is able to create stable O/W creams and lotions with both low and high polarity oils. In order to demonstrate this, NatraGem E145 has been compared with two commonly used emulsifiers, Polysorbate 60 and Steareth-20. Two simple emulsions with different oil phase polarities have been createdwith each emulsifier and subjected to three months stability at 40°C. NatraGem E145 is able to produce stable emulsions that are comparable to Steareth-20 and more stable than Polysorbate 60. HLB 14.5 (experimentally determined)". Supplier - Croda. [author's comment: this material is not found in nature but it is a half-way path on the route to naturals as there is no other totally natural solution available].

Sodium Cocoyl Glutamate.

Sodium salt of coconut fatty acid, used as emulsifier and lightly sudsing surfactant. Function: Cleansing/ surfactant.

Sodium Oleate.

9-Octadecenoic acid, sodium salt, (Z)-. FDA PART 186 - Indirect food substances affirmed as generally recognized as safe. Subpart B--Listing of Specific Substances Affirmed as GRAS Sec. 186.1770 Sodium oleate. Sodium oleate is the sodium salt of oleic acid, a monounsaturated fatty acid. This anionic surfactant and emulsifier is a component of commercial soaps. As a fatty acid, sodium oleate is not generally found free in nature, but rather occurs as a component of more complex naturally occurring lipids. In its isolated form, however, the substance exudes a tallow-like scent and is a white crystalline solid at room temperature. Function: Cleanisng/emulsifying/ surfactant/ viscosity controlling.

Sodium Olivoyl Glutamate

Glutamic acid, N-olive oil acyl derivs., sodium salt. Olivoil Emulsifier is a new emulsifier based on Olivoil Glutinate, which is a new non ethoxylated (PEG free) vegetable derived surfactant combining the very exceptional fatty acid profile of olive oil with the beneficial hydrolysed wheat proteins, achieving a perfect compatibility with the human body (high tolerance for the skin) and with the environment (high biodegradability). By combining this new natural derived molecule with olive oil waxes, it is possible to achieve an amphoteric system adapted to the formulation of oil in water emulsions. Traditional ethoxylated emulsifiers are Chemicals which are highly irritant for the skin. Olivoil Emulsifier, based on vegetable materials, combines high tolerance for the human body and the environment to high efficiency to form stable and fine emulsions. Emulsions based on Olivoil Emulsifier don't interfer with the hydrolipidic balance of the skin and leave a soft and smooth feeling while bringing emoliency. This gentle efficiency is of particular interest in the formulation of moisturizing and anti-aging emulsions, and skin care products in general. Function: Cleansing.

Sorbitan olivate

It is most often provided as a mixture. Cetearyl Olivate, Sorbitan Olivate is a new non ethoxylated, mild emulsifier. It gives emulsions a pleasant silky touch, moisturising properties and good spreadability, because of its origin from olive oil. The gel network stabilizes the emulsion, absorbs on the skin and reduces the transepidermal water-loss without solubilizing the skin proteins and lipids. CO/SO is free from soap and ethoxylated ingredients. While the Cetearyl Ester derivative stabilizes the liquid crystals, Sorbitan Olivate enhances the emolliency properties and provides an easier dispersion of powders (UV filters and pigments are easily dispersed at high percentages). Furthermore CO/SO combines the liquid crystal structure with the oleic component derived from olive oil, which is responsible for its easy skin penetration, with a soft, silky and smooth after-feel. The substantivity of its composition is very similar to human sebum and provides retention of the skin moisture and increases the active ingredients' resistance to water and/or sweat. CO/SO works by forming liquid crystals in emulsions which increase the hydration capacity on our skin. In this structure the amount of intralamellar water can reach 70% percent, and all this water is immediately available for the skin as soon as the cream is spread and the network is broken. This structure is also responsible for the cooling effect one feels while spreading.

Sorbitan Stearate.

Sorbitan stearate is an oil soluble emulsifier used to prevent the oil and water phases from separating in a cream or lotion. The Food and Drug Administration (FDA) allows Sorbitan Stearate to be added to food as a multipurpose food additive. Sorbitan Stearate is also on FDA's list of and flavoring substances and adjuvants that may be added to food.

Stearyl Citrate.

Stearyl citrate is used as an oil-soluble chelating agent. Stearyl group is lipophilic and carboxyl groups are active in chelating. This function is useful for various oil stability for long shelf-life. It is used as a surface lubricant in the manufacture of paper contacting fatty acids and water. It is used as a plasticizer for food contact plastics. It is used in polymeric coatings. Used in foods E484, Stearyl citrate, Stearyl citrate emulsifier, rated as SAFE. Function: Emollient/ skin conditioning.

Sucrose Dilaurate

The Food and Drug Administration (FDA) has approved the use of Sucrose Fatty Acid Esters as direct food additives. Sucrose oligoesters (average degree of esterification from four to seven) are also approved for use as direct food additives. The safety of many fatty acids, including Stearic Acid, Lauric Acid Myristic Acid and Coconut Acid, has been assessed by the Cosmetic Ingredient Review (CIR) Expert Panel. The CIR Expert Panel evaluated the scientific data and concluded that the fatty acids were safe for use in cosmetics and personal care products. Please search this website for the specific fatty acid for more information about the CIR reviews. Function: Emulsifying/ surfactant/ skin conditioning/ emollient.

Sucrose Laurate.

Sucrose laurate. The Food and Drug Administration (FDA) has approved the use of Sucrose Fatty Acid Esters as direct food additives. Sucrose oligoesters (average degree of esterification from four to seven) are also approved for use as direct food additives. Sucragel AOF is based on sweet almond oil and is recommended for use with vegetable oils to obtain clear oily gels. It can also be used to make milks, lotions and creams by adding an oil phase followed by a water phase. It is considered by many to be a natural emulsifier. . INCI: Prunus Amygdalus Dulcis (Sweet Almond) Oil & Glycerine & Aqua & Sucrose Laurate. Supplier: Alfa Chemicals

Sucrose Palmitate.

The Food and Drug Administration (FDA) has approved the use of Sucrose Fatty Acid Esters as direct food additives. Sucrose oligoesters (average degree of esterification from four to seven) are also approved for use as direct food additives. The safety of many fatty acids, including Stearic Acid, Lauric Acid Myristic Acid and Coconut Acid, has been assessed by the Cosmetic Ingredient Review (CIR) Expert Panel. The CIR Expert Panel evaluated the scientific data and concluded that the fatty acids were safe for use in cosmetics and personal care products. Function: Emulsifying/ surfactant/ skin conditioning.

Sucrose Polystearate.

100% vegetable origin, and considered a natural emulsifier. Plantec Natural Emulsifier HP10 is a non-ionic emulsifier, PEG-free. This emulsifier is characterised by a high HLB (HLB: 9±1), it is adapted for use in O/W emulsions. It is an association between sucrose esters and olive unsaponifiables. Sucrose esters are natural molecules well-known for their exceptional sensitive skin feel and their non-irritating activity. Olive unsaponifiables, the active part of olive oil, present a high affinity with human skin lipids, helping to hydrate and restore the cutaneous barrier. Due to the waxes naturally present, olive unsaponifiables increase the emulsifying power. Supplier: CRM International - Plantec: Natural Emulsifier HP10: Sucrose Polystearate (and) Cetearyl Alcohol (and) Olea Europaea (Olive) Oil Unsaponifiables.

Sucrose Stearate.
The Food and Drug Administration (FDA) has approved the use of Sucrose Fatty Acid Esters as direct food additives. Sucrose oligoesters (average degree of esterification from four to seven) are also approved for use as direct food additives. The safety of many fatty acids, including Stearic Acid, Lauric Acid Myristic Acid and Coconut Acid, has been assessed by the Cosmetic Ingredient Review (CIR) Expert Panel. The CIR Expert Panel evaluated the scientific data and concluded that the fatty acids were safe for use in cosmetics and personal care products. Please search this website for the specific fatty acid for more information about the CIR reviews. [CAS: 25168-73-4; EINECS: 246-705-9]. Function: Emollient/ emulsifying/ surfactant/ skin conditioning. The actual or estimated LD_{50} value: 20,000 mg/kg body weight. AICS status (NICNAS Australia): AICS Compliant. Oral LD_{50} value (rat): 20,000 mg/kg. Comedogenic value: 0.

Conclusion

The choice of natural emulsifiers will remain the greatest challenge to the formulator.

Chapter 6

Natural Preservatives

Natural preservation is one of the most followed topics in the industry and attracting the attention of even the most traditional of formulating chemists. This chapter looks at the theoretical development of a natural preservative system. The traditional methods of preservation, many taken from the food industry, are summarized and the latest raw materials offered within the industry are evaluated.

Warning: Parabens are very natural, they do occur in nature and a vast sum of money has been spent proving how wonderfully safe and trustworthy this group of preservative have been. They were the preservative of choice for almost all dermatologists, because they rarely produced adverse skin reactions and are well tolerated by even the most sensitive of skins. Pack copy that claims "no parabens" implies that parabens may have some harmful or potential adverse effects and this is not true. Claims on packaging that are not true infringe the Trade Descriptions Act and numerous Codes of Advertising Practice. Prosecution for such an unfair and unjustified claim is therefore a real possibility and the Trading Standards in the UK may well take a case to court. [A similar case did set a precedent when "no lanolin" was claimed and the manufacturers successfully filed a damages suit].

Introduction

There was a time when the mention of natural preservation would have been greeted with skepticism, however, today the views of many scientists has changed, and the use of plant materials and natural molecules is seen as a distinct possibility in their formulations. This author has been writing and updating the benefits of natural preservatives for more than a decade [Dweck, 1994; Dweck, 1995; Dweck, 2003; Dweck, 2005]. Costs for these new materials is high, but more than compensated for by the additional marketing claims such as "preservative-free" or "contains no synthetic preservatives." The paper reviews the most commonly used methods of preservation that are available to the formulator. The food and beverage industry may be called upon for many of these examples.

Legal Position

No preservative may be used that does not appear in Annex V Part 1 or 2 of the Regulation of the European Parliament and of the Council on cosmetic products (recast) 2008/0035 (COD) dated 10 November 2009 (finally as 1223/2009 on 30 November 2009).

Regulation (Ec) No 1223/2009 of the European Parliament and of the Council of November 30, 2009, on cosmetic products (recast) [*Journal of the European Union*] has recently consolidated the original Directive and all of its amendments shown above. Annex VI has now become Annex V. This legislation will be adopted by the ASEAN countries and South Africa. In the United States there is no restricted list of preservatives and one should look to the CIR (Cosmetic Ingredient Review) panel in order to find the preservative restrictions.

However, these natural materials are not legislated as preservatives when used for their beneficial effect on the skin that may coincidentally have a positive effect on the total preservative requirement of the formulation. Of course, no material appearing in Annex II may be considered.

The food industry often uses a preservation technique known as the "hurdle approach" where a number of different materials and factors that might eliminate organisms on their own if used at a high level are used in combination, but at significantly lower levels. This gives rise to a series of hurdles over which most organisms are unable to cross. The idea of using a whole variety of these hurdles to slowly weaken each organism, but at individual levels that would be ineffective, is an almost alien concept to the cosmetic and toiletry industry.

Sugar
High levels of sugar can preserve against spoilage organisms. This is a technique used in homemade jams, preserves, sweet pickles and marmalades. It is also an important factor in the preservation of boiled sweets and chocolates. Increasingly, it will be noticed that many products now have to be kept in the refrigerator of freezer once opened, because sugar has been replaced by an artificial sweetener that may be cheaper and healthier to eat, but has no preserving qualities.

Honey
Honey in its undiluted form is also a natural preservative and, indeed, there are many learned papers citing honey as a viscous barrier to bacteria and infection.

Alcohol
Not all organisms are detrimental. The production of alcohol from sugar by yeast is an industry in its own right. A wine that is carefully produced using sterile equipment and fermented to 13% by volume will just about resist further infection from external organisms once that ferment has completed. The time during the fermentation of the must is when the must is most vulnerable to infection. The naturally produced fermentation-grade alcohol may be concentrated by distillation and used as a natural preservative in toners, aftershaves and colognes. Alcohol at a level of 15% is effective, but 20% is more assured.

The denaturant present in the alcohol is not natural and there is a need to return to quassin, the bitter substance present in Quassia (*Picraena excelsa*), which used to be acceptable as a denaturant.

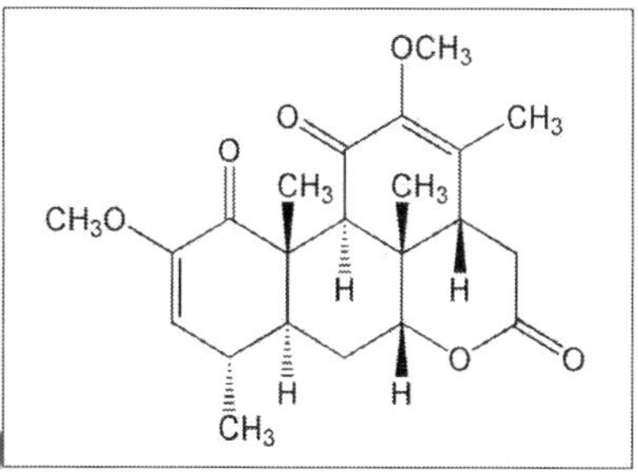

Quassin

Sadly, the Customs & Excise might not see the antique solution to denaturing with quite the same enthusiasm, especially since those in the industry lobbied so strongly to be able to use Bitrex. (INCI: Denatonium benzoate or N-(2-((2,6-dimethylphenyl) amino)-2-oxoethyl)-N,N-diethyl-, benzoate).

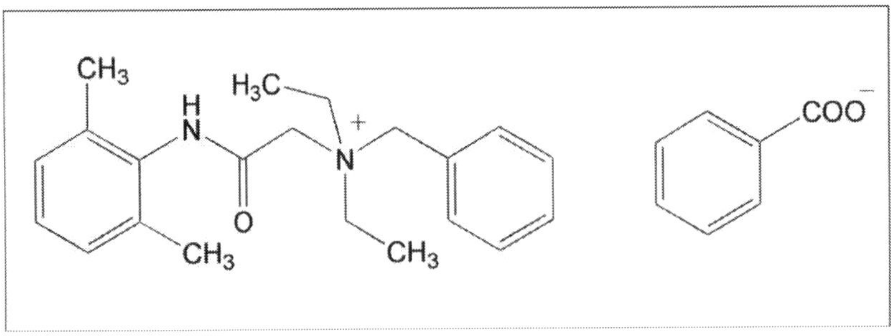

Bitrex or Denatonium Benzoate

Heat

Heating, cooking, steaming and pasteurization are other natural forms of preservation that will sterilize products, especially where that product is designed as a one-shot use product—for example, a vial or a sachet. Alternatively, once opened, the product can be stored in the refrigerator or freezer to prevent microbiological degradation.

Cold

Placing a product in the cold merely stops the clock on microbiological growth and this is perfectly fine, provided the product was fairly clean microbiologically when it was placed in the cold. The discovery of prehistoric animals and mummified people in various frozen wastes show that this method of preservation is effective.

Desiccation

Removing water from a product, or making it totally dehydrated, will greatly reduce the possibility of spoilage, however, it must be recognized that the presence of spore-bearing organisms could become active once that water is reintroduced.

Anhydrous

In a similar vein, one could make products with materials that do not contain any traces of water and deliberately design and formulate a totally anhydrous product. However, creams that can be finished by the consumer, by introducing water to the blend of oils, fats and waxes, are prone to the same restrictions as the desiccated products.

Salt

The use of extreme levels of salt as used by the ancient mariners to preserve their meat is effective and it may be very likely that the preservation of the Egyptian mummies was, in part, achieved with the 40-day treatment in *natron* (a local salt that osmotically drained the tissues of water).

Acid pH

The preservative activity can be boosted by operating at as low a pH as possible. Natural acidity could be obtained from one of the many of the alpha hydroxy acids (AHAs) that are obtained from citrus species, where the major components are citric and malic acids.

Alkaline pH

High values of pH also inhibit the growth of organisms and it is not necessary to preserve bar soaps, although it might be necessary to put a fungicide in the cardboard carton that surrounds a naked soap, as the moisture that leaves the soap can be the cause of mold and fungal growth in the card.

Chelating Agents

In addition to formulating at low pH, chelating agents such as phytic acid extracted from rice bran could be added to enhance the activity of the natural preservative. There are a number of suppliers for this material.

Another option is to use a naturally produced material such as sodium gluconate. Sodium gluconate is the sodium salt of gluconic acid, produced by the fermentation of glucose. It is a white crystalline powder, very soluble in water. Non-corrosive, nontoxic and readily biodegradable (98% after two days), sodium gluconate is an effective chelating agent especially in alkaline and concentrated alkaline solutions.

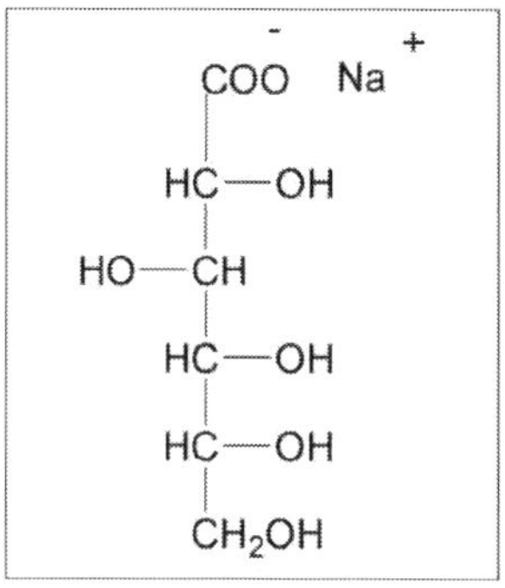

$COO^- \; Na^+$
HC—OH
HO—CH
HC—OH
HC—OH
CH_2OH

Sodium Gluconate

It forms stable chelates with calcium, iron, copper, aluminum and other heavy metals. It is as effective as other chelating agents, such as ethylenediaminetetraacetic acid (EDTA) and related salts. Aqueous solutions of sodium gluconate are resistant to oxidation and reduction, even at high temperatures. However, it is easily degraded biologically (98% after two days) and thus presents no wastewater problem. It is used in the food industry.

Chelating agents interfere with the cellular membranes that surround all organisms and weaken them by depriving them of the trace elements that they need for cellular function. Extremely high levels of chelating agent have been used as preservatives on their own.

Antioxidants
Antioxidants such as natural tocopherol and ascorbic acid will further aid in preservation, as well as in reducing the potential rancidity. It has to be reminded that ascorbic acid is extremely unstable in water and is particularly sensitive to copper, iron or nickel, and will brown quite quickly in aqueous systems.

Glycerin
High levels of vegetable glycerin, up to 15–20%, will also have a preservative effect, similar to that effect obtained by the use of high levels of sugar. There is a downside to these high levels and that is increased stickiness.

Emulsion Form
It has been argued that the formula comprised of a water-in-oil emulsion, where the oil is the continuous phase, is far less likely to be subject of attack by spoilage organisms. This might be true, but it certainly does not exclude the use of a preservative system. It does, however, form another link in the hurdle approach to preservation.

Emulsifier Type
A material was marketed many years ago called Lauricidin, which was glyceryl laurate and was said to mimic the sterile and protective action found in mother's milk. It had been found that the properties which determine the anti-infective action of lipids are related to their structure: e.g., free fatty acids and monoglycerides. The monoglycerides are active; diglycerides and triglycerides are inactive. Of the saturated fatty acids, lauric acid has greater antiviral activity than either caprylic acid (C_8), capric acid (C_{10}), or myristic acid (C_{14}). Lauric acid is one of the best "inactivating" fatty acids, and its monoglyceride is even more effective than the fatty acid alone.

The system has been shown to work, but the formulating is difficult and not always predictable. It might be a good solution perhaps for those with a large research department and plenty of human resources.

Dr. Straetmans makes a series of natural solutions, among the emulsifiers is glyceryl oleate pyroglutamate (Dermosoft GMO P-30).

Plant Self-preservation
Plants that are living and connected to their root systems remain vibrant and resistant to attack by yeast, mold and bacterial attack. This is because every plant contains its own preservative system that keeps it fresh and vibrant. Smell a rose and then cut it and smell the same rose after as short a time as one hour. The smell of the rose will have changed drastically as the chemical composition of the fragrant parts degrades with great rapidity.

In fruits, the seeds develop until maturity and then the fruit falls from the plant. Immediately the system that had protected the fleshy pulp surrounding the fruit from attack by yeasts, molds and organisms ceases to function and now disintegrates. It is this breakdown of the fruit that provides nutrients for the seed to germinate and prosper.

If it were not for the failure of the natural preservation system, plant material would not rot. Plants that contain high levels of essential oil, such as pine needles and firs, take a very long time for clippings to rot down—hence the advice that they should not be used for composting.

The chemicals present in all parts of the plant protect it from the environment. However, examples can be seen where tampering with the plant leads to a reduction in the efficacy of this natural mechanism.

It is concluded that the chemical constituents within each plant clearly differ in composition. Furthermore, in certain species, there may be a chemical or group of chemicals present in the plant that are capable of killing microorganisms. This chemical composition varies according to whether the plant is alive or dead, and in certain/most plants will vary according to season.

In many cases, when these plants are extracted, it is found that the extracts are capable not only of resisting certain spoilage organisms, but also in some cases can actively destroy them. The time and speed of extraction of the fresh plant is often critical if the preservative activity is to be retained.

Commercial Products

There are natural preservatives available on the market that are not legal, strictly speaking, since they have no entry in Annex VI as a permitted preservative. However, the use of a plant for its marketing claim, or for other functional benefits smudges the issue. One may use a number of plant derivatives as fragrance components and coincidentally achieve a lower overall preservative requirement for the product in which they are used.

The debate over the safety of para-bens (and the ludicrous report that they cause breast cancer) shows the ignorance of phytochemistry, since parabens abound in nature. There are many cases where plants may contain paraben-type compounds in addition *Methylparaben* to other functional actives, and the difficulty is to decide whether the botanical is being used as a preservative or for other legitimate and perfectly legal benefits.

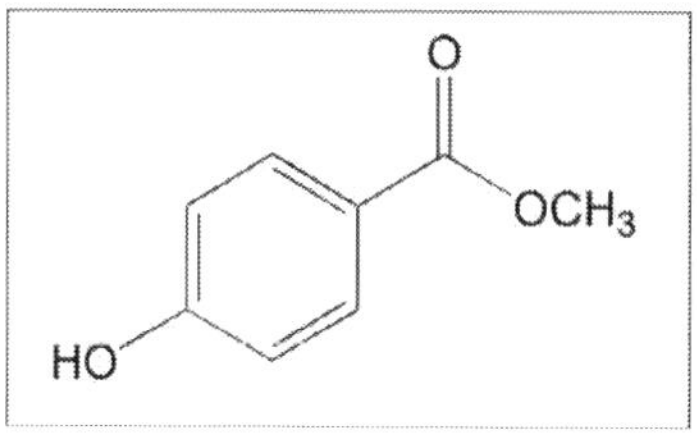

A Review of the Solutions Already Present on the Market

In days of old, wine and water was stored in silver vessels because it had been observed that the keeping time vastly improved when compared to earthenware jugs and pots. This is somewhat surprising; one might have expected that the glazes on those pots, often rich in lead, might have further aided preservation.

Silver Chloride

The modern preservative is comprised of silver chloride (20%) deposited on a substrate of titanium dioxide. It does appear in Annex VI, but is prohibited for use in products for children under three years of age. It is not allowed in oral products and those products intended for application around the eyes and lips. It is limited to 0.004% when calculated as silver chloride.

John Woodruff (Creative Developments Limited) wrote, "*A silver chloride titanium dioxide composite is the basis of the JMAC range of antimicrobial products produced by Johnson Matthey in association with Microbial Systems International.*

JMAC is a silver chloride/titanium dioxide composite and JM ActiCare is a suspension of particles of a silver chloride/titanium dioxide composite in a water/sulfosuccinate gel that improves its activity against yeasts and molds. Although recommended for most types of leave-on and rinse-off products, JM ActiCare is especially useful for preserving products containing finely dispersed particulates such as sunscreen preparations based on microfine inorganic oxides and makeup preparations.

When incorporating JM ActiCare into the formulation it is important not to add it to the oil phase as its activity is seriously affected if the particles become coated with oil. It is stable across the pH range 3–10, but it is affected by xanthan gum that binds the silver, and by some AHAs, but not lactic and glycolic acids. Materials such as ascorbic acid and sodium metabisulfite that may reduce the silver chloride need careful evaluation, and strong cationic materials may also be detrimental."

Nature Identicals

There are a number of materials already allowed in the legislation that occur naturally in nature. These include benzoic acid (limit 0.5% as the acid) and benzyl alcohol (limit 1%). They can be obtained naturally from natural sources such as balsamic resins, but the price is expensive.

Geogard™ ECT from Lonza has a composition of benzyl alcohol, salicylic acid, glycerin and sorbic acid is a typical "natural" blend that offers a broad spectrum activity over a broad range of pH values 3 – 8.

Rokonsal BSB from ISP is composed of benzoic acid and sorbic acid in a mixture of glycerin and benzyl alcohol. It is used at pH <5.

Benzoic acid is moderately good against Gram+ve bacteria, yeast and molds, but moderately poor against Gram-ve bacteria, while benzyl alcohol is good to very good against Gram+ve, moderately poor against Gram-ve, poor versus fungi and moderately poor against yeast.

Sorbic acid (and its salt, potassium sorbate) is found in nature (originally from *Sorbus aucuparia* or Rowan berry) but can be purchased synthetically and used up to 0.6%. It is moderately effective against all bacteria and good against fungi and yeasts.

Salicylic acid. Occurs in the form of esters in several plants, notably in wintergreen leaves (*Gaultheria procumbens*) and the bark of sweet birch (*Betula lenta*). It is quite poor against bacteria, yeast and molds and needs to be used in acidic situations, but it does seem to have quite a good synergistic effect when used with other preservatives.

Formic acid (systematically called methanoic acid) is the simplest carboxylic acid. It is an important intermediate in chemical synthesis and occurs naturally, most famously in the venom of bee and ant stings. In nature, it is found in the stings and bites of many insects of the order Hymenoptera, mainly ants. Its name comes from the Latin word *formica*, meaning "ant," referring to its early isolation by the distillation of ant bodies. It is moderately effective against bacteria, but is poor against yeasts and molds and once again operates best in pH range 4-6.

Phenylethyl Alcohol. The *Cosmetic Ingredient Review* (USA) has determined that phenethyl alcohol is safe for use in personal care products at concentrations <1%. It is a component of rose oils and has good antimicrobial activity. It is normally produced synthetically because the price of natural material is very expensive. PEA prevents the growth of Gram-negative organisms by disrupting the structure of lipids in the Gram-negative membrane. It also can hamper protein synthesis. PEA is present in this medium at 0.025%; if the concentration was higher. it would affect both Gram-positive and Gram-negative cells.

Piperonal (Heliotropine). Piperonal (heliotropine, protocatechuic aldehyde methylene ether) is an aromatic aldehyde that comes as transparent crystals, $C_8H_6O_3$, and has a floral odor. It is used as flavoring and in perfume. It can be obtained by oxidation of piperonyl alcohol. It is also a minor natural component of the extract of vanilla. Heliotropine is an extract of sassafras used for manufacturing perfumes and soaps. It is a derivative of safrole, a naturally occurring aromatic obtained from botanical sources such as *Cinnamomum petrophilum* and *Sassafras albidum*. Heliotropine (piperonal) is an aldehyde form of piperonic acid, which is one of the sharp-tasting constituents in black pepper and used as an insecticide.

Conarom H-3 from ISP provides this material but in combination with PPG-2 methylether and phenylpropanol. This might disappoint the purist "natural" chemist.

β-Caryophyllene. Caryophyllene, or (−)-β-caryophyllene, is a natural bicyclic sesquiterpene that is a constituent of some essential oils, especially clove oil and the oil from the stems and flowers of *Syzygium aromaticum*. It is usually found as a mixture with isocaryophyllene (the *cis* double bond isomer) and α-humulene (obsolete name: α-caryophyllene), a ring-opened isomer. Caryophyllene is notable for having a cyclobutane ring, a rarity in nature. Caryophyllene is one of the chemical compounds that contributes to the spiciness of black pepper.

Germacrene-D. Germacrene is a subset of volatile organic hydrocarbons called sesquiterpenes and is typically produced in a number of plant species for its antimicrobial and insecticidal properties, though it also plays a role as insect pheromones. The two prominent molecules are germacrene A and germacrene D.

Borax. Fungicide and mold remover and multipurpose cleanser, but must be used with extreme care.

Lactoperoxidase and glucose

A two-pack system consisting of lactoperoxidase, glucose oxidase and glucose does not appear in Annex VI, but has found a good following among the green brigade. It is very fiddly to work with and has to be premixed just prior to addition to the finished batch. Its mechanism is said to mimic the conditions that keep a cow's udder free of infection while it is suckling its calf.

This material was formerly called Myavert C, but is now renamed Biovert.

The "Illegal" Preservatives

Citrus Seed Extracts
There are other dodges used by the 'green' formulators in their quest to avoid the preservatives listed in Annex VI that do not occur in nature. Citrus fruits have always been a useful source of alpha hydroxy acids, of fragrant essential oils and useful astringents.

Everything in the fruit is useful: the juice for its vitamin C claims, the peel for its fragrant essential oil and claims of "zest", the flowers' yield of exquisite and very expensive essential oil called neroli, and the seeds yield of an antibacterial, which is either naringenin, hesperedin or hesperitin, depending on the citrus species chosen.

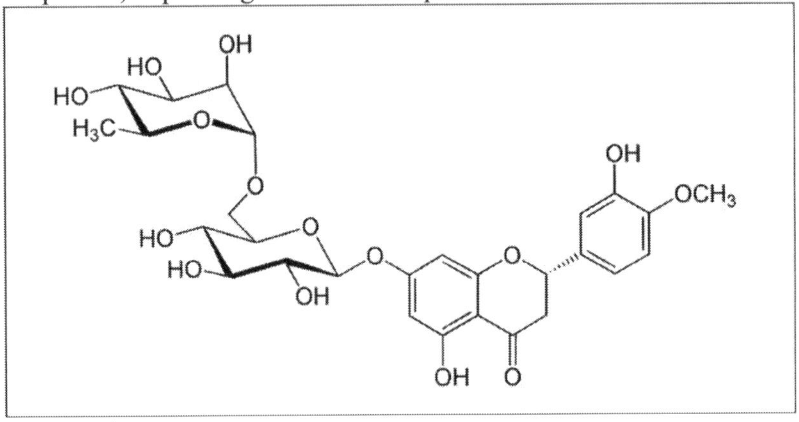

Hesperidin

This can be slipped in with other citrus components and is conveniently lost among the myriad of exotic ingredients. As if by magic, the need for a preservative has disappeared. It is not simple to formulate with these types of materials and you have to do a lot of experiments, since not all systems are compatible, but success can be achieved.

The introduction of a natural preservative with the INCI name Citrus Grandis (Grapefruit) Fruit Extract called Citricidal, which was available from a number of sources, gave great hope for the future of natural preservation.

Hesperitin

However, reports started to circulate that the material had been contaminated with quaternary materials, and that it was this material that was giving the extract its preservative qualities. The original thought, that the molecules above were responsible for the action, was dashed when a feature was found on the Internet at a site that has since disappeared:

"*Grapefruit Extract (GSE) is made by first converting grapefruit seeds and pulp into a very acidic liquid. This liquid is loaded with polyphenolic compounds including quercetin, helperidin, campherol glyceride, neohelperedin, naringin, rutinoside, poncirin, etc. The poly phenols themselves are unstable but are chemically converted into more stable substances that belong to a diverse class of products called quaternary ammonium compounds.*"

The Web article then went on to say, "*Some quaternary compounds, benzethonium chloride and benzalkonium chloride, for example are used industrially as antimicrobials but are toxic to animal life ...*"

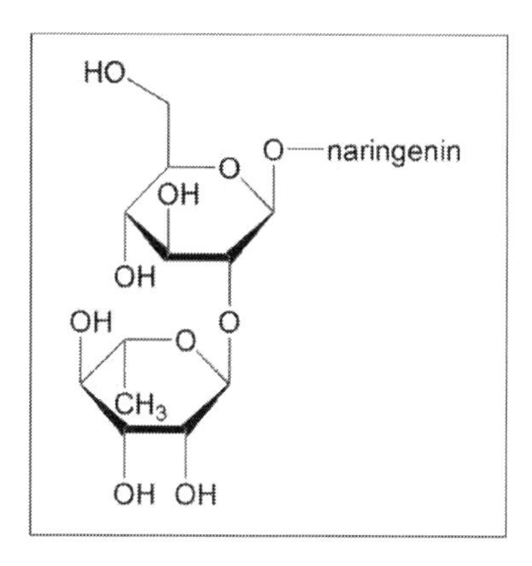

Naringin

Clearly this material is no longer natural but is best classified as a natural derivative, and should not have the INCI name ascribed to it, since this is inaccurate.

Tree Lichen Extracts

The tree lichen (*Usnea barbata*) contains usnic acid that is a fairly powerful agent against yeast and molds. Biostat (Variati & Co.), is the copper salt of usnic acid, extracted from lichen in ethoxyglycol. Evosina Na2 GP is the sodium salt in propylene glycol and both materials are claimed to reduce body odor.

It comes as no surprise therefore, that when this extract is used at a reasonable concentration that these spoilage organisms are not able to grow. The traditional use of this material for infections of the feet is well justified.

Usnic acid and usnate salts

Usnic acid and its salts are available from a number of sources (e.g. A&E Connock) and as Deo-Usnate from Cosmetochem. A word of caution: There are some individuals who are susceptible to irritation from this material and it is felt that this is because it has some similarities to the potential allergens found in Oak Moss and Tree Moss that are now listed in the 26 potential allergens.

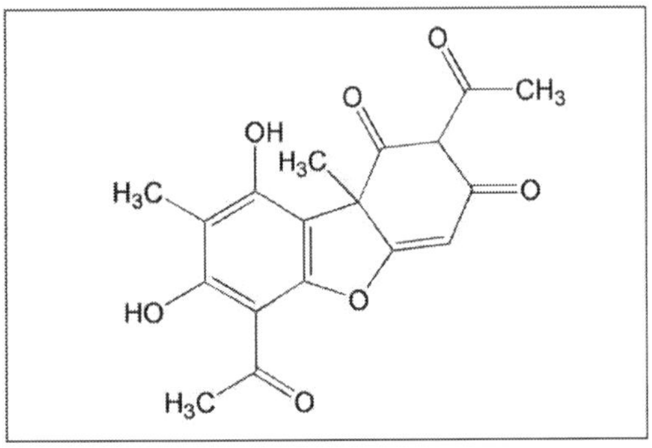

Usnic acid

Japanese Honeysuckle Extracts

A plant preservative that is based on the Japanese Honeysuckle *Lonicera japonica*) is available that is described as being a complex mixture of esters of lonicerin and other materials. The commercial material from Campo is called Plantservative WSr, WMr (INCI: *Lonicera Caprifolium* Extract). Lonicerin is luteolin-7-O-galactoside [Chen *et al.*].

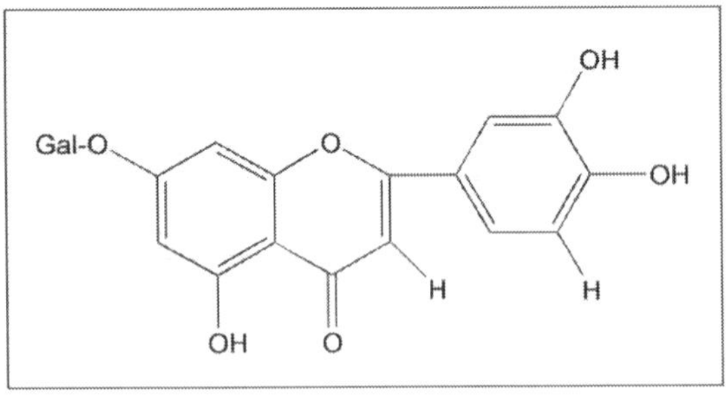

Lonicerin

It has been reported [Lee *et al.*] that *Lonicera japonica* has anti-inflammatory activity and though not as potent as the normal benchmark of prednisolone, it would nonetheless be effective in treating inflammatory disorders. This factor makes the preservative very attractive, since it has benefits for its soothing properties and also has antimicrobial activity. There are not many preservatives that would have this dual benefit. Literature has not been searched to see whether other luteolin derivatives have been found to have antimicrobial properties, but the flavonoids are certainly well respected for their anti-inflammatory activity wherever they are found in plant materials.

Formosan Hinoki Tree
Hinokitiol is a white crystalline acidic substance first isolated from the essential oil of Formosan Hinoki (*Chamaecyparis taiwanensis* Masamune et Suzuki) by Nozoe in 1936. This substance was also found in the essential oil of the Aomori Hiba tree (*Thujopsis dolabrata* Sieb. et Zucc).

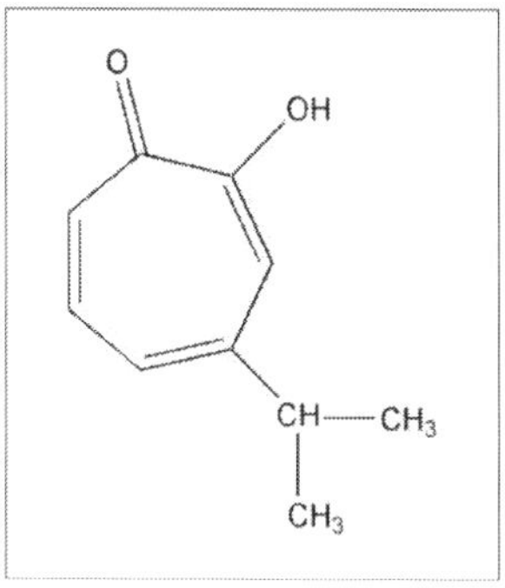

Hinokitiol

Though the natural form of hinokitiol is no longer available, the nature identical form is still made. It may be a surprise to learn that this material is listed as a hair conditioning agent in the Cosmetic, Toiletry, and Fragrance Association (CTFA) *Ingredient Dictionary*. It is one of those unfortunate events where a conditioning effect also had a preservation contribution. It is also unusual in that it has a 7-membered ring and is quite unlike any other preservative normally encountered. It is available from S. Black, A&E Connock and Nikko with CTFA names Hinokitiol or Chamaecyparis Obtusa Powder or Oil (Ichimaru Pharcos), although it is not known if the last materials, presumably all natural, have preservative action.

The *Parfum* (Fragrance)
Another clever idea is to look at essential oils and then isolate one or two of the components that coincidentally have antimicrobial activity. Since these components came from an essential oil, they must be perfumery-based materials and so can be listed as *parfum* or fragrance.

One isolated material present in the Dr. Straetmans range of such preservatives was once revealed as being anisic acid which quite clearly is a paraben structure. Another material revealed was levulinic acid [$CH_3COCH_2CH_2COOH$] or 4-oxopentanoic acid.

Levulinic acid, or 4-oxopentanoic acid, is a white crystalline keto acid prepared from levulose, inulin, starch, etc., by boiling them with dilute hydrochloric or sulfuric acids. It is soluble in water, ethanol, and diethyl ether, but essentially insoluble in aliphatic hydrocarbons. Conversion comes from transforming paper mill sludge. Biofine Corp. show that a multipurpose chemical called levulinic acid, which normally is produced from refined petroleum, can be produced from paper mill sludge. It has also been isolated from *Opuntia ficus-indica*.

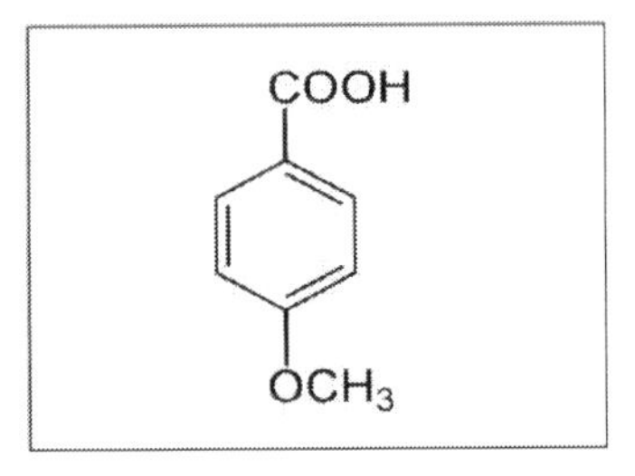

Anisic acid

Anisic acid is found in Aniseed (*Pimpinella anisum*) among many sources, and levulinic acid has been found as a by-product in the production of diosgenin from Wild Yam (*Dioscorea villosa*).

Natural Preservatives Based on *Parfum*

Dermosoft 1388 from Dr. Straetmans. (INCI: Aqua, Glycerin, Sodium Levulinate, Sodium Anisate) is a combination of fragrances of natural origin with a mild inherent odor and bio-stabilizing properties at the same time. It is useful for stabilizing surfactant products as well as emulsions and can be used in cold processing.

Dermosoft 250 from Dr. Straetmans. (INCI:Phenylpropanol). Found in a number of natural oils.

Dermosoft GMC, GMCY, 700 from Dr. Straetmans. (INCI: Glyceryl Caprylate) is a multifunctional cosmetic raw material that acts as fatting, wetting, moisturizing, and co-emulsifying agent. It is very skin-friendly. Also it is effective for the biological stabilization against yeast and bacteria and can be incorporated into the oil- or water-phase. It is a good replacement for Triclosan.

Naticide (INCI: *Parfum*) from Sinerga. One of the first "natural" preservatives. Another product from Sinerga called Naticide has also been shown to have excellent results at around 1.0%, and despite continual pestering, it's not able to be determined what plants have been used as the source of the active materials. (INCI name: *Parfum*).

Conarom H-3 from ISP. (INCI: PPG-2 Methylether, Phenylpropanol, Piperonal) has a vanillalike odor and has good antimicrobial activity.

Conarom P from ISP. (INCI: Phenylethanol (Fragrance), Caprylyl Glycol, Trideceth-8) has a roselike odor and a preservative action.

Conarom E from ISP. Blend of PPG-2 Methylether, Piperonal and Phenylethanol (Fragrance).

Bio-Botanica has two natural preservatives that are also extracted from plant actives, one called Neopein and the other Biopein, which are composed of Origanum Leaf Extract, *Thymus vulgaris* (Thyme) Extract, *Cinnamomum zeylanicum* Bark Extract (not present in thc Biopcin), *Rosmarinus officinalis* (Rosemary) Leaf Extract, *Lavandula angustifolia* (Lavender) Flower Extract and *Hydrastis canadensis* (Golden Seal) Root Extract.

Natrulon™ GPS 341 from Lonza is a multifunctional cosmetic ingredient composed of a unique blend of natural and nature-identical ingredients. The material imparts a delicate fragrance to mildly enhance the attributes of a personal care product. In addition, it offers antimicrobial capabilities to add an additional level of protection in order to maintain the product's integrity. The material is water soluble, compatible in a diverse range of personal care products and can be used at low concentrations. It contains: 1,3 Propanediol, Polyglyceryl-10 Oleate, Vanillin and Caprylic Acid. It might not satisfy the purists of natural formulation but is nonetheless a welcome addition to the arsenal of 'natural' preservatives.

Suprapein from Bio-Botanica is a third generation development based on a blend of active materials from essential oils. *Origanum vulgare* Leaf Extract, T*hymus vulgaris* (Thyme) Extract, *Cinnamomum zeylanicum* Bark Extract, *Rosmarinus officinalis* (Rosemary) Leaf Extract, *Lavandula augustifolia* (Lavender) Flower Extract, *Citrus medica limonum* (Lemon) Peel Extract, *Mentha piperita* (Peppermint) Leaf Extract, *Hydrastis canadensis* (Golden Seal) Root Extract and *Olea europaea* (Olive) Leaf Extract. It is classified by the U.S. Food and Drug Administration as an extract. Usage Rate: 0.45% (tests confirm the product usage to be safe up to 2.5%)

All of these materials are well-known for their antimicrobial activity and give rise to a host of active molecules: carvacrol, thymol, cinnamaldehyde, eugenol, cineole, camphore, α-pinene, rosmarinic acid, berberine, hydrastine, linalyl acetate and linalool.

This area could be exploited far more, because there are many other essential oil components that have antibacterial properties.

Multiex Naturotics ME05 from BioSpectrum is a blend of materials that provides a mixture of natural preservatives. *Magnolia biondii* Bark Extract, *Propolis* Extract, *Camellia sinensis* Leaf Extract, *Thujopsis dolabrata* Extract (source of hinokitiol), *Citrus grandis* Fruit Extract, *Chamomilla recutita* Extract, *Salix alba* Bark Extract (salicylic acid source).

Perillic Acid

A material that was presumably first found in *Perilla frutescens* or the Japanese Shiso oil is perillic acid. The perillaldehyde present has already been found effective against *Acnes propionibacterium* and *Staphylococcus aureus* [Balacs].

This material is made commercially by the conversion from limonene using a biotechnology process. It has been found to have good activity against Gram+ve and Gram-ve bacteria. Perillic acid is available from Dr. André Rieks.

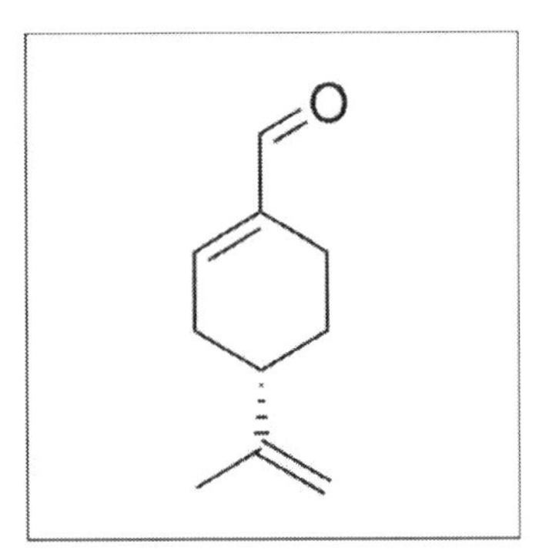

Perillaldehyde

Melaleucol
A refinement of *Melaleuca alternifolia* (Tea Tree) Oil, which contains 1,8-cineole, α-terpineol, α- and γ-terpinene, terpinen-4-ol and sesquiterpenes, has been found to deliver good performance against a wide range of spoilage organisms. The smell is quite clearly the characteristic antiseptic smell of tea tree, but this might be acceptable when one considers that it has also been shown to have mosquito-repelling effects as a bonus.

Wasabi (*Wasabia japonica*) or Japanese Horseradish
Active Concepts have a material called ACB Wasabi extract that is rich in allyl isothiocyanate. Used at 1–3%, it is said to have a significant contribution to the antimicrobial activity and at the same time has antioxidant activity.

Totarol (Mende DEK).
Totarol is extracted from recycled *Podocarpus totara* heartwood using supercritical CO2. It is a natural extract enriched in the aromatic diterpenoid, Totarol: $C_{20}H_{30}O$. It also contains other bioactives related in structure to totarol. Totara can live for up to 1000 years—totarol and related compounds in Totarol protect the tree from microbial attack. It has potent antibacterial activity against Gram-positive and Gram-negative bacteria as well as possessing potent antioxidant properties. It is also highly active against *Propionibacterium acnes*, even at a dilution of 0.005%.

Natamycin - E235
Nataseen™ from Siveele

Natamycin is a natural, toxicologically safe food additive. It is used for inhibiting yeasts and moulds in a wide range of foods. It has no effect on bacteria. It is available as a 50% mix with lactose, glucose or salt (NaCl) or at a higher concentration (>50%-95%). This product can be used in alcoholic beverages, baking, cheese, dairy products, and the culinary and meat industries.

Zinc and Copper Salts
The effectiveness of copper and zinc salts in reducing body odor is now thought to be because of their action in inhibiting the development of the bacterial enzymes that cause malodor.
Zincotrat is the trade name given by Vevy Europe to a solution of zinc citrate in propylene glycol citrate, which has astringent and deodorant properties.

Biostat (Variati & Co.), is the copper salt of usnic acid, extracted from lichen in ethoxyglycol.

Tegodeo JP 100 and Tegodeo HY 77 (Th Goldschmidt) are two differently solubilized forms of zinc ricinoleate. Neither have bactericidal or fungicidal properties, but Goldschmidt states that the zinc salts chemically combine with the bacterial decomposition products of perspiration, thus inhibiting odor development. JP 100 is recommended for nonaerosol and predominantly aqueous-based products such as roll-on deodorants and deodorant sticks, whereas HY77 is used in predominantly alcoholic products such as sprays.

Ajidew ZN-100 from Ajinimoto (Zinc PCA) The zinc salt of L-pyrrolidone carboxylate (Ajidew ZN-100) from Ajinomoto is a natural humectant associated with the Natural Moisturizing Factor of the skin that helps to reduce the appearance of excess sebum on the skin and scalp, leaving a clean, refreshed sensation. Ajidew Zn-100 is ideal for the widest range of personal care and beauty products. It also exhibits good antimicrobial activity.

Materials with Primary Effects and Secondary Antimicrobial Activity

Lexgard O from Inolex (Caprylyl Alcohol). Lexgard O is a humectant that can also boost or aid the preservation of cosmetic formulations. It can be used as an alternative to parabens or other preservatives that may be undesirable. It can be used alone or in combination with certain ingredients to achieve completely preservative-free claims. Lexgard O melts at approximately 30°C to 35°C and contributes a smooth, dry sensory characteristic to topical skin care products.

Lexgard Natural from Inolex "*Lexgard Natural is an all-natural multi-functional ingredient system for preservative-free and self-preserving cosmetics. Lexgard Natural is composed of high purity monoesters of caprylic acid (C8 acid) and undecylenic acid (C11 acid). The former is well established for its biostatic activity against bacteria and yeast. The latter is known for its activity against fungus. This combination provides remarkable broad-spectrum effects. Formulations with Lexgard Natural efficiently pass challenge tests required in the cosmetics industry.*"

Lexgard GCP "*is a biostatic glyceride and fragrance blend that can aid in the preservation of cosmetic formulations. It can be used as an alternative to parabens or other preservatives that may be undesirable.*"

Glyceryl Caprylate. Lexgard GMCY is a multifunctional, high purity monoester of glycerin and caprylate fatty acid that combines excellent skin compatibility with remarkable selective effects against microorganisms. It is especially useful for high quality skin care preparations as an antimicrobial agent with re-fatting, moisturizing and wetting agent properties. It may be used to control the growth of Gram-positive and -negative bacteria, yeast and mold in products designed to be self-preserving.

Another increasingly popular constituent of preservatives is ethylhexylglycerin that although it looks fairly natural does not occur freely in nature. Available in Lexgard E

Geogard Ultra™ Gluconolactone & Sodium Benzoate from Lonza
Geogard Ultra is a mixture Gluconolactone (GDL) that is a naturally derived product of fermentation, is a proven moisturizer and has no formulation color issues, and is a synergist and formula potentiator. Sodium benzoate is a food-grade preservative and Calcium gluconate is a processing aid that might also function as a chelating agent. This is a broad spectrum material that functions well in the pH 3-6 range.

Geogard™ 111 A. This is Dehydroacetic Acid a naturally derived material that is effective against yeast and mold in the pH range 2 – 6.5. There is a related product Geogard™ 111 S that is sodium dehydroacetate that has similar effects. A related product Geogard™ 221 is dehydroacetic acid, benzyl alcohol in aqueous solution that has broad spectrum activity in the pH range 2 to 7.

Natrulon GPS IV from Lonza is another blend. Glucono delta lactone is a naturally derived product of corn fermentation. Organisms produce glucose oxidase enzyme, then oxidize glucose to gluconic acid. Water removed, results in GDL and is used in some personal care applications as a moisturizer and exfoliator. It has no formulation color issues, comes from a non-GMO source. Sodium erythorbate is natural and comes from a non-GMO corn source. The corn sugar fermentation is a natural process. It is a stereoisomer of vitamin C.

Packaging Considerations

The base and the additives that could be added to products in order to reduce or eliminate spoilage organisms have been considered. The last piece of the jigsaw is the packaging. Wide-neck jars with shives (the plastic discs that cover the necks) are probably the worst news for the microbiological integrity of a product. Those covers allow water to condense on the surface and then enrich the organisms. The cardboard seal in the lid is another microbial sponge just waiting to act as a growth medium.

Tubes are far better, which is why they are more widely used in the pharmaceutical industry. The nozzle offers a smaller and more discrete surface for contamination. There are now tubes that have non-return valves, so that once pressed, the tube cannot relax to permit the ingress of air. Notice how tubes for eye products have long tapering nozzles with a small pin hole for delivery of product. This is good microbiological sense.

The new generation of pots does not allow the consumer to insert fingers of high contamination. There are pots with nozzles and sealed flat surfaces that have airless pistons which follow the product to completion. The product is offered to the consumer at a push and those days of scooping out are over, especially with the 7th amendment demanding a use-after-opening period to be declared.

It could be argued that, in these sealed and hygienic environments, the need for a microbial challenge test is over, since the consumer and the air will never enter the product during its active life. The preservative requirement will be a fraction of that required for a wide-necked cream jar.

The most secure pack is a single-application pack, the sachet, the blister pack and the single shot capsule. These are technologies that come to us from the fast food and pharmaceutical industries. You use it all or throw away the residue—it is the perfect preservative-free environment and the worst example of wastefulness.

Conclusion

There is a move toward preservative-free, which is being achieved by many means. It is hoped that this overview has provided an insight into some of the techniques available.

Chapter 7

Minerals

Introduction

While in Dominica in 2009, I visited one of the many volcanic pools in the area and also purchased a tourist sample of the mud extracted out of this hot spluttering ferment. The local residents have a superb knowledge of their local flora and ethno-pharmaceutical remedies, but they also placed great store by the muds and silts that are prevalent in the area.

Examination of the sample showed that the local residents were eager to show that the sulfur mud was antiaging, good for arthritis, pimples, rashes and many kinds of skin disorders. Although it's uncertain about the claim to preserve a youthful skin, the claim that it would be good for pimples is well-substantiated from a search of the literature, since sulfur has long been known as good for acneic skin. The presence of mineral in the mud would also be beneficial in any preparation seeking to alleviate distressing problematic skin conditions.

Additional minerals are found in the chapter on natural thickeners and bulking agents.

Sulfurous Muds

In the 5th century B.C., the beneficial properties of the sulfurous springs were already well-known because they were used to treat muscle and articular diseases such as arthritis and also used to improve skin complaints. This use continued through Roman times and became a part of the hot springs and bathing culture.

Sadly, the sulfur paste sold by the side of the sulfur pool was not available in large quantities commercially, so we searched our literature for a possible alternative and discovered that a source was available from Provital called Thermal Alluvium with Sulfur.

The literature explained that the mud had settled after long periods of time, in sediments near thermal springs or at the bottom of the sea and was essentially composed of clay and small quantities of other compounds. These compounds are responsible for the different characteristics of the mud, such as color, odor and other special properties, depending on its place of origin.

Muds are formed because of the erosion of several rocks, especially feldspar and mica. Once transformed into powdered materials, they are easily carried away by water flows and deposited in sediments at the river sides or at the bottom of the seas and lakes.

Although the composition and morphology of these muds are very different, they mainly contain aluminium, iron and magnesium hydrated silicates (crystalline and amorphous forms) that include nacrite, of thermal formation, up to montmorillonite, as well as the stable kaolinite. In some cases these deposits contain hydrated alumina and ferric oxide. The size and distribution of the particles, which are of colloidal size and have a plain or flattened shape, give them a characteristic slip, whereas the very fine quartz sand, feldspar, mica, powdered opal, fossils fragments and high density minerals such as sulfates and carbonates, give them a certain degree of abrasiveness. The thermal alluvium is picked up in sediments closed to a sulfurous thermal spring, when the most granulated hard particles are eliminated. It contains approximately 7% of elemental sulfur. The main properties of the mud are:

***Absorbing power*:** It has a very high adsorption and absorption capacity because of its small colloidal particle size and so has a large surface area, which gives it excellent qualities as an antitoxic and decongestive. Due to its strongly absorbing power, mud is of great use to treat greasy acne-prone skin. The mud exerts a proven action on sores, eczema, skin burns and other dermatological conditions, promoting rapid wound healing.

C***icatrizing action*:** It has an high aluminum silicate content and so helps to cicatrize wounds, eczema and other disturbances of the skin very quickly. Improved speed of healing leads to a reduction in possible scarring and blemishes. The mud is a natural bio-catalyst, especially efficient on the connective tissue, because of its high silicon content. Silicon is essential for the correct formation of collagen fibers. In this context, oligoelements also play a relevant role.

***Stimulating capacity*:** Mud absorbs several radiations (solar, magnetism, radioactivity, etc). It releases these radiations, but in smaller doses and at a lower frequency, resulting in a very concrete potential of organic stimulation, which influences the body vigor and metabolism, invigorating the cells reconstruction and accelerating all the organic processes.

***Antimicrobial effect*:** According to Provital's brochure, this highly powerful absorption inactivates the microorganism action. This action seems to be strengthened by a certain natural antibiotic capacity of mud.

The occlusion that results of mud application on the skin raises skin temperature and promotes microcirculation in the conjunctive tissue, all of which produce a certain degree of skin moisturization. These effects would confirm the claims made on the small purchased sample in Dominca.

Clays

Beraca has a similar product called Amazonian White Clay. This clay is an oleaginous soil that has a great number of shades, depending on the site of origin. It is a natural product, very rich in mineral salts that take the toxins to the skin surface and help remove them. It is rich in iron, aluminum, potassium, calcium, boron and sulfur. The clay was quite popular natural therapy, well-known locally for its beneficial effects in headache and stress illnesses. The Brazilian Indians used geotherapy often, as the clay may protect them against skin problems and also protect them from biting insects. Clays are full of nutrients that combat free radicals and they are believed to channel positive energy. This product is said to be effective when it is sun-dried to energize it. The Amazonian white clay can be applied in natural treatments for relaxation, skin regeneration, hydration and cleansing.

Silts

Active Concepts has a sea silt that forms in calm estuaries and coastal regions with low tides. Often rich in minerals, clay, algae and other nutrients, sea silt is said to both fortify and rejuvenate the skin. During the manufacturing process, sea silt is carefully extracted with water and butylene glycol to retain all of its nutritive benefits. ABS Sea Silt Extract is a healthy addition to cosmetic and personal care products.
Gaiamare (through Gattefosse) has a Fossil Mud that is silt 40%, magnesium aluminium silicate 35% and aqua 25%. The product is extracted by washing to a more than 700 meters depth zone of Catalonia where the earth presents a great concentration in salts and minerals that confer to these products and their derivates wide properties. It is a natural saline mud that contains an elevated portion of halite (NaCl), calcite ($CaCO_3$), anhydrite ($CaSO_4$), quartz (SiO_2) and philosilicateated fractions of muscovite and rests of nontronite. It is combined with red clay and is used as a basis for cosmetic and therapeutic preparations due to its thermoactive, remineralizing and exchanging salts properties.

There are a number of marine muds that would happily fit into this category.

Dead Sea Mud

Alban Muller has Dead Sea muds and salts that are of great potential in a variety of personal care applications.

The thermal baths of the Dead Sea are contemporary to Great King Solomon. Their reputation of efficacy spread throughout the Antique world. Cleopatra imported salts from the Dead Sea for her beauty care.

The Dead Sea is situated 400 meters below sea level. The water of the rivers that flow into it cannot flow out. It concentrates, evaporates, then progressively deposits treasures such as mineral salts and organic sediments eroded from the Judean Mounts. The salinity of the Dead Sea is thus 10 times higher than that of any other sea. Today, many countries encourage their citizens suffering from psoriasis or arthritis to make a stay at the thermal baths of the Dead Sea as the mud and salts bring relief from these distressing conditions. The much sought-after salts and mud are also quite famous for their remineralizing and purifying properties.

The mud is highly recommended for seborrheic skin or skin suffering from acne, and Dead Sea mud is very efficient for skin care, to deep cleanse and also to be used to relieve painful joints. The black mud, rich in minerals, is composed of layers of clay sediments constituted as millennia passed. The concentration in minerals is high. The mud contains natural exfoliating agents with an action that purifies and tones the skin. As a result, the skin is more supple and smoother.

In a mask, the mud can be used to balance and smooth the skin. The skin is cleaned in depth and as a result ends up nourished and better hydrated. In scrubs, it provides a light lifting to leave the skin fresh and revitalized. On the hair, in a mask or in a shampoo, it is used to clean the hair, get rid of excess sebum and rebalance the scalp. In products designed for body, it is said to favor cellular exchanges and slimming and helps to improve the elasticity of the skin. As the figure gets thinner, the skin breathes. Ideal for anti-stress cures or after birth, around the joints and to relieve pains.

The salts of the Dead Sea bring a calming and relaxing effect. In a bath, they enable the body to feel well and are an ideal solution for body tiredness, muscular or nervous tension. In cosmetics, the salts purify, cleanse and remineralize the skin for a progressive come back to a healthy and supple aspect. Their beneficial aspect is furthermore improved when associated with warm applications of Dead Sea mud.

Rassoul Mud

Rassoul or Ghassoul was created at the piedmont of the Atlas Mountains in Morocco, known as "the country where the sun sets." It was formed in deposits during the Jurassic period in the Mesozoic era, 208 to 144 million years ago. This detoxifying clay is mined from 2.5 miles within the Atlas Mountains. Its purity is renowned throughout history going back as far as 2,500 B.C. Rhassoul contains lithium as well as magnesium and other trace elements, such as iron, potassium, copper and zinc. The mud has been used since the 12th century. Moroccans used Rassoul as a daily therapeutic source for skin cleansing and purification, dermatitis, sensitivities, smoothing rough or scaly skin, seborrheic skin, and scalp and hair treatments. The underground mining of Rassoul originally commenced with ancient North African civilizations.

The Societé du Ghassoul et de ses Dérivés (available through Unifect) exploit the only known deposits of Ghassoul clay in the world, located 200 km from Fez in Morocco.

Ghassoul, used for more than 12 centuries by people from all North African and Middle Eastern countries, takes its name from the Arabic word for "washing"—*rassala*.

The material has been awarded Ecocert organic certification for the cosmetics sector. [*www.ghassoul.org*]

The topic of minerals and the number of new raw materials seem to be growing in popularity, so here are examples of commercial sources of these materials.

Bamboo Silica

This is one of the natural sources of plant silica.

Lessonia has Bamboo 200, 500, 1000 and Bamboo 100. Bamboo Exfoliator is a solidified organic silica extracted from the nodes of specific bamboo stems, called *tabasheer* in the traditional Indian medicine. Bamboo 100 is especially ideal for microdermabrasion since it improves skin refinement and complexion clearness.

Another exfoliant based on bamboo silica is available from Libiol. The bamboo exfoliant is obtained from the bamboo tabasheer stems collected in the form of an exudate that crystallizes at ambient temperature. The product is presented in the form of a very hard white mass. After crushing, one obtains a crystalline powder very rich in mineral salt. It is a source of silica (SiO_2)

since this particular variety of bamboo contains nearly 70% of it among the other minerals found. The bamboo exfoliant is particularly interesting for the manufacture of exfoliating creams or gels to remove the dead cells and debris from the skin surface.

Bamboosilk from Naturactiva. Certain female bamboo contain a white secretion in their joints called tabasheer. It consists almost of pure hydrated silica and has been used in Indian phytotherapy for its remineralizing properties in treatment of arthritis, osteoporosis, and for strengthening hair and nails. Eastern medicinal tradition describes Tabasheer as alkaline, cold in nature and sweet in taste. It is considered as an aphrodisiac and tonic in China where it is called *tian zhu huang*. Bamboosilk imparts to powder formulations a variety of properties that include: a high sebum absorption capacity, a mattifying aspect and a silky smooth feel. It also helps to improve the ease of application and the spreadability for body lotions, absorbs excessive perspiration, reduces lipstick exudation, adds volume in mascaras and improves nail polish wear.

Lava Silica

Rhyolite (Si/Al) 75 & 200 (Lessonia). Extremely abrasive, rhyolite is a low density lava rich in silica (71%), formed in volcano magma from the Lipari Islands. Processed into a very fine powder, Rhyolite Exfoliator allows a tonic exfoliation. Such a mineral is ideal to regenerate skin and to smooth the epidermis without affecting the hydrolipidic surface.

Marine Sources of Minerals

Lithothamnium (Ca/Mg) (Lessonia). Lithothamnium is a red coral plant highly rich in calcium (30–35%) and magnesium. Its marine origin makes it a revitalizing mineral treasure gathering more than 32 trace elements in a readily available form. Its hardness also ensures energetic exfoliations.

Spring Sea Water (from Soliance) is pumped from a natural reservoir more than 22m in depth which is constantly supplied with sea water. It is naturally rich in minerals and trace elements because of its contact with the sand and the network of granitic faults of its original environment. The marine station is located in North Brittany, France. It optimizes water exchanges and improves the cohesion between the dermis and the epidermis by increasing the hemidesmosomes number and maintains endogenous skin hydration by stimulation of epidermal lipids synthesis. A huge range of marine derivatives and similar spring water are available from Codif.

There comes a time when the cosmetic scientist wants to splash out on something totally unreasonable, pure fantasy and totally indulgent. It is a champagne, truffles and absurdly romantic opportunity to add a wow factor that is not going to come cheaply! It is the new realm of total luxury. Most of the words are from the suppliers who have waxed lyrical on these new concepts.

Precious Stones

Amber. Exfo-Amber from Provital is fossilized resin coming from forests of *Pinus succinifera*, ground to a particle size smaller than 400 μm. This plant product has been developed as a physical exfoliant agent to accelerate skin desquamation and therefore improve skin appearance. Amber has been credited with numerous beneficial properties both physical and spiritual. The Ancient Romans and Greeks used it to cure ailments such as asthma, rheumatism and internal problems. Its purported healing powers have extended to epilepsy, jaundice, kidney and bladder complaints and even the plague. It has also been used as an aphrodisiac and as a protection against witchcraft. Amber is a stone of health and wellness, stone of happiness and sun, its orange (amber) color is considered exciting, effusive and invigorating.

Amethyst: Soliance supplies an amethyst in its Lithocosmetic range of materials. The amethyst is approved by ECOCERT for ecological and organic cosmetics.
Metaphysical characteristics: Tourmalines are said to enhance one's understanding, increase self-confidence and amplify one's psychic energies. They also are said to neutralize negative energies, dispel fear and grief, and to help in concentration and communication.

Aquamarin is available as Aquamarina from Soliance in its Lithocosmetic range of materials.. From the light blue of the sky to the deep blue of the sea, aquamarine shines over an extraordinarily beautiful range of mainly light blue colours. This stone is the most common of gem beryl, which is beryllium aluminium silicate. Aquamarine occurs in pegmatite and forms a much larger and clear crystals than emerald. It is found in Brazil, India, Russia and USA and mined in a number of exotic places including Nigeria, Madagascar, Zambia and Mozambique.
Metaphysical characteristics: Aquamarine has a soothing effect on 'just married' couples, assisting them in working out their differences and insuring a long and happy marriage. It is also said to re-awaken love in long-married couples and to induce the making of new friends. Aquamarine works against nerve pain, glandular problems, toothache and disorders of the neck, jaw and throat. It soothes eye, ear and stomach problems and relieves cough. Moreover, it protects from sea perils, including seasickness and eases depression and grief.
Cosmetic characteristics: Aquamarina is a powder of micronized powder of aquamarine that belongs to the Lithocosmetics Range composed of cosmetic precious stones. Aquamarina triggers microcirculation through a physical phenomenon. It improves the complexion by fighting against the effects of stress. Activate microcirculation, improves skin complexion and tone and relaxes

Citrine. Citrina is supplied by Soliance and is described as the relaxing gemstone cosmetic powder. Citrine is a yellow member of the quartz mineral group. It is one of the most affordable gemstones, thanks to its durability and availability. It is often confused with the costly yellow-orange topaz gemstone.
Metaphysical characteristics: Citrine acts against digestion problems and eye disease. Citrine is also called the "stone of the mind" because it is believed to increase psychic power. It increases self-esteem, protects from negative energy, opens the mind and clarifies thoughts. Citrine also has the power to calm and sooth depression, to give joy and love.
Cosmetic characteristics: Citrina is a citrine micronized powder and improves the complexion by fighting against the effects of stress. Citrina also stimulates the immune system thus protecting sensitive skin.

Diamond. Active Concepts supplies a diamond material. For centuries, diamonds have been used as alternative medicines. It has been said that diamonds are capable of increasing energy and brain activity while improving glaucoma. Although these claims have never been scientifically validated, diamonds have come to symbolize generosity, prosperity and love. Diamonds are the strongest mineral on earth, and are often used not only in jewelry, but also for industrial equipment to cut hard surfaces.

The popularity of diamonds is most likely attributed to their beauty. When polished and cut, the sparkle and shimmer of a diamond may seem rather hypnotic to some. Although most people are familiar with the "4 C's" of cut, color, clarity and carat, many often confuse carat as being a measurement of size when it is actually a measurement of weight. The term carat is derived from the Greek word *keration* which is the fruit of a carob. Carob seeds were noted to have uniform weight and were originally used as weights on precision scales. The confusion with the word carat does however seem logical, as larger diamonds tend to weigh more than smaller diamonds. Many of us may want a large diamond due to the value; but from an aesthetic sense, they may be more appealing due to the greater distance the light must travel. The result is an increased prism effect, which is registered as a greater sparkle and shimmer than seen in smaller diamonds.

But why would you ever think to use diamonds in cosmetics? Skin is a living organ that is constantly changing. It consists of the dermis and the epidermis. The epidermis is comprised of several layers, with the outermost layer of the epidermis consisting of dead cells. These cells tend to build up in time and without proper exfoliation, skin appears dull and dry. By regularly exfoliating skin can be kept looking smooth and revitalized while evening out skin tone.

AC Diamond Dust (IMCD in United Kingdom) consists of diamonds that are ground into a fine powder for use as an exfoliant in personal care and cosmetic products. This can be used not only in foot and elbow exfoliants, but also in products designated for exfoliating more sensitive areas such the face, neck, shoulders and lips. They also supply a diamond extract.

Diamond is also supplied by Soliance and is described as the purity gemstone cosmetic powder.
Metaphysical characteristics: Diamonds are the traditional emblem of fearlessness, virtue, power and wealth. Today, diamonds symbolize eternity and are often seen on engagement rings. Diamond is a great assistance for all brain diseases. It is also beneficial in the stomach area and can strengthen the owner's memory. The wearer of diamond is protected from the bad influence of evil spirits and snakebites. Diamond is an antidote to poison and is capable of detecting poison.
Cosmetic characteristics: Diamond is a micronized powder and this stone shows healing properties and stimulates immune defenses. Diamond can be useful to treat bone diseases.

Emerald. Active Concepts also supplies an emerald. Green—the color of love, money and envy—is also one of the most sought after colors in gemstones. Emerald is the green variety of the mineral beryl ($Be_3Al_2(Si0_3)_6$) and its green to bluish green color is attributed to trace amounts of chromium and iron present. Considered to be the most valuable of all the gemstones due to its rarity, emeralds are typically formed under hydrothermal reactions and tend to have slight water inclusions. Emeralds also have a hardness of 7.5 on the Mohs scale where 10 is considered to be the highest.

Regarded as the traditional birthstone for May, the emerald also represents the astrological signs of Cancer and Taurus. It is believed by many that emeralds have the power to ward off evil spirits, cure epilepsy and dysentery while also aiding poor eyesight. Typically regarded as a symbol of pride, emeralds are also the traditional gifts given for the 20th, 35th and 55th wedding anniversaries in several different cultures. Emeralds are commonly recommended in chakra and crystal therapies, and are attributed with improving the wearer's physical, emotional and mental balance. Emeralds are also considered to be a source of inspiration, while improving patience and providing healing properties.

The Incas and Aztecs of South America revered the gemstone as holy. Emeralds were also a favorite of ancient Egyptian pharaohs between 3,000 and 1,500 B.C. Not surprisingly, the emerald mines were referred to then as "Cleopatra's Mines." The Vedas, holy scriptures written in India many centuries ago, mention the gemstones as having healing properties, while enhancing well-being and ensuring good luck. Not surprisingly, artifacts with emeralds are highly sought after and prominently displayed in museums and private collections. An example is the cup of Emperor Jehangir (made entirely of emerald) which has been displayed at the New York Museum of Natural History next to the "Patriciaç," a Colombian emerald that weighs 632 carats.

When added to cosmetic and personal care products, AC Powdered Emerald may provide mystical properties that have been associated with the gem. Commonly used for healing, and as an "energy medicine tool," gemstone products are also popular in holistic healing treatments such as chakra and crystal therapies.

Garnet (Si/Fe) 100, 200 & 300 (Lessonia). Semiprecious stone, the Almandite Garnet used in Garnet Exfoliator is known for natural hardness and abrasive characteristics. It is especially rich in iron (30%), essential for the body's well-being. These minerals, associated with silica and aluminium, act in synergy for a powerful regenerating and energizing experience.

Jade. Active Concepts supplies Jade, which has been known to exhibit various therapeutic properties, both for the body and the mind. People in many ancient cultures, such as the Mayas and Aztecs, used to wear a belt of Jade around their waists because it was thought to relieve kidney problems. Jade is also known as being one of the toughest natural stones, and the people in these cultures used it to make many of their everyday tools. The peaceful features of this stone give it the qualities to be deemed the birthstone for the month of March as well as for the zodiac sign Virgo.

In gemstone therapy Jade is believed to improve creativity and mental agility while also having a balancing and harmonizing effect. This stone is known in China as the "stone of heaven" and is known to embody the Confucian virtues of wisdom, justice, compassion, modesty and courage. The qualities were so important that the Aztecs buried their loved ones in Jade masks so that they would have a peaceful death.

Jade is actually two distinct stones and although it is typically thought to be a green color, it also comes in shades of white, yellow, pink, purple and red. Nephrite is the more common of the two stones and can be found in China, New Zealand, Russia, Guatemala and the Swiss Alps. This stone has been the provincial gemstone of British Columbia since 1968 with more than 150 communities depending on the mining industry in Canada. Jadeite is the less common of the two stones and can be found in China, Russia and Guatemala, while the most valuable type is found in Burma. The value of the Jade stone is determined according to the intensity of color, vivacity, texture, clarity and transparency.

When applied to cosmetic formulations, AC Powdered Jade may provide the mystical properties that have been associated with the gem. Commonly used for healing, gemstones are also popular in holistic healing treatments, and jade may be able to bring about a sense of balance and harmony.

Moonstone: Active Concepts supplies a moonstone material that many different cultures have associated with possessing mystical powers. Ancient Romans believed that an image of Diana, the goddess of the moon, was encapsulated within the sacred stone; and until the 16th century it was believed that the appearance of moonstone changed with the different phases of the moon. It has also been said that wearing the gem will bring one victory, health and wisdom In India it is considered to be sacred and is typically displayed on a yellow cloth that will bring good fortune, which is brought on by the spirit that resides in the stone Recognized as a symbol of fertility in many Arab countries, the stone has also been sewn into women's garments.

Many believe that moonstone may improve one's intuition and sensitivity toward others, while also improving one's emotional and daydreaming tendencies. Considered a stone for lovers, it has been said that it will also increase feelings of tenderness and the desire to protect loved ones. Often cut into a cabochon shape, the shimmering opalescent stone was very popular during the Art Nouveau period and was commonly used in jewelry and ornamental decorations by the famous goldsmith René Lalique and his contemporaries. Many pieces containing moonstone that were created during the Art Nouveau period are now displayed in museums and private collections.

Moonstone is a type of feldspar that naturally occurs in many igneous and metamorphic rocks. An important group of silicate minerals, feldspar comprises approximately half of the earth's crust; and under certain geological conditions, gem-quality stones such as moonstone, labradorite, amazonite and sunstone form. These gem-quality feldspars are actually aluminosilicates that contain aluminum, silicon, oxygen sodium and potassium. Although Sri Lanka is considered to be the primary location for mining the highest quality moonstone, the gems are also found in the Alps, Malagasy, Burma and India.

When added to cosmetic and personal care products, AC Powdered Moonstone may provide mystical properties that have been associated with the gem. Commonly used for healing, and as an "energy medicine tool," gemstone products are also popular in holistic healing treatments.

Rose Quartz. Active Concepts supplies a form of rose quartz. Historically, holistic healers and therapists around the world have employed the natural energy of gemstones to help correct physical, mental, and spiritual abnormalities. Such imbalances are thought to cause everything from dry skin to insomnia. Today, with a renewed interest in the natural living and wellness market where health and beauty habits increasingly overlap, it is not surprising that the ancient practice of using gemstones has found a place in the modern world.

A very interesting gemstone, rose quartz is one of the most desirable varieties of quartz. It is found in Madagascar, India, Germany and several localities in the United States. Much rose quartz was originally extracted from a famous site near Custer, South Dakota, but now, most of the world's supply comes from Brazil. So amazing are the rose quartz crystals that modern mineralogists from around the world dismissed the first ones discovered as fakes.

In ancient Egypt, women wore facial masks carved of pure rose quartz to soften and beautify the skin. Hair was dry shampooed using fine powdered clays and gemstones. The powders were combed through to absorb the grease and dirt. The gem has since become the established stone of all things feminine and goddesslike. Rose quartz can be categorized as one of nature's most energizing minerals.

In its powdered form, it may be used to buff away dead skin cells and impurities revealing the soft, fresh and youthful skin beneath. Its light-reflective crystals illuminate the skin, softening flaws and boosting radiance. Its restorative and energetic properties sweep away impurities, improving dull or flaky skin and opening hair follicles, while protecting against future damage. AC Powdered Rose Quartz creates a stronger, more potent bio-energy field in formulations.

Ruby. Rubisa from Soliance is described as the modulating gemstone cosmetic powder. Ruby is the red variety of the corundum mineral. For thousands of years Ruby has been considered one of the Earth's most valuable gemstones. It has got all it takes for a precious stone: a wonderful color, excellent hardness and an overwhelming brilliance. Slight traces of the color, creating elements such as chrome, iron, titanium or vanadium, are responsible for the red color synonymous of passion and power.
Metaphysical characteristics: Ruby protects people from afflictions, sorrow and disasters. This stone strengthens the spirit, prevents egoism and authoritarianism. The wearer of the ruby enjoys wealth and property and becomes fearless.
Cosmetic characteristics: Rubisa is a ruby micronized powder that is a strong stimulator and good regulator of blood circulation. It also has a toning effect.

Active Concepts supply a ruby extract. Often considered to be the color of love and the color of passion, red is also the unmistakable color of rubies. These rare gems are the most valuable and most sought after compared to other gems. With a hardness of 9.0 on the Mohs scale, second in hardness only to diamonds, rubies are actually the mineral corundum that was formed in the presence of chrome. When cut into cabochons rubies can show an optical phenomenon referred to as asterism, where the reflected light creates a star shape on the stone. Used in holistic medicine, rubies are thought to treat a variety of ailments. AC Ruby Extract may be added to aqueous systems and emulsions and may convey feelings of passion along with a noble sense of refinement.

Sapphire. Soliance (IMCD in the United Kingdom) has a product called sapphira, which is described as the regulating gemstone cosmetic powder. Sapphire, "gem of heaven," has been cherished for thousands of years. The ancient Persians believed that the earth rested on a giant sapphire and its reflection colored the sky. From the deep blue of evening skies to the bright and deep blue of a clear and beautiful summer sky, sapphire exists in all the shades of blue

skies. This splendid gemstone, however, also comes in many other colors, displaying the bright fireworks of sunset colors: yellow, pink, orange and purple.
Metaphysical characteristic: Sapphire symbolizes loyalty and faithfulness, while at the same time expressing love and desire. Blue sapphire is used to transmute angers and impulsiveness into patience and waiting. Sapphire wakes up the conscience as well as qualities of wisdom, intuition, interior peace, knowing and simplicity. Sapphire calms and alleviates pain.
Cosmetic characteristics: Sapphira is a sapphire micronized powder that belongs to the Lithocosmetics range composed of cosmetic precious stone. Sapphira is excellent for the skin, the hair and the nails. It can be used to fight against skin eruptions and excessive perspiration.

Active Concepts also supply a sapphire extract. Characterized as 'gems of the sky', in the past some believed that the sky was an enormous blue sapphire in which the earth was embedded. The varied shades of blue sapphires are associated with the virtues of loyalty, trust and friendship as well as the feelings of confidence, sympathy and harmony. Formed when aluminum oxide crystallized into corundum in the presence of chrome and iron under intense heat and pressure, sapphires can actually be found in several different colors including pink, yellow and orange. Thought to be associated with calming properties useful for relaxing the mind in gemstone therapy, AC Sapphire Extract may provide an air of mysticism to any formulation.

Topaz. Active Concepts supplies a topaz that is a very hard gemstone which shares the traits of a diamond. Topaz exists in various shades of color from brown, yellow, orange, pink and red. They are mostly found in Brazil, Sri Lanka, Pakistan, and Russia. Pale topaz, which has been enhanced by irradiation to become blue, is mostly found in Brazil, Sri Lanka, Nigeria and China. Topaz has been associated with Jupiter, the Roman god of the sun. Being the largest planet in the solar system, the planet Jupiter is known as a celestial metaphor for the king of gods. Jupiter is also associated with health complications. The combination of bad transits and weak Jupiter at birth causes complaints like jaundice, hernia, liver problems, skin disease, and cerebral congestion. In such cases, it is believed that wearing topaz may act as a remedial measure against diseases governed by Jupiter. The Greeks, on the other hand, focused on the power of topaz to increase strength and make its wearer invisible in times of emergency. Additionally, they use the powder to cure insomnia.

Today, this elegant stone is used in cosmetics in its powder form to help stimulate blood flow circulation and improve dull skin. When used in conjunction with plant extracts, AC Powdered

Topaz hydrates and protects, infusing the skin with visible life and radiance, bringing a healthy glow to the surface. It can also be used in skin rejuvenation techniques, such a microdermabrasion, to correct topical flaws.

For hair care, the crystals can act as tiny mirrors focusing positive light on the follicle, where it may stimulate nascent hair.

Tourmaline. Soliance supplies Tourmalina, which is described as the energizing gemstone cosmetic powder. It is approved by ECOCERT for ecological and organic cosmetics. Tourmaline is a mineral found in granite pegmatic and metamorphic rock. It has been characterized since Ancient Egypt by a wide range of different shimmering colors.

Metaphysical characteristics: Tourmalines are said to enhance one's understanding, increase self-confidence and amplify one's psychic energies. They also are said to neutralize negative energies, dispel fear and grief, and to help in concentration and communication. Tourmaline gives off very focused directional energy. It is supposedly good for moving energy between the chakras. Pink tourmaline is said to inspire love, spirituality and creativity, and to give wisdom and enhance one's willpower. It is one of the foremost heart charka healers. It treats stress and nervous conditions.
Mechanism of action: Tourmaline is a piezo-electric crystal that emits energy in the far infrared region (4–20 µm). These radiations have unique interaction with biological systems by resonance absorption. The skin absorbs IR radiations at 9µm so tourmaline crystals can transmit the energy and induce improved blood flow. Far infrared rays improve microcirculation and are involved in better healing as they increase collagen synthesis. Niwa (1993) proved that tourmaline is capable of activating leukocyte functions without promoting oxidative injury (lipid peroxidation from unsaturated fatty acid was markedly inhibited).
Cosmetic characteristics: Tourmalina is a micronized powder of tourmaline. Tourmalina triggers microcirculation through a physical phenomenon: far infrared energy emission. It improves the complexion by increasing skin temperature.

Active Concepts supplies a tourmaline. Tourmalines are unique gemstones that are not only found in a variety of colors, but they also slightly change color depending on the type of light. According to ancient Egyptian legends tourmalines are found in so many different colors because they passed over a rainbow on their long journey up from the center of the earth. Although this is just a legend, tourmalines are still referred to today as the 'gemstones of the rainbow'. The gemstone is also pyroelectric meaning that is capable of generating an electrical potential. When exposed to changes in temperature the crystal becomes polarized as the positive and negative charges move to the opposite ends of the gemstone. AC Tourmaline Extract is an aqueous extract of tourmaline, which may be used in cosmetic and personal care product to create a sense of luxury.

Precious Minerals

***Corindon* (Al) Special Microdermabrasion** (Lessonia). Fashionable skin-freshening and skin-renewing technique. Corundum is the crystalline form of aluminium oxide. It is naturally clear but can have various colors depending on the crystals it contains. The transparent specimens are used as gems (e.g. ruby, sapphire).

Hema'Tite (Gattefossé). Hematite is capable of acting right inside the cells to stimulate enzymatic activity of prolylhydroxylase, thus increasing fibroblast production of pro-collagen.

Rhodo'lite (Gattefossé). The anti-stress stone par excellence, rhodochrosite encourages intuition and creativity by emitting high-frequency vibrations that are actual fields of energy which surround the bearer. When placed directly on the skin, it helps regulate emotions and calm restlessness and anguish. In lithotherapy, it is used to purify the skin and to improve general physical well-being. In the face of everyday pressures imposed by the urban and professional environment, rhodochrosite prepares the mind and renews strength and energy. By aiding memory and mental acuity, it stabilizes emotional and energy networks.

Mala'Kite (Gattefossé) The effects of malachite have been demonstrated at several levels of the antioxidant defense mechanisms in the cells. Through its effect on the superoxide radical, this mineral acts directly on the reactive species by inhibiting the superoxide anion at a rate of more than 95%. At the same time, malachite acts at a more complex level in the body's defense system. Its effect on hydrogen peroxide occurs via reduced glutathione—or more precisely, through the enzyme in charge of regenerating it. This indirect activation effect allows elimination of one of the main factors limiting the reaction between hydrogen peroxide and reduced glutathione: the reduced glutathione itself.

Oli'Vine (Gattefossé). Olivine is a stone extract, rich in magnesium. It stimulates the cells metabolic activity, acting both on cellular respiration and on the synthesis of ATP. It has powerful anti-stress action. Energizing mineral complex, Olivine is intended for the treatment of dull and tired skin. Olivine will also enable the skin to adapt to and resist environmental stress.

Zin'Cyte (Gattefossé) or smithsonite (zinc carbonate) can offer advanced cell protection that encourages natural DNA repair and shields against permanent cell lesions. Metallothioneins have a genoprotective effect that helps prevent DNA lesions. They can maintain the nuclear redox potential thus making DNA repair easier through the exchange of zinc atoms. The metallothionein induction capability of smithsonite makes it a select ingredient for daily application formulation such as day creams and foundations. It provides the best, constant protection for cells in the daily fight against environmental stress. When added to detoxifying serums, zinc carbonate mineral helps fight against the effects of certain pollutants. Its special genoprotective action makes it particularly effective in protecting DNA, the building block of skin cells.

Pearl Sources of Minerals

***Pearl* (Ca/Mg) 200 & 500** (Lessonia). The mother-of-pearl is alternatively formed with layers of aragonite and conchiolin organized around a nucleus. Thus, both mineral and organic, it especially activates microcirculation and gives brightness to the face; with sparkling particles, it gives cosmetic scrubs a deep precious image that is pure luxury for the skin.

Cosmepearl Freshwater Pearl Powder (Cosmetochem): an exciting cosmetic ingredient to add a touch of luxury to personal care products. This is a fine powder made from micronized freshwater pearls coated with jojoba oil. They contain calcium carbonate in the form of aragonite crystals and trace elements and natural moisturizers in the form of a wide range of amino acids and protein complexes. A water pearl abrasive is also available.

Crodarom have a pearl extract that is described as being antiinflammatory, stimulating invigorating, antibacterial and detoxifying. It re-energizes skin tissue and strengthens hair and skin.

There are also pearl powders available from A&E Connock, Sino Lion and Eurocostech.

Precious Metals

Gold
Grant Industries have Colloid PMG-WP which is a colloidal suspension of ultra- fine particles of gold in water. a prestigious, pure natural 24 carat gold colloidal formula that is standardized to about 1,000ppm, The ruby red to reddish purple color called "purple of cassius" is resistant to UV degradation. It is a mixture of Aqaua (Water), Hydrolyzed Wheat Protein and Gold.

Silver
The silver preservative from JMAC could no doubt be used (subject to the legal restrictions, if such a claim were wanted.

Precious Metal Bio-Chelates:

ACB Bio-Chelate Copper & ACB Bio-Chelate Copper Powder:
Copper helps to strengthen the bond between collagen and elastin, allowing the skin to remain strong and durable. A lack of copper in the skin is detrimental, causing it to become fragile and brittle. By adding copper to the skin, it may help to generate collagen and allow skin to regain its natural strength.

ACB Bio-Chelate Gold:
Gold is used as an anti-inflammatory agent and when this product is applied to the skin, it may help to reduce redness and inflammation.

ACB Bio-Chelate Silver:
Silver, when applied topically to the skin, is able to work as an anti-inflammatory, which allows it to heal the uncomfortable effects which may be caused by ultra dry, red skin.

ACB Bio-Chelate Malachite Powder:
Malachite or copper carbonate hydroxide is a stone that historically has been fashioned into jewelry and artwork. The name is derived from the Greek word *molochitis* meaning "mallow-green stone". More recently malachite has been recognized for having antioxidant benefits and for protecting the skin from environmental damage. ACB Bio-Chelate Malachite Powder consists of ground malachite that is fermented with *Saccharomyces* and spray dried into powder form. During the fermentation process a biotransformation takes place to create a complex consisting of malachite coupled with *Saccharomyces* peptides. Useful in skin and hair care products, ACB Bio-Chelate Powder is intended to deliver the benefits of malachite with the soothing properties of yeast peptides.

Conclusion

The growth of the use of minerals has provided a brand new and exciting portfolio of raw materials for the cosmetic industry, and it looks forward to a wealth of further ideas from these novel sources.

Author acknowledgments: I would like to thank all the suppliers mentioned in this chapter for the use of their brochures to compile material and apologize to those suppliers not mentioned because space did not permit a complete review.

Chapter 8

Botanical Extracts

The original volume had a very long chapter that became so large that it had to be removed from this volume and is now available as Handbook of Natural Ingredients and has nearly 4,000 entries on over 600 pages.

Chapter 9

Essential Oils

A

Abies alba. Fir, Silver. Fir has traditionally been used to help reduce symptoms of arthritis, rheumatism, bronchitis, coughs, sinusitis, colds, flu and fevers. It has been found to be a useful antiseptic, anticatarrhal, antiarthritic and stimulating. The buds are antibiotic, antiseptic and balsamic. The bark is antiseptic and astringent. The leaves are expectorant and a bronchial sedative. The resin is antiseptic, balsamic, diuretic, eupeptic, expectorant, vasoconstrictor and vulnerary. Both the leaves and the resin are common ingredients in remedies for colds and coughs, either taken internally or used as an inhalant. The leaves and/or the resin are used in folk medicine to treat bronchitis, cystitis, leucorrhea, ulcers and flatulent colic. The resin is also used externally in bath extracts and rubbing oils for treating rheumatic pains and neuralgia. Oil of Turpentine, which is obtained from the trunk of the tree, is occasionally used instead of the leaves or the resin. The oil is also rubefacient and can be applied externally in the treatment of neuralgia.

Abies balsamea. Fir, Canada. The resin is a very effective antiseptic and healing agent. It is used as a healing and analgesic protective covering for burns, bruises, wounds and sores. It is also used to treat sore nipples and is excellent for a sore throat. The buds, resin, and sap are used in folk remedies for treating corns and warts. The resin is also antiscorbutic, diaphoretic, diuretic, stimulant and tonic. It is used internally in propriety mixtures to treat coughs and diarrhea, though taken in excess it is purgative. A warm liquid of the gummy sap was drunk as a treatment for gonorrhea. A tea made from the leaves is antiscorbutic. It is used in the treatment of coughs, colds and fevers. Widely used medicinally by various North American Indian tribes as an antiseptic healing agent applied externally to wounds, sores and bites, it was used as an inhalant to treat headaches and was also taken internally to treat colds, sore throats and various other complaints.

Abies grandis. Fir, Grand. Fir has traditionally been used to help reduce symptoms of arthritis, rheumatism, bronchitis, coughs, sinusitis, colds, flu and fevers. It has been found to be a useful antiseptic, anticatarrhal, antiarthritic and stimulating. The buds are antibiotic, antiseptic and balsamic. The bark is antiseptic and astringent. The leaves are expectorant and a bronchial sedative. The resin is antiseptic, balsamic, diuretic, eupeptic, expectorant, vasoconstrictor and vulnerary. Both the leaves and the resin are common ingredients in remedies for colds and coughs, either taken internally or used as an inhalant. The leaves and/or the resin are used in folk medicine to treat bronchitis, cystitis, leucorrhea, ulcers and flatulent colic. The resin is also used externally in bath extracts, rubbing oils, etc., for treating rheumatic pains and neuralgia. Oil of Turpentine, which is obtained from the trunk of the tree, is occasionally used instead of the leaves or the resin. The oil is also rubefacient and can be applied externally in the treatment of neuralgia.

Acorus calamus. Calamus (Sweet Flag). In Arabia and Iran it is used as an aphrodisiac.In Japan, the leaves were used as a bathing agent to make "Sweet Flag bath water." It is an aquatic perennial, which emits a smell rather like that of mandarin oranges. Used for treating rheumatism, fever and lumbago.

Aloysia triphylla. Lemon Verbena. The essential oil is used in aromatherapy in the treatment of nervous and digestive problems and also for acne, boils and cysts. The essential oil obtained from the leaves (yield 0.5%) is extensively used in perfumery. There is evidence that the use of this oil can sensitize the skin to sunlight and has been largely replaced by the lemongrass, *Cymbopogon* spp. The dried leaves retain their fragrance well and are used in potpourri. The plant is an insect repellant and repels midges, flies and other insects. The essential oil is an effective insecticide at 1–2%.

Amomum subulatum. Cardamom large. Larger or Greater Cardamom or Nepal Cardamom. Medicinally, the seeds are credited with stimulant and astringent properties. It is used in gastrointestinal and genito-urinary complaints. It is correctly described by the Arabian physicians under the name Hil-Bawa.

Amoracia rusticana. Horseradish. Horseradish is listed by the Council of Europe as a natural source of food flavoring (category N2). This category indicates that horseradish can be added to foodstuffs in small quantities, with a possible limitation of an active principle (as yet unspecified) in the final product. In the United States, horseradish is listed as GRAS (generally regarded as safe). Horseradish is commonly used as a food flavoring. Horseradish possesses antiseptic, circulatory and digestive stimulant, diuretic, and vulnerary properties. It has been used traditionally for pulmonary and urinary infection, urinary stones, edematous conditions and externally for application to inflamed joints or tissue. It is a powerful and severe irritant and should be used with great caution.

Anethum graveolens. Dill Weed. Dill is a sedative herb and a good remedy for sleeplessness, acting as a mild tranquillizer. Flatulent pain in infants. Chewing dill seeds will help to sweeten the breath. Carminative and local anodyne. The essential oil in the seed relieves intestinal spasms and griping, helping to settle colic.

Angelica archangelica. Angelica. In the form of an ointment, it has a soothing effect on skin complaints, arthritis and rheumatism. A decoction of the root can also be used for scabies or itching and for wounds; as a compress in gout. The tea is a good eye tonic.

Aniba Rosaeodora. Rosewood Wood Oil. Rosewood is an evergreen tree now controlled as an endangered species by the Government of Brazil. It has a reddish bark and heartwood. Rosewood has a warm and woody scent, while at the same time being both floral and fruity. Rosewood or Bois de Rose does not come from roses, but from the cayenne Rosewood tree, which grows in the Amazon region and French Guiana. It is reputed to be an important component of Chanel No.5. The literature does not give any references to its use in skin care, though it is a valued essential oil in aromatherapy and is used for its relaxing properties, since it is felt to have a steadying and balancing effect on the nerves. Rosewood oil is rich in linalool, a chemical which can be transformed into a number of derivatives of value to the flavor and fragrance industries.

Anthemis nobilis. Chamomile, Roman. Roman Chamomile flower [Syn. *Chamaemelum nobile*] has certain uses similar to those of Matricaria flower (German Chamomile), although some of its constituents are markedly different and it is much less investigated pharmacologically and clinically. Anti-inflammatory and sedative effects of volatile oil have been demonstrated in rats.

Apium graveolens. (Celery) Oil**.** Family: Umbelliferae. Traditional use: As a salad vegetable or made into a tea, celery can be helpful also in clearing up skin problems. In Folk medicine, the oil is used as a diuretic in dropsy and bladder ailments, as a nervine and antispasmodic, and in rheumatoid arthritis. Seeds are used as carminative, stomachic, emmenagogue and others. Chemistry: Apigenin, apiin, isoquercitrin, apigravin, apiumetin, apiumoside, bergapten, celerin, celeroside, isoimperatorin, isopimpinellin, osthenol, rutaretin, seselin, umbelliferone, 8-hydroxy-5-methoxypsoralen, furanocoumarins. Volatile oil: limonene, selenine, phthalides, eudesmol, santalol, 3-n-butyl phthalide, sedanenolide, choline, ascorbate, dihydrocarvone, linoleic acid, myristic acid, myristicic acid, myristoleic acid, oleic acid, palmitic acid, palmitoleic acid, petroselinic acid, stearic acid. Traditional use: In Folk medicine, the oil is used as a diuretic in dropsy and bladder ailments, as a nervine and antispasmodic, and in rheumatoid arthritis.

Artemisia absinthium. Wormwood. Oil of wormwood is used in liniments for sprains, bruises and lumbago, and in fomentations for rheumatism and inflammations. Excessive use is dangerous and pure wormwood oil is a strong irritant that should be used only as a liniment. It is of value for gout and rheumatism, when used as a bath additive. It is not advised for general topical use.

Artemisia dracucnculus. Tarragon (Estragon). The name tarragon is derived from *dracunculus*, meaning a little dragon. It is a member of the Wormwood family and, like most of them, it possesses a strong and individual flavor more popular in France than in Britain. The dried herb is more aromatic than the fresh, and French tarragon has a sharper flavor than Russian. Its oil is used in perfumery.

Artemisia herba-alba. Artemisia Herba-Alba Oil is the volatile oil obtained from the whole plant, *Artemisia herba-alba*, Compositae. White wormwood, Desert wormwood. Most of the more than 250 species of Artemisia are dry-land inhabitants playing an important role in the desert-like vegetation in Central Europe, North Africa and Asia. The so-called artemisia oil, or armoise oil, is obtained by steam distillation of white wormwood, a dwarf shrub growing wild in various areas of North Africa and the Middle East. White wormwood exists as various chemotypes and the composition of the oil may vary widely. Moroccan oil (Marrakesh-type) has a light herbaceous odour characteristic of thujone. This species of sagebrush is widely used in folk and traditional medicine for its antiseptic, vermifuge and antispasmodic properties. Artemisia herba-alba was reported as a traditional remedy of enteritis, and various intestinal disturbances, among the Bedouins in the Negev desert. In fact, essential oil showed antibacterial activity, as well as, antispasmodic activity on rabbits. The oil contained 10 components with percentages higher than 10%. The main components were cineole, thujones, chrysanthenone, camphor, borneol, chrysanthenyl acetate, sabinyl acetate, davana ethers and davanone. The oil also contains a number of irregular monoterpene alcohols, e.g. artemisia alcohol, and the anthelmintic sesquiterpene lactone α-santonin has been isolated from the flowering branches. Artemisia oil is used in fairly large amounts in fine perfumery, e.g. for Chypre notes, one great example being Kouros Fraicheur (Yves Saint Laurent 1993). Investigations on the medicinal properties of A. herba-alba extracts reported anti-diabetic, leishmanicidal, antibacterial and antifungal properties.

Artemisia pallens. Flower/ Leaf/ Stem Oil. The leaves and flowers yield an essential oil known as oil of Davana. Several species yield essential oil and some are used as fodder, some of them are a source of the valuable anthelmintic drug santonin. Davana blossoms are offered to Shiva, the God of Transformation, by the faithful, and decorate his altar throughout the day. Davana has been widely used In Iraqi and Indian medicine for the treatment of diabetes mellitus. Oral administration of an aqueous/ methanolic extract from the aerial parts of the plants was observed to reduce diabetes in glucose–fed hyperglycemic and alloxan-treated rabbits and rats. Davana oil is soothing to rough, dry, chapped skin, skin infections and cuts.

Artemisia vulgaris. Artemisia. Used in fomentations for skin diseases and ulcerative sores. The entire plant is often made into a decoction and used as a wash for all sorts of wounds and skin ulcers. The boiled leaves are used as a poultice to allay headaches and nervous twitching of the skin and muscles. The dried leaves cut into small fragments are used to help induce more rapid scarring of unhealed wounds. Practitioners also use the leaves in eczema, herpes and purulent scabies. Wormwood extract is the main ingredient in absinthe, a toxic liquor that induces absinthism, a syndrome characterized by addiction, gastrointestinal problems, auditory and visual hallucinations, epilepsy, brain damage, and increased risk of psychiatric illness and suicide. Thujone-free wormwood extract is currently used as a flavoring, primarily in alcoholic beverages such as vermouth. Also known as armoise.

Azadirachta indica. Neem. The medicinal and antimicrobial activity of plant extract has been known for generations. The earliest use of a plant being used as human medication is found on an Egyptian papyrus dated about 1550 B.C. (*The Ebers Papyrus*—ACD). Almost every part of the neem tree is used in traditional medicine in India, SriLanka, Burma, Indochina, Java and Thailand. The stem, root bark, and young fruits are used as a tonic and astringent and the bark has been used to treat malaria and cutaneous diseases. The tender leaves have been used in the treatment of worm infections, ulcers, cardiovascular diseases and for their pesticidal and insect-repellant actions. It is used to reduce dental caries and inflammation of the mouth when used as an ingredient in dental preparations. It is a naturally occurring oil (from seeds of *A. indica*) with pronounced antimicrobial properties. It is not really an essential oil, but is often mistaken for one!

B

Backhousia citriodora. Lemon Myrtle. The essential oil distilled from the leaf has strong antibacterial, antifungal and antiviral properties. It has a fine, rounded lemon scent with a somewhat spicey undertone. Its antibacterial qualities are more powerful than tea tree oil. The antimicrobial and toxicological properties of the Australian essential oil, lemon myrtle,

(*Backhousia citriodora*) were investigated. Lemon myrtle oil was shown to possess significant antimicrobial activity against the organisms Staphylococcus aureus, Escherichia coli, Pseudomonas aeruginosa, *Candida albicans*, methicillin-resistant *S. aureus* (MRSA), *Aspergillus niger*, *Klebsiella pneumoniae* and *Propionibacterium acnes* comparable to its major component—citral.

Barosma betulina. Leaf oil. Buchu essential oil is known by the common name round leaf buchu. It has historically been used as a flavoring agent and an herbal remedy. Buchu essential oil and extract of the leaves are used as flavouring for teas, candy, and a liquor known as buchu brandy in South Africa. The extract is said to taste like blackcurrant. Buchu essential oil has a minty, camphorous and sweet berry aroma. It is a clear, colourless or slightly yellow, mobile liquid. It has a fresh top note. Buchu essential oil originates from the South Africa and is extracted by steam distillation of the fresh flowering tops of the shrubs. Buchu essential oil is a member of the Rutaceae family.

Betula alba. Betula Alba Oil is the volatile oil obtained from the Birch, *Betula alba* L., Betulaceae. Birch tar oil resembles oil of Cade in its properties, and is used for external application in the form of ointment (10%) or soap (10%) for eczema, psoriasis, and other skin conditions. Mixed with other essential oils it is used to keep away mosquitoes. The crude tar from Birch is used in pharmaceutical preparations for dermatological diseases. Traditional use: The oil is also applied externally in skin complaints. The oil contains methyl salicylate and is a good substitute for wintergreen oil as a rheumatic liniment. A decoction is good for bathing skin eruptions. The dried leaves are described as having healthful benefits for skin chaffing.

Betula lenta. Birch, Sweet.

Boronia megastigma. Boronia Megastigma Flower Oil is an essential oil obtained from the flowers of the plant, *Boronia megastigma*, Rutaceae. *Boronia megastigma* is a species of shrub in the citrus family known by the common name brown boronia. This is one of several species of Boronia cultivated for its intense, attractive scent. It is the main Boronia source of essential oils, The two main aroma compounds of the oil of this species are β-ionone and dodecyl acetate. The oil is used in perfumes and as a food additive that enhances fruit flavors. It is described as having a woody and violet note. Others say it is reminiscent of cassie, violet warm sweet woody fresh fruity green.

Boronia absolute oil is solvent extracted with petroleum ether from the flowers. The shrub is native to Tasmania where it is harvested commercially. The concrete is a thick dark green oil. The absolute is a somewhat lighter green oil. Boronia Absolute Oil has a rich tenacious floral undertone of warm, woody, sweet character. Fresh, fruity, green and sweet. It is used in perfumery and is one of the most expensive oils. Boronia absolute is extracted from blossom primarily for use as a food additive. A major component is b-ionone and B. megastigma is one of the commercial, natural sources of this compound.β-ionone, dodecyl acetate, α-pinene, β-pinene and limonene in the oil extract were also present to varying degrees depending on location of crop. Therapeutic benefits: Uplifting. Calming to the nervous system.

Boswellia carterii. Frankincense. Fresh, woody, balsamic, slightly spicy and fruity fragrance. Externally, it served in the treatment of stiffness, blood vessels, joints, and various wounds. It is also used in inflammatory conditions, pain in the legs, infections, stomach problems, pressure in the ear and to stimulate birth. The oil was used as an ingredient in embalming liquids and in mummification. It is also used to treat various diseases of the eyes, toothaches, etc. The smoke was considered helpful for women's problems, to eliminate odors in the home, clothing, or body. It was known as a multipurpose disinfectant. Mixed with pomegranate juice it found use as an astringent.

Boswellia sacra. Frankincense. *Boswellia sacra* is a tree in the Burseraceae family. It is the primary tree in the genus Boswellia from which frankincense, a resinous dried sap, is derived. Some literature identifies *B. sacra* as growing in Oman and Yemen, and *B. carterii* as growing in Somalia. The latest scientific opinion is that these are both the same species and should correctly be called *B. sacra*. The trees start producing resin when they are about 8–10 years old

Boswellia serrata. Frankincense, Indian. Indian frankincense is a gum resin from *Boswellia serrata* of Burseraceae used in Ayurveda and Western medicine for the antinflammatory effects of boswellic acids. *B. serrata* is listed in the USDA Database/Plants Profile as Indian frankincense, which was not considered true frankincense by traditional standards. It produces a soft, odorous resin that hardens in a year. As a result, it is used as incense solely by the natives.

Bulnesia sarmientoi. Bulnesia Sarmientoi Wood Oil is the volatile oil obtained from the wood of *Bulnesia sarmientoi* (Zygophyllaceae). Also called Pala Santo, Ibiocaí, and Verawood. It is a tree that inhabits a part of the Gran Chaco area in South America, around the Argentina-Bolivia-Paraguay border. Its wood is often traded as Argentine lignum vitae or Paraguay lignum vitae, since it has properties and uses similar to the "true" lignum vitae trees of genus Guaiacum, which are close relatives. Palo santo is appreciated for the skin-healing properties. From its wood, also, a type of oil known as oil of guaiac (or guayacol) is produced, to be used as an ingredient for perfumes. Fractional distillation of guaiacwood oil leads to guaiacwood alcohols which are a blend of isomers (guaiol and bulnesol) generally called commercially Guaiol. Natives of the Chaco region employ the bark to treat stomach problems. Its resin can be obtained by means of organic solvents. The oil is also obtained by steam distillation of the ground wood and sawdust. It has a soft, "precious-woody" odor. Two sesquiterpene alcohols, guaiol and bulnesol, make up about 85 % of the oil. The oil is highly prized by the inhabitants of the Dry Chaco due to its healing properties for skin wounds. Small pieces of the wood are also used as a form of natural incense in spiritual rituals. During indigenous marriage ceremonies, and in the absence of witnesses, the couple must plant a seedling of Palo Santo as a symbol of linking their destinies. Contains: β-guaiene, guaioxide, elemol, germacrene-B, eudesm-5-en-11-ol, γ-eudesmol, α-eudesmol and (-)-hanamyol.

C

Callitris intratropica. Blue Cypress, Northern Australian Cypress Pine. Traditional use: The essential oil is normally used, which has antibacterial activity. Folklore: Blue Cypress oil is obtained by steam distillation of the bark and wood. Callitris Introtropica Wood Oil is the volatile oil obtained from the wood of *Callitris introtropica*, Cupressaceae. The INCI name is incorrectly spelt. The oil is extracted from *Callitris intratropica* pine. It consists of a mixture of around 190 constituents. The oil contains 8 major constituents which are closely related in structure. These can be further divided into two subgroups; the alcohols (56.5% of the oil) and the alkenes (10% of the oil). The composition of the oil: The commercial oil contains sesquiterpenes such as β-elemene, α-guaiene and δ-selinene, and sesquiterpene alcohols such as guaiol (26%) and β-eudesmol (6.3%). The blue colouring may be ascribed, at least in part, to the presence of guaiazulene (1.6%) although other complex structures with a resemblance to the azulene moiety are present.

Camphor. Camphor is a waxy, white or transparent solid with a strong, aromatic odour. It is a terpenoid with the chemical formula $C_{10}H_{16}O$. It is found in wood of the camphor laurel (Cinnamomum camphora), a large evergreen tree found in Asia (particularly in Borneo and Taiwan) and also of Dryobalanops aromatica, a giant of the Bornean forests. It also occurs in some other related trees in the laurel family, notably *Ocotea usambarensis*. Dried rosemary leaves (Rosmarinus officinalis), in the mint family, contain up to 20% camphor. It can also be synthetically produced from oil of turpentine. It is used for its scent, as an ingredient in cooking (mainly in India), as an embalming fluid, for medicinal purposes, and in religious ceremonies. A major source of camphor in Asia is camphor basil. Camphor is readily absorbed through the skin and produces a feeling of cooling similar to that of menthol, and acts as slight local anesthetic and antimicrobial substance. There are anti-itch gels and cooling gels with camphor as the active ingredient. Camphor is an active ingredient (along with menthol) in vapor-steam products, such as Vicks VapoRub. Although touted as a cough suppressant, it has no effects on respiratory tract function. A recent publication in Pediatrics suggests the topical application of VapoRub may improve symptoms of colds and sleep quality when compared to a control.

Cananga odorata. Ylang-ylang. Ylang-ylang is extremely effective in calming and bringing about a sense of relaxation. It is antispasmodic, balances equilibrium, said to help with sexual disabilities and frigidity and has been used traditionally to balance heart function. Ylang-ylang in the Malayan language means "flower of flowers." The scent is very sensual, sweet and reminiscent of almonds. It is mentally relaxing and soothing. It is useful in treating insomnia, anger, anxiety and low self-esteem. It is said to relax facial muscles, and a massage with ylang-ylang helps to ease tension headaches.

Canarium commune gum oil**.** Elemi oil is extracted from the resin of *Canarium luzonicum* (also known as *C. commune*) of the Burseraceae family and is also known as Manila elemi. The tree exudes a pale yellow resin when the tree sprouts leaves. The resin solidifies on contact with the air and the resin stops flowing when the tree loses its leaves. The Elemi tree is known locally as 'Pili" and the gum is exported from Manila in two qualities: 'Primera' which is cleaned gum and 'Secunda' which is still crude and unclean. The ancient Egyptians used elemi oil in the embalming process. These days is used in incense, soaps and varnish. Elemi oil is extracted from the gum by steam distillation. A resinoid and resin are also produced in small quantities. The main chemical components of elemi oil are terpineol, elemicine, elemol, dipentene, phellandrene and limonene. Gum Elemi (soft) *Canarium commune* from the Philippines. Elemi produces a bright lemony, woody fragrance with a hint of fennel, frankincense and grass. Elemi is known to be clarifying and cleansing with energizing properties. It stimulates mental ability and works well for morning meditation, tai chi or yoga exercises. It creates a spirit of hopefulness and is said to relieve depression. Traditionally, people use elemi with substances that are refreshing and cleansing such as mastic, lemongrass, and sweet grass.

Carum petroselinum Parsley Oil. Carum Petroselinum Seed Oil is a volatile oil obtained from the seeds of the Parsley, *Carum petroselinum* L., Apiaceae. Traditional use: Fresh crushed Parsley can be used to relieve painful insect stings and bites. Can be used to sooth and heal abrasions and cuts. It has slight bleaching properties, and it has been used in the past to lighten freckles. Can help combat oily skin. Folklore: In France, a popular remedy for scrofulous swellings is green Parsley and snails, pounded in a mortar to an ointment, spread on linen and applied daily. In Greek mythology, Parsley is said to have sprung up from the blood of Archemorus. It contains apiol, terpene and pinene.

Cedrus atlantica. Cedarwood (Atlas Cedar). Good for stress related disorders. Said to soothe acne, eczema, arthritis and rheumatism. One of the most ancient oils traditionally used as a fixative in the perfume industry. Soothing woody aroma. It is an oil that is helpful for oily skin and itchy scalp. Add to a fragrance jar in a wardrobe to repel moths. A very calming oil for respiratory problems. The oil is widely used for insect repellent activities and Turkish carpet shops are walled with cedarwood boards to deter the moths.

Cedrus Deodara. Cedrus Deodara Wood Oil is the volatile oil obtained by steam distillation of the stumps of the Deodar Cedar, *Cedrus deodara*, Pinaceae. It is the national tree of Pakistan and the name deodar comes from the Sanskrit word 'devdar' which means 'timber of the gods'. The wood is strong, durable and fragrantly scented and as it has religious associations is typically used for construction in temples and palaces. The curative properties of Deodar are well recorded in Indian Ayurvedic medicines which include many skin diseases such as eczema and psoriasis. The inner wood is aromatic and used to make incense as well as being distilled into essential oil. Insects avoid this tree and the essential oil has insect repellent properties. It has antifungal properties. Cedar oil is often used for its aromatic properties, especially in aromatherapy. It has a characteristic woody odor.

Chamaemelum nobile. Chamomile, Roman. May help with insomnia, muscle tension, cuts, scrapes and bruises. It is useful against infestions and is used extensively in Europe for skin disorders. Soothing and calming, especially on nervously excited children.

Chamomilla recutita (German Chamomile). Soothing, calming, restoring chamomile brings peace to the most troubled of skin. German Chamomile is an exceptionally useful and versatile member of the Compositae (daisy) family. Both the oil and the aqueous extract, which are prepared from thc flowers, have beneficial properties. The water-soluble flavonoids (particularly apigenin and apigenin-7-glucoside) present in the extract give it anti-inflammatory, sedative, wound healing, sedative and soothing properties. It is listed in all of the major pharmacopoeias, including the British Pharmacopoeia, the British Herbal Compendium and the British Herbal Pharmacopoeia.

Chenopodium ambrosioides. Wormseed. The essential oil is derived from the aerial parts and is called Oleum Chenopodii. The seeds also contain a toxic essential oil and an overdose can result in poisoning and death. The active ingredient in wormseed is ascaridol, a volatile oil. Wormseed, as the name suggests, is still used as an anthelmintic; however there are some risks involved since the oil is very toxic.

Cinnamomum camphora. Camphor (White). Camphor is well-known for its analgesic and infection-fighting abilities when used in combination with eucalyptus oil. The U.S. Food and Drug Administration's *Over-The-Counter (OTC) Drug Review Ingredient Status Report*, (December 1991) listed camphor as a Category I ingredient for fever blisters and as a counterirritant in the External Analgaesics Monograph. Camphor is also listed as an antitussive ingredient in the cough and cold monograph. A nasal product indicated for the relief of nasal irritations and nasal congestion due to colds consists entirely of a blend of essential oils cajeput, eucalyptus and peppermint. Methyl salicylate, or oil of wintergreen, is listed as a counterirritantin the external analgaesic monograph.

Cinnamomum camphora. Camphor (Yellow), Ho Oil. The essential oil has been used as an anthelmintic, antirheumatic, antispasmodic, cardiotonic, carminative, diaphoretic, sedative and tonic. It has been used externally in liniments for treating joint and muscle pains, balms for chilblains, chapped lips, cold sores and skin diseases. It is often used as an inhalant for bronchial congestion. Some caution is advised, excessive use causes vomiting, palpitations, convulsions and death. It is possible that the oil can be absorbed through the skin, causing systemic poisoning. The essential oil is used in aromatherapy.

Cinnamomum camphora linalooliferum. *Cinnamomum Camphora Linalooliferum* Leaf oil is an essential oil obtained from the leaves of the Camphor Tree, *Cinnamomum camphora* var. *linalooliferum*, Lauraceae. Syn. Ho Leaf Oil. The chemical variants (or chemotypes) seem dependent upon the country of origin of the tree. The tree is native to China, Japan, and Taiwan. It has been introduced to the other countries where it has been found, and the chemical variants are identifiable by country. i.e., *Cinnamomum camphora* grown in Taiwan and Japan, (often commonly called "Ho Wood") is normally very high in linalool, often between 80 and 85%. In India and Sri Lanka the high camphor variety/chemotype remains dominant. It has many extractives and physically modified derivatives such as tinctures, concretes, absolutes, essential oils, oleoresins, terpenes, terpene-free fractions, distillates, residues, etc. A slightly spicy, sweet, woody and floral scent that is sometimes called Ravintsara (not the same as Ravensara) but as this variety is from China, it is best called Ho Leaf. The tree grows in Madagascar as well as China but produces two very different oils. It is said to have similar properties to Rosewood. It is a gentle oil that can help relieve fatigue, soothe the mind and balance the emotions. It is also supportive to the immune system and has antiseptic, antibacterial and antifungal properties. It is also believed to be a mild aphrodisiac. Although their chemical compositions may be nearly identical, Ho Wood and Rosewood do not smell alike at all. Rosewood has a sweet, soft, rose scent (just as one would expect), while Ho Wood is astringent, sharp and camphor-like. It is said Ho Leaf is often used as an environmentally friendly alternative to Rosewood essential oil. The information is conflicting, for when compared they appear dissimilar. Aphrodisiac, antiseptic, antibacterial are not confirmed by our own research yet.

Cinnamomum cassia. Cassia. It yields Cassie oil for barber shops, was one of the holy annointing oils mentioned in Exodus as being used by Moses on sacred occasions. The dried bark is used. Cassia is chiefly used to scent potpourri and to flavor chocolate; but in China it is given as an antiseptic and as a digestive tonic, and it flavors other medicines. Essential oil, basically cinnamon essential oils, comes in two forms—cinnamon bark, which is steam distilled from the inner bark of the tree; and cinnamon leaf, steam distilled from the leaves and twigs. Cinnamon leaf essential oil has many uses in modern practice, and in the *British Herbal Pharmacopoeia* as a specific for flatulent cholic and dyspepsia with nausea.

Cinnamomum cassia. Cinnamon. Cinnamon is a cooking spice that has many medicinal therapeutic properties that include anticancer effects, however, its inherent allergic and irritant properties can lead to contact stomatitis in sensitive individuals.

Cinnamomum zeylanicum. Cinnamon. Essential oil, basically cinnamon essential oils, comes in two forms–cinnamon bark, which is steam distilled from the inner bark of the tree; and cinnamon leaf, steam distilled from the leaves and twigs. There is also a cinnamon cassia, but this is from a slightly different subspecies and is discusssed under "Cassia." Cinnamon bark oil is mostly used for fragrance and flavoring, and is not to be used in aromatherapy, as it may be harmful on the skin. However, cinnamon leaf essential oil has many uses in modern practice, and is current in the *British Herbal Pharmacopoeia* as a specific for flatulent cholic and dyspepsia with nausea. The refreshing and cooling quality of the bark is due to the presence of methyl amyl ketone. Cinnamon oil has antifungal, antiviral, bacteriacidal and larvicidal properties.

Cistus ladaniferus** **oil. Cistus is a small, sticky shrub native to the Mediterranean and Middle East. It is a pale yellow liquid with a very strong, sweet, dry-herbaceous, musky aroma.The odour is deep, sweet balsamic, faintly but persistently herbaceous with an amber-like, rich backnote. The latter is characteristic and rarely found in other perfume materials. In the right quantities, and if the fragrance is perceived as pleasing, sweet-floral scents lull the senses into a dreamy, relaxed state. Labdanum (cistus oil) is obtained as a gum by boiling the leaves of Rock Rose (Cistus) and then distilling to obtain the oil. The gum can be combined as an incense of frankincense, myrhh and galbanum. Cistus is a very useful oil. Labdanum is used as a fixative in perfumery with the aroma seen as an alternative to ambergris. Its odor effect is generally perceived as warming and restorative. It acts as an antiseptic. Cistus essential oil can be used on the skin in a carrier, alone or included with other essential oils. It has an uplifting aroma but should be used sparingly as it can overpower other essential oils. It is considered one of the ancient spiritual oils with a history of being used in incense. It is used extensively in oriental style perfumes and can be found as a flavoring in food, alcohol and soft drinks

Cistus ladaniferus resin. It is often called Labdanum gum and is the resinous matter derived from the plant *Cistus ladaniferus* and other species of cistus by boiling the leaves and twigs of this plant in water. It is a natural oleoresin and differs slightly from other oleoresins as it contains more waxes and less volatile oil than most of the other natural oleoresins. The plant grows wild in most countries around the Mediterranean but the production of labdanum is found mainly in Spain. The gum is skimmed off the surface of the water and mixed with other resinous matter, which sinks to the bottom of the boiling water. It is a small shrub, the white flowers of which have only a very faint odor. The flowers as such are not exploited in perfumery. Labdanum gum is a dark brown, more or less solid mass. When fresh, labdanum is plastic but not pourable. The odor of Labdanum is sweet, herbaceous balsamic, somewhat amber like and slightly animalistic, rich and tenacious.

Citrus aurantifolia. Lime. Acts like lemon and the other citrus oils, beautiful tangy, fresh stimulating aroma. Warning: Do not use lime on the skin in direct sunlight; however, if the essential oil of lime is distilled rather that expressed, then it does not have a phototoxic effect. The aroma enhances and enlivens the mood and energizes, and can help relieve fatigue and stimulate mental activity and memory.

Citrus aurantium amara. Orange, Bitter. A volatile oil obtained by expression from the fresh peel of the bitter orange. Dried peel is official in the *British Pharmacopoeia* as a bitter tonic. It may have applications as a topical antifungal agent; oil of bitter orange was effective in curing patients with treatment-resistant fungal skin diseases. In vitro tests show that limonene from citrus peels may have relevant anticancer, antitumor, and cell-differentiation promoting activities.

Citrus aurantium amara. Petitgrain. It is made from the twigs and buds of the orange tree, similar properties and aroma to neroli (see *C. aurantium dulcis*), but not quite as sophisticated or floral. Bitter orange is regulated by the U. S. Food and Drug Administration (FDA). Although *C. aurantium* is listed in its Poisonous Plants Database, *C. aurantium* orange oil extract, peel, flowers, and leaf are listed in its food additive database, an inventory often referred to as "Everything" Added to Food in the United States (EAFUS). In frozen concentrated orange juice, the volume of bitter orange that may be added cannot exceed 5%. Petitgrain oil is extracted from the leaves of the tree but was once extracted from the green unripe oranges, when they were still the size of cherries, hence the name Petitgrain or "little grains."

Citrus aurantium bergamia. Bergamot. It is used for oily skin, acne, seborrhea of the scalp, herpes, psoriais, ulcers and wounds. Bergamot is an antidepressant and gentle relaxant and has a refreshing, uplifting quality.

Citrus aurantium dulcis. Neroli. The essential oil extracted from the fragrant flower of the bitter sour, or Seville orange tree, also known as *Citrus bigaradia*. The therapeutic properties are effective in treating the nervous system. For insomnia the oil has an almost hypnotic effect and a few drops into a warm bath prior to bedtime will help give a good night's sleep. The Queen of all essential oils.

Citrus aurantium dulcis. Orange, Sweet. The volatile oil obtained by expression from the ripe peel of the fruit *Citrus sinensis* (L.) Osbeck (Fam. Rutaceae)/*C. dulcis*.

Citrus clementina. Citrus Clementina Peel Oil is the volatile oil obtained from the peel of Citrus clementina, Rutaceae. A clementine is a variety of mandarin orange (C. reticulata), named in 1902.The exterior is a deep orange color with a smooth, glossy appearance. Clementines separate easily into seven to fourteen moderately-juicy segments. They are very easy to peel, like a tangerine, but are almost always seedless. Clementines are, thus, also known as seedless tangerines. Like all citrus fruits, limonene is a major constituent of oils in clementines. Other constituents include linalool, α-terpineol, α-pinene, β-pinene and myrcene. Traditional uses: Clementines can keep cats from nibbling at houseplants, just rub the peel from the clementine on the leaves.

Citrus grandis. Grapefruit. Grapefruit oil is a good astringent when used as a facial toner. It has a cooling, refreshing and stimulating effect on lifeless skin and jaded senses. When inhaled, grapefruit is an antidepressant and helps relieve anxiety. It can be effective in treating symptoms of premenstrual syndrome and menopause. It is also a good addition to air freshner preparations and proves to be especially effective against kitchen smells and odors.

Citrus medica limonum. Lemon. Refreshing, revitalizing and stimulating. This oil is used wherever a fresh, awakening and invigorating property is needed in foam baths, shower gels or massage oils. Lemon oil is stimulating, calming, carminative, astringent, detoxifying, antiseptic, disinfectant, sleep inducing and has antifungal properties.

Citrus reticulata. Mandarin, Petitgrain. A sweet citrus, fruity essential oil with the typical oriental orange smell. Used in aromatherapy for acneic skin types. Tangerine (*Citrus reticulata*) is a citrus fruit that is well-known for being sweet and easy to peel. Tangerine contains vitamin C, folate, and β-carotene. Several laboratory studies have shown that tangerine may have antioxidant properties. Tangerine peel has also shown antineoplastic activity *in vitro*. However, there is currently insufficient available evidence in humans to support the use of tangerine for any medical indication.

Citrus reticulata. Tangerine. Tangerine oil is extracted from *C. reticulata* (also known as *C. nobilis*, *C. madurensis*, *C. unshiu* and *C. deliciosa*) of the Rutaceae family and is also known as European mandarin, tangerine, naartjie and true mandarin. The odor is the traditional oriental, slightly spicey orange reminiscent of Christmas. The oil is described as calming, sedating, anti-inflammatory, and is said to help with anxiety, dizziness and nervousness. It has a warming freshness.

Citrus sinensis. Orange, Sweet. See *C. aurantium dulcis* [Syn. *C. sinensis*]. The essential oil of orange is warm, radiant, sweet, uplifting and best described as alive. It is uplifting to the mood, while promoting relaxation and being calming. It is also calming and brightening to dull complexions.

Citrus tangelo. Tangelo. No data. The tangelo is another mandarin-, tangerine-, clementine-type orange fruit.

Citrus tangerina. Citrus tangerina. Tangerine. Tangerine oil is extracted from *C. reticulata* (also known as *C. nobilis*, *C. madurensis*, *C. unshiu* and *C. deliciosa*) of the Rutaceae family and is also known as European mandarin, tangerine, naartjie and true mandarin. The odor is the traditional oriental, slightly spicy orange reminiscent of Christmas. The oil is described as calming, sedating, anti-inflammatory, and is said to help with anxiety, dizziness and nervousness. It has a warming freshness.

Commiphora erythraea. Opopanax. Opopanax is used as a fixative and fragrance for high class perfumery. Aromatherapy uses have been suggested to be similar to myrrh oil (to which it is related). Opopanax is often adulterated. Myrrh is commonly used in Chinese medicine for rheumatism, arthritis and circulatory problems. It is employed in perfumery. Not to be confused with Cassie (*Acacia farnesiana*), which is also called Opopanax.

Commiphora myrrha. Myrrh. Myrrh has been used as a medicine and for ceremonial and religious purposes. In traditional medicine, myrrh is used for embalming, leprosy, bronchitis, diarrhea, dysentery, typhoid, mouth ulcers, inflammation, viral hepatitis, female disorders, wounds, coughs, tumors etc. It is also used to some extent in Ayurveda and Unani medicine, although more preference is given to the related resin known as guggulu obtained from *Commiphora mukul* Engl. The Commission E approved myrrh for topical treatment of mild inflammations of the oral and pharyngeal mucosa. *The British Herbal Pharmacopoeia* indicates myrrh tincture as a mouthwash for gingivitis and ulcers. Myrrh is also an important drug in Chinese Traditional Medicine. It is used in East Africa as a decoction of myrrh resin to treat stomach ache.

Copaifera officinalis. Balsam Copaiba resin**.** Indigenous peoples in the Amazon region have long applied balsam copaiba as a healing agent on sores and ulcers and have also used it as an insect repellent. Folklore: Amazonian Indians have long known the Copaiba balsam. They use it to treat wounds, cure ulcers and cure the navel of newborn babies.

Coriandrum sativum. Coriander. The oil is said to have antiinflammatory and sedative properties. Coriander has been used as a flavoring and medicine since ancient times. Seeds have been found in the tombs of Pharohs, and the Roman legions carried coriander as they progressed through Europe, using it to flavor their bread. Externally it has been used as a lotion to treat rheumatic pains.

Croton glabellus. Croton Glabellus Bark Oil is the volatile oil obtained from the bark of Croton glabellus L., Euphorbiaceae. Cascarilla Bark, Cocomcoli, Eleuthera Bark, Kouli, Quina Aromatica, Quina Falsa, Quina Morada, Sassafras, Serosee, Sweetwood Bark. The constituents of the bark include cascarillin, tannins, resins and an essential oil which together produce the pungent, bitter and astringent properties of extracts of the bark. The bitter-tasting diterpenoid which is Cascarillin A is possibly the ingredient which makes cascarilla extracts effective treatment for stomach indigestion. The essential oil is an international commodity, which is produced by steam distillation and by hydro-diffusion from the bark. The specifications for the steamdistilled essential oil show its major constituents to be α- and β-pinene followed by myrcene and β-selinene. In perfumery terms, cascarilla bark essential oil is said to provide a fresh spice woody black pepper sweet anise odor. Another source describes it as of balsam woody spice terpene herbal.

Cuminum cyminum. Cumin. Cumin is an aromatic, astringent herb that benefits the digestive system and acts as a stimulant to the sexual organs. Orally, cumin is used as an antiflatulent, stimulant, antispasmodic, diuretic, aphrodisiac, for stimulating menstrual flow, treating diarrhea, colic and flatulence; also for chest conditions and coughs, as a pain killer and to treat rotten teeth. The essential oil obtained from the seed is antibacterial and larvicidal. In spices, foods, and beverages, cumin is used as a flavoring component. Cumin oil is used as a fragrance component in cosmetics (maximum use level 0.4% in perfumes). The chemical composition of the essential oil was examined by GC and GC–MS; 37 components, representing 97.97% of the oil, were identified. Cuminal (36.31%), cuminic alcohol (16.92%), γ-terpinene (11.14%), safranal (10.87%), p-cymene (9.85%) and β-pinene (7.75%) were the major components. Another source provided the analysis α–pinene (29.2%), limonene (21.7%), 1,8-cineole (18.1%), linalool (10.5%), linalyl acetate (4.8%), and α-terpineole (3.17%) were the major components of the essential oil from *C. cyminum* L., and the oil showed a strong inhibitory effect on fungal growth.

Cupressus funebris. It is produced by the steam distillation of the wood of Chinese Weeping Cypress, which is an evergreen shrub growing in Kweichow and Shaansi, Guizhou, Kansu, and-Szechwan provinces. The oil and wood are also used to prepare incense in China.The analysis is given as: α-cedrene 27.1%, β-cedrene 8.8%, thujopsene 30.0%, α-himalchene 0.5%, β-himalchene 0.5%, curcumene 0.4%, α-selinene 3.0%, cuparene 3.3%,cedrol 12.0%, widdrol 4.9%, methyl carvacrol 0.6%. The oil from *C. funebris* is pale yellow or yellow-green, with an odor that is almost invariably smoky, woody-, bordering on crude, with an almost cresylic character. In comparison to cedarwood oil it appears smooth woody, slightly oily pencil sharpenings, and somewhat slightly sweet in comparison, whereas the Chinese oil is altogether more powerful.

Cupressus sempervirens. Cypress. Funeral Cypress, Italian Cypress, Mediterranean Cypress, Mediterranean Cyprus. It is used for circulatory conditions where its effects as a venous tonic make it helpful for varicose veins, hemorrhoids, bruising and thread veins. It helps decongest the circulatory and lymphatic systems and may help with edema, cellulite and water retention. It has been found useful for muscular stiffness, arthritis, rheumatism, cramps and post-sports treatment. It is also used externally as a foot deodorant. It is anti-infectious, antibacterial and antimicrobial.

Curcuma zedoaria. Zedoary. It is a highly aromatic species related to turmeric. Known as zedoary, it is native to India and Indonesia. During the sixth century it was brought to Europe by Arab traders and had some success in medicine and as a source of perfume, reaching the height of its popularity in the Middle Ages. Aromatic, stimulant. Used in a similar manner to ginger. The rhizome is used to treat certain types of tumors in China.

Cymbopogon citratus. Lemongrass. With its lemony aroma, it is an excellent general skin tonic and antiseptic. It is also believed to soothe fevers, and to help relieve migraine. Said to normalize overactive oil glands, and so good for acne and open pores. Makes a good skin tonic.

Cymbopogon flexuosus. Lemongrass. It is known as cochin lemongrass or British Indian lemongrass and has similar properties to *C. citratus*. The aroma definitely stimulates the brain and facilitates the logical thinking process. It is useful where clear thinking and powers of concentration are required and will refresh a weary traveler.

Cymbopogon martini. Palmarosa. This oil is floral, fresh and sweet with the characteristic smell of geranium oil, but with a rose-overtone, hence the alternative name rose geranium. It helps with skin problems such as candida, rashes, scaly and flaky skin. It is antimicrobial, antibacterial, antifungal and antiviral. It is also used as an insect repellent and a carminative. It has been used on cuts, as an insect repellant and to relieve headaches. Contains a high level of geraniol. It has antimicrobial properties and has been said to help with rashes and other skin problems. In aromatherapy it is used to aid in clarity of mind.

Cymbopogon martinii var.sofia. Gingergrass. Very close relative of the delightful Palmarosa oil, but with a hint of ginger and a different chemotype known as Gingergrass. Said to be anti-inflammatory and to have insect repellant properties.

Cymbopogon nardus. Citronella. It is extracted from an Asian grass and used for its fragrancing properties in the food industry, as well as in perfumes and body care. It also has good insect-repelling properties.

Cymbopogon schoenanthus. Lemongrass. Geranium grass, Lemon grass, Camel grass, Fever grass, West Indian lemon grass is a herbal plant of southern Asia and northern Africa with fragrant foliage. The volatile oil obtained by steam distillation of *C. schoenanthus* (Guatamalan/light type, also known as *C.citratus*) as opposed to *C. flexuosus* (Indian/Cochin type).

Cymbopogon winterianus. Citronella (Java type). Citronella oil has been used as a flavoring for foods and beverages. In traditional medicine, the oil has been used as an aromatic tea, as a vermifuge, diuretic and antispasmodic. Perhaps the most widely recognized use for the oil is as an insect repellant. It is sometimes incorporated into perfumes and soaps. Citronella candles have been promoted as an effective way to repel mosquitoes. Sometimes referred to as *Andropogon nardus*.

D

Dianthus caryophyllus It is the volatile oil obtained from the flowers of the Girofle (carnation), *Dianthus caryophyllus* L., Caryophyllaceae.. Obtained by ethanolic extraction of the concrete, itself produced by solvent extraction of the flowers of *Dianthus caryophyllus*. Herbal remedies are only prepared from the aerial parts. The flower petals have a strong smell of cloves and are candied, used as a garnish in salads, for flavouring fruit, fruit salads etc. They can also be used as a substitute for rose petals in making a syrup. The petals should be removed from the calyx and their bitter white base should be removed. Carnation flowers are an aromatic, stimulant herb that has been used in tonic cordials in the past to treat fevers in the past. It has been prescribed in European herbal medicine to treat coronary and nervous disorders. The plant has been used as a vermifuge in China.

Dipteryx odorata. It is the seed extract of tonka beans, which are also known as *tonquin* or *coumarouna beans* or *coumarins*, has a vanilla-like fragrance. It is used as a fixative and aromatic ingredient in hundreds of products, including acne products, antiseptics, deodorants, hair care, soaps, suncare and toning lotions. Tonka Bean (Syn. *Coumarana odorata*) seeds are used. It is a member of the Family Leguminosae, and is a forest tree native to Brazil and British Guiana and called there 'Rumara'. The odor of the coumarin, which distinguishes the Tonka Bean, is found in many plants, especially in Melilotus, sweet vernal grass, and related grasses. One pound of beans has yielded 108 grains of coumarin, which is the anhydride of coumaric acid. In addition to its use in perfumery as a fixative, coumarin is used to flavour castor-oil and to disguise the odour of iodoform. Thc fatty substance of the beans is sold in Holland as Tonquin butter. One reference refers to coumarin as Tonka bean camphor. Coumarin is the odourous principle of Tonka seed (Tonka or Tonquin bean); it may be prepared synthetically. Coumarin is used as a fixative in the perfumery industry. It has been used as a flavoring agent to mask unpleasant odors but because of its toxicity this use is now considered undesirable and it is no longer used as a food additive. It is reported to be an immunostimulant and has been tried in the treatment of malignant neoplasms. Coumarin is reported to be a macrophage stimulant and has been tried with cimetidine in the treatment of melanoma and other tumors.

E

Ellettaria cardamomum. Cardamom. Good for nervous exhaustion and mental fatigue. Used as a flavoring, and also for its carminative and stimulating properties. Properties are antiseptic and stimulative. Used for its aphrodisiac effect, also has an uplifting effect, helping to clear the mind of noise and confusion. Makes an excellent bath oil— light, refreshing and stimulating.

Eriocephalus punctulatus. Chamomile, Cape. It has a fine fruity fragrance, with roman chamomilelike notes. It has been suggested that the oil can be used as an alternative to roman chamomile, and experts have suggested it could add a whole new dimension to the flavor and fragrance industry. The leaves are used in baths for its relaxing and invigorating scent. Used in pillows, the scent encourages pleasant dreams. Anti-inflammatory and soothing.

Eucalyptus citriodora. Eucaplyptus. There are a number of different Eucalyptus varieties (over a hundred) and they all have subtly different aromas. *Eucalyputus citriodora* is also known as Lemon Scented Gum and was found to be effective against the inhibition of *Candida albicans*. As the name suggests, it has the typical aroma that would be expected from eucalyptus, but with the added highlight of a deliciously refreshing citrus note.

Eucalyptus globulus. Eucalyptus. Characteristic aromatic camphoraceous odor and a pungent camphoraceous cooling taste. Eucalyptus oil taken by mouth for catarrh and used as an inhalation often in combination with other volatile substances. Eucalyptus oil has also been applied as a rubefacient. Has local antiseptic, expectorant, deodorant and refreshing effects.

Eucalyptus radiata. Eucalyptus. Anti-infectious, antibacterial, antiviral, anticatarrhal, expectorant, anti-inflammatory. *E. radiata*, with its richness in alcohols, is more suitable for infectious conditions, be they viral or bacterial. It is prefered to *E. globulus* for higher infections.

Eucalyptus smithii. Eucalyptus. Anticatarrhal, expectorant, excellent liquefaction of the secretions, digestive stimulant (internally), anti-infectious, antiviral, antibacterial, parasiticide, antirheumatic, analgesic, antineuralgic, febrifuge, stimulant and calming, balancing.

Eugenia caryophyllus. Eugenia Caryophyllus (Clove) Bud Oil**.** In its pure form, clove oil has an overpowering and heady fragrance. It was traditionally used to stop the pain of an aching tooth and was used in dental practice for its antiseptic and analgesic properties until quite recently. It is a powerful anti-inflammatory agent in the case of wasp stings and has antibacterial properties. In aromatherapy, it is found to be invigorating, stimulating and warming and is often included in a mix of essential oils to confer these properties. LD_{50} (oral, rat) 1,370 mg/kg; mod. toxic by ingestion; severe skin irritant; heated to decomp., emits acrid smoke and fumes. Clove and clove oils are used safely in foods, beverages, and toothpastes.

Eugenia caryophyllus Eugenia Caryophyllus (Clove) Flower Oil. Clove and clove oils are used safely in foods, beverages, and toothpastes. Toxicity has been observed following ingestion of the oil, but this rarely occurs.

F

Ferula galbaniflua. (Galbanum) Resin Oil. An incision made in the root exudes a juice which hardens into the tears of galbanum. It is a stimulant, expectorant in chronic bronchitis, antispasmodic. It is used externally as a plaster and in ointments for inflammatory swellings. It has been used internally for hysteria, rheumatism, chronic affections of mucous air passages. It is mentioned in the bible for making incense (Exodus XXX,34)

Foeniculum vulgare dulce. Fennel, Sweet. Antispasmodic, antiseptic and stimulating to the cardiovascular and respiratory systems. Topically, fennel is good for conjunctivitis, and blepharitis (as eyewash). Useful as an oil when rubbed onto affected parts to relieve rheumatic pains. The essential oil is used in tinctures as a gargle and eyewash and in carminative preparations. Shown to be anti-inflam-matory in rats and mice studies. The two forms of fennel are very similar and not always distinguished.

Foeniculum vulgare* var. *vulgare. Fennel, Bitter. Fennel has been used as a flavoring, a scent, an insect repellent, as well as an herbal remedy for poisoning and gastric conditions. It has also been used as a stimulant to promote lactation and menstruation. However, there is a lack of clinical evidence to support the use of fennel for any indication. The two forms of fennel are very similar and not always distinguished.

Foeniculum vulgare. Fennel. Ingestion of the volatile oil may induce nausea, vomiting, seizures and pulmonary edema. The principal hazards with fennel itself are photodermatitis and contact dermatitis. Some individuals exhibit cross-reactivity to several species of Apiaceae, characteristic of the so-called celery-carrot-mugwort-condiment syndrome. Rare allergic reactions have been reported following the ingestion of fennel. A serious hazard associated with fennel is that poison hemlock can easily be mistaken for the herb.

Fucus Vesiculosus Thalle Oil.

G

Gaultheria fragrantissima. Wintergreen, Fragrant. It has been described as a fresh, cleansing minty aroma, but the smell is better known as the characteristic smell of Germolene and the typical smell of many embrocations. It is the methyl salicylate that gives it the characteristic smell. Used for its muscle relaxing and decongestant properties. The oil is employed in a rub used externally for rheumatic and muscular pains and to flavor dental preparations. The leaves yield around 1.25% of an essential oil; this is a wintergreen substitute and it is used in perfumery, as a hair oil and medicinally.

Gaultheria procumbens. Wintergreen. An essential oil (oil of wintergreen obtained from the leaves contains methyl salicylate, which is closely related to aspirin and is anti-inflammatory). The oil is analgesic, anti-inflammatory, aromatic, astringent, carminative, diuretic, emmenagogue, stimulant and tonic.

Grass-Hay herb oil. Grass-Hay Herb Oil is an essential oil obtained from the herbs of various species of grass or hay (Gramineae). It is described as sweet herbal tea clary sage hay. This is a complicated area but hay absolute and hay oils including many 'foin' qualities are traditionally obtained from certain grasses growing in the South of France including *Lolium perenne* L. and other *Lolium* spp. including *Lolium italicum* L., *Phleum pratense* (Timothy grass), *Poa pratensis* L. (Meadow grass), *Cynosurus cristatus* (Crested Dog's-Tail), *Anthoxylum odouratum* L. and *Melilotus* spp. amongst others.

Guaiacum officinale Wood Oil**.** Guaiacum is stated to possess antirheumatic, anti-inflammatory, diuretic, mild laxative, and diaphoretic properties. Traditionally, it has been used for subacute rheumatism, prophylaxis against gout, and specifically for chronic rheumatism and rheumatoid arthritis. It is recommended that guaiacum is avoided by individuals with hypersensitive, allergic, or acute inflammatory conditions.

H

Hayflower Oil. Traditional use: The blossoms of wild hay of the mountains are an herbal aid for healing detoxifying baths, ointments and poultices. The hay flowers can be used for foot and hand soaks and in the bath. They have a good effect on the skin and are said to help remove toxins and reduce inflammation. Folklore: Hayflowers are *Graminis flores* or a mixture of the Poaceae species. They are found throughout Central Europe and harvested throughout the summer. The flowers and complete inflorescences of various poaceae herbs are collected—used in medicinal bath preparations and soothing compresses for the treatment of rheumatism, lumbago and chilblains.

Helichrysum angustifolium. Helychrysum. The plant is known as Everlasting or Immortelle and has a fresh, earthy, almost herbaceous aroma. It is said to be good for stretch marks and particularly useful in cases of damaged and problematic skin conditions. Said to reduce skin redness, skin bruises and is good for acneic skins.

Helichrysum angustifolium. Immortelle. See ***H. angustifolium***.
Hibiscus abelmoschuus. Ambrette. It is aromatic, insecticide. It is known as an insecticide, being dusted over woollens to protect them from moths etc. It is antispasmodic, nervine, stomachic. An emulsion made from the seeds is said to be useful for spasmodic problems. An emulsion made from the milk can be used for itchy skin.

Humulus lupulus (Hops) Cone Oil**.** It is sedative, visceral antispasmodic, bitter digestive tonic, locally antiseptic and healing. Some extracts are used as emollients in skin preparations. Folklore: It was used by the ancients as a tonic. Hops were introduced to England by the Dutch in the 16th century. The common name is said to originate in Old English hopen or hoppan, meaning "to climb." Its most known use is in beer brewing. Function: Fragrance ingredient/ masking. Extracts and oil are used as flavoring in nonalcoholic beverages, frozen dairy desserts, candy, baked goods, gelatins, and puddings, with the highest average maximum use level of 0.072% reported for an extract used in baked goods.

Hyssopus officinalis. Hyssop. The essential oil is spicy, fresh, warm and woody. The scent uplifts the mood, provides direction and suggests purity and clarity of spirit. It has the ability to inspire, increase concentration and focus on difficult tasks. Hyssop is believed to bring quick relief to the pain and bruising of a black eye.

I

Inula graveolens. Inula. This is a powerful respiratory oil and mostly used for clearing congestion and easing breathing. It is said to be good for chest infections. The common name stinkweed, or stinkwort, is not particulalry pleasant. Some aromatherapists use it for backache and muscle cramps.

Inula helenium. Alantroot. See ***I. graveolens***.

Inula helenium. Elecampane. See ***I. graveolens***.

Iris florentina Root Extract. Iris Florentina Root Extract is an extract of the roots of the Orris, *Iris florentina* L., Iridaceae. Orris root is a term used for the roots Iris germanica, Iris florentina, and Iris pallida. Once important in western herbal medicine, it is now used mainly as a fixative and base note in perfumery, as well as an ingredient in many brands of gin. In the manufacturing of perfumes using orris, the scent of the iris root differs from that of the flower. After preparation the scent is reminiscent of the smell of violets. After an initial drying period, which can take five years or more depending on the use, the root is ground, dissolved in water and then distilled. One ton of iris root produces two kilos of essential oil, also referred to as orris root butter, making it a highly prized substance, and its fragrance has been described as tenaciously flowery, heavy and woody.. Orris root has been used in tinctures to flavour syrups; its taste is said to be indistinguishable from raspberry.

Iris germanica Root Extract. Iris Germanica Root Extract is an extract of the roots of the German Flag, Iris germanica L., Iridaceae. Bombay Sapphire gin contains flavoring derived from particular bearded iris species. Rhizomes of the German Iris (I. germanica) and Sweet Iris (I. pallida) are traded as orris root and are used in perfume and medicine, though more common in ancient times than today. Today Iris essential oil (absolute) from flowers are sometimes used in aromatherapy as sedative medicines. The dried rhizomes are also given whole to babies to help in teething. Gin brands such as Bombay Sapphire and Magellan Gin use orris root and sometimes iris flowers for flavor and color. For orris root production, iris rhizomes are harvested, dried, and aged for up to 5 years. In this time, the fats and oils inside the roots undergo degradation and oxidation, which produces many fragrant compounds that are valuable in perfumery. The scent is said to be similar to violets. The aged rhizomes are steam-distilled which produces a thick oily compound, known in the perfume industry as "iris butter".

Iris pallida Root Extract Iris Pallida Root Extract is an extract of the roots of the Pale Flag, *Iris pallida*, Iridaceae. Orris root is an excellent natural perfume base note and fixative. It is a thick, waxy oil and will need to be gently heated in a warm water bath to make it easier to work with. Sweet, woody raspberry, fruity peach with a fatty floral nuance.

J

Jasminum grandiflorum. Jasmine. The Hindus string the flowers together as neck garlands for honored guests. The flowers of one of the double varieties are held sacred to Vishnu and are used as votive offerings in Hindu religious ceremonies. An oil made by boiling the leaves of this eastern Jasmine is used to annoint the head for complaints of the eye, and an oil obtained from the roots is used medicinally to arrest the secretion of milk. This prized essence is calming, relaxing; it also raises the libido and is used to overcome frigidity.

Juniperus communis. Juniper. Stimulates circulation. Good for acne and eczema. Said to be good for anxiety and stress. Some references say good for hangovers. Cleansing. Juniper when applied externally is a good penetrator of the skin and is useful in cases of rheumatism, sciatica and dermatitis.

Juniperus mexicana. Juniper. Juniperus tetragona is probably the correct name for this species. The oil has antifungal and antibacterial activity.

Juniperus virginiana Wood Oil**.** "Cedar Wood Oil Virginian". Juniperus Virginiana Wood Oil is an essential oil obtained from the wood and twigs of the Red Cedar, *Juniperus virginiana* L., Cupressaceae. It is not an INCI name but a perfumery name. Virginian cedarwood oil is the volatile essential oil produced by the steam distillation of the chopped wood, stumps, logs, wood shavings or sawdust of the slow-growing Eastern Red Cedar, growing in the S.E. of USA, Canada and Japan. The oil is yellow to yellow-brown (sometimes red-brown), it is often solid or semi-solid at room temperature due to precipitated cedrol. Redistilled oils, where sold, are almost colourless. The oil has a woody, oily, somewhat dry odor but cleaner, and markedly smoother than Atlas or Himalayan oils although less powerful. There is a distinct pencil-sharpenings note. Dry-down is still smooth woody-oily, retaining the pencil-sharpenings aspect. Virginia cedarwood is considered by many perfumers to have a smoother and finer (but less powerful) odor than Texas oil. The major odor impact compound of cedarwood oils is (+)-α-cedrol, which occurs at up to 15% in the oil; the major portion of the oil being composed α- and β-cedrenes and thujopsene which are weak woody odorants. The sesquiterpene hydrocarbon content is 55 and 65%. Cedarwood oil Virginia shows a higher cedrol content than the Texas oil, and higher β-caryophyllene and γ-eudesmol, whereas Chinese oils show higher thujopsene contents. Cedarwood oil Virginia is used in insecticides, polishes and cleaning products, soaps and liniments.

L

Laurus nobilis. Bay, Sweet. Bay leaf oil is used externally for sprains, bruises, etc., and was sometimes dropped into the ears to relieve pain. Use with care and never more than 1.25%. The oil is used as a food flavoring as well. Mildly narcotic, said to be good for promoting hair growth and ridding the scalp of dandruff.

Lavandula angustifolia. Lavender. A few drops in a foot bath will banish fatigue. Applied to the body it will act as a strong stimulant and may relieve various neuralgic pains, sprains and rheumatism, while in France it is used to treat painful bruises. Sedative to restless children and sleepless adults, a few drops on the pillow will bring sleep more easily.

Lavandula hybrida. Lavandin, abrialis. It is a lavender oil derived from a hybrid plant that combines the properties of Aspic and true lavender. It is said to be more effective than any of the other lavender types in reducing skin redness. Good for muscle aches and sprains and said to improve skin circulation.

Lavandula hybrida. Lavandin, grosso. This is a hybrid between *L. officinalis* and *L. latifolia* and has a more herbaceous smell (some would say harsher) than some lavenders. It has the same skin calming and sedative properties of all lavenders.

Lavandula hybrida. Lavandin, sumian. This is another hybrid and in this case the lavender is more camphoraceous with an exciting woodiness. It is more rounded and smoother than some lavandins. This soft oil has the same calming and sedative properties possessed by all lavender oils.

Lavandula hybrida. Lavandin, super.

Lavandula latifolia. Lavender, Spike. It yields a higher quantity of oil than the *L. angustifolia*, but the quality of the oil is perhaps inferior in odor. The oil is still sedative and has all the properties associiated with lavenders.

Leptospermum petersonii. Lemon Tea Tree. It has antiseptic, antimicrobial, carminative and sedative properties. It is used in aromatic blends to combat coughs and colds. It is also successfully used as a powerful insect repellent. Used for years by the Maori's the oil has qualities similar to Tea Tree Oil. Calms and reduces stress and tension, helps relieve aches and pains, relaxes tight muscles. Healing to the skin.

Leptospermum scoparium. Manuka. Manuka oil is antibacterial, antifungal, used in cases of acne and believed to be anti-inflammatory. It is used in skin and hair care preparations. It is used for the prevention of body and foot odor, foot care and also oral hygiene products.

Lippia citriodora. Verbena. The alternative name is *L. citriodora* and this essential oil should not be used as a fragrance ingredient according to the latest legislation Council Directive 76/768/EEC (CAS No. 8024-12-2).

Lippia javanica. *Lippia javanica* (wild verbena). The herb is steam distilled in South Africa. Also known as Lippia Oil. The color is a light yellow to orange. The odor is a combination of mint and vanilla. It has floral notes with a slightly green base. Another describes it as pleasant, fruity and sweet smell with a peppery top note somewhat reminiscent of tagette oil. Traditionally: Adds a beautiful fragrance to cupboards and will keep away moths and insects. The major components were carvone (61.07%) and limonene (28.21%) in the oil of *L. javanica* representing Swaziland population. The essential oil composition of *L. javanica* major components: linalool (1.8-68.8%), myrcene (0.5-54.0%), limonene (0.4-39.9%), 2,6-dimethylstyrene (trace-26.9%), neral and geranial (trace-29.3%) and geraniol (trace-15.8%). Cultivated L. javanica from 2 sites contained myrcene (33.1-52.4%) and linalool (1.7-19.1%) as the main components.

Lippia rehmanni. Lippia. Similar to *L. citriodora*, check the legal status. Specific data not available.

Lippia Sidoides. Lippia Sidoides Leaf Oil is the volatile oil obtained from the leaves of *Lippia sidoides*, Verbenaceae. *Lippia sidoides* leaf oil in Northeastern Brazil is widely used in local phytotherapy. This plant is mainly used as a general antiseptic due to its strong action against many microorganisms. Research conducted at the Federal University of Ceará found that a mouth rinse containing *Lippia sidoides* essential oil was safe and effective in the treatment of gingivitis and may provide an inexpensive therapy for treating gum infections.

Liquidambar styraciflua. Liquidambar Styraciflua Oil is the volatile oil obtained from the exudate of the American Styrax, *Liquidambar styraciflua* L., Hamamelidaceae. Styrax gum. Crude gums of American and Asian styrax should not be used as fragrance ingredient. Only extracts or distillates (resinoids, absolutes and oils), prepared from exudations of *Liquidambar styraciflua* L. var. *macrophylla* or *Liquidambar orientalis* Mill., can be used and should not exceed a level of 0.6% in consumer products. This is equivalent to 3% in a fragrance compound used at 20% in the consumer product. Major constituents: terpinen-4-ol, α-pinene and sabinene. The high terpinen-4-ol and low 1,8-cineole content make the oil of some pharmacological interest. The hardened sap, or gum resin, excreted from the wounds of the Sweetgum, for example the American Sweetgum (*Liquidambar styraciflua*), can be chewed on like chewing gum and has been long used for this purpose in Southern United States. The sap was also believed to be a cure for sciatica, weakness of nerves, etc. Aztecs applied a hot gum extract to the cheek for toothache. Indians chewed it, as a preservative for their teeth. Much used for horses as a vulnerary, called 'Gum Elimy'. Indians used liquidambar resin to cure fevers and heal wounds. It was a resin formerly used for scabies. The oil is called copalm in Mexico, used for chronic catarrh, and for affections of lungs, intestines, urinary tract.

Litsea cubeba. This material is also known as May Chang, because its origins lie in traditional Chinese medicine. It should not be confused with Cubeb oil, which is a powerful pepper oil extracted from *Piper cubeba*. The aroma is a subtle blend of Lemongrass, Melissa with a hint of Lemon, it is fresher and cleaner smelling than those individual essential oils alone, and is used for its deliciously natural, refreshing fragrance.

M

Matricaria chamomilla. Chamomile, German. Scientific Name(s):
M. recutita L., Family: Asteraceae (daisy). Synonyms: *Chamomilla recutita*, *M. chamomilla* and *M. suavoelens*. It is antispasmodic, anti-septic/wound healing, sedative, anti-inflammatory, tonic, soothing for conjunctivitis/sore eyes. Has been used for helping skin conditions such as dermatitis, boils, acne, rashes, and eczema as well as for hair care, burns, cuts, toothaches, teething pains, inflamed joints, menopausal problems, insomnia, migraine headaches and stress related complaints.

Melaleuca alternifolia. Tea Tree. The smell of tea tree is well-known, a combination of medicinal, eucalyptus and woody citrus notes. The oil is antibacterial, used for cold sores, acne skin washes, and decongestant baths when a cold is coming on. Good for spots and other skin infections.

Melaleuca cajuputi. Cajuput. The leaves possess antibacterial, anti-inflammatory and anodyne (pain relief) properties, and used for burns, colds, influenza and dyspepsia. Cajeput oil is produced from the leaves by steam distillation. The oil is a common household medicine, especially in Southeast Asia, used internally for the treatment of coughs and colds, against stomach cramps, colic and asthma. It is used externally for the relief of neuralgia and rheumatism, often in the form of ointments and liniments, and for the relief of toothache and earache. It is also applied in treating indolent ulcers. The oil is reputed to have insect-repellent properties. It is a sedative and relaxant and used as a flavoring in cooking and as a fragrance in soaps, cosmetics, detergents and perfumes.

Melaleuca ericifolia. *Melaleuca ericifolia* Smith ("Rosalina oil") from northern NSW contains linalool (35-55%), 1,8- cineole (18-26%), α-pinene (10%) and aromadendrene (4%). It is believed to be mildly antimicrobial. Its pleasant odor is more acceptable than *Melaleuca alternifolia*. It is suggested for topical use for acne, boils, tinea and herpes. Melaleuca Ericifolia Leaf Oil is the volatile oil distilled from the leaves of the Tea Tree, *Melaleuca ericifolia*, Myrtaceae. Another analysis suggests the composition to be 1,8 cineole (≤26%), linalool (44%), α-terpineol (4%), α-pinene, aromadendrene, viridiflorene (2.5%).

Melaleuca leucadendron. Cajuput. see ***M. cajuputi*** [Syn: *M. leucadendron*].

Melaleuca quinquenervia. Niaouli is a close relative of the Tea Tree and is typically found in Australia. It has similar properties to *Melaleuca alternifolia* (see below), but is more often used for its fragrance.

Melaleuca viridiflora. Niaouli. This oil has a sweet, fresh fragrance with a hint of tea tree. It is strongly antiseptic and so useful for treating acne, boils and other skin irritations. It is used as a chest rub and also has analgesic properties.

Melissa officinalis. Lemon Balm. Melissa Officinalis Leaf Oil is a volatile Despite extensive traditional medicinal use, melissa oil was initially prohibited by the International Fragrance Association (IFRA)'s 43rd amendment,[19] but this restriction appears to have been revisited and relaxed in the 44th amendment and is now restricted. Folklore: It is a well-known monastery herb and monks and nuns prepare from it a fragrant cologne and healing salves. Lemon balm contains eugenol, which kills bacteria and has been shown to calm muscles and numb tissues. It also contains tannins that contribute to its antiviral effects, as well as terpenes that add to its soothing effects. Traditionally this herb has been used as a sedative, and as an antispasmodic. It is mentioned in the treatment of herpes. Melissa is extremely potent and should be used with caution

Mentha arvensis. Cornmint/Peppermint. Cornmint oil, field mint oil, Japanese mint oil, marsh mint oil. Internal use of mint oil for flatulence, functional gastrointestinal and gallbladder disorders, catarrhs of the upper respiratory tract, and external use for myalgia and neuralgic ailments. Mint oil is official in the *Indian Pharmacopoeia* as a carminative. It is official in the *Chinese Pharmacopeia* as an aromatic, flavoring agent, and carminative, for application to the skin or mucous membrane, and to relieve pain or discomfort. In Germany, it is taken internally as a carminative or cholagogue, inhaled as a secretolytic, and applied externally for its cooling property. Menthol, derived from mint oil, is widely used as an antipruritic component of over-the-counter preparations to treat burns and sunburn, poison ivy rash, athlete's foot, and as a counterirritant in external analgesic preparations.

Mentha citrata. Mentha citrata (Ehrh.) (syn. Mentha × piperita L. var. citrata (Ehrh.) Briq.; syn. Mentha x aquatica var. citrata (Ehrh.) Benth. Mentha odorata is a herb. It is also known as Bergamot mint, Eau-de-cologne Mint, Horsemint, Lemon Mint, Lime Mint, Orange Mint, Pineapple Mint, Su Nanesi, Water Capitate Mint, Water Mint, Watermunt, Wild Water Mint, and in Central America Yerba Buena. This herb has a characteristic lemon odour when crushed. It is sometimes used to make a tasteful tea similar to lemonade, with medicinal value. The leaves and flowering plant have analgesic, antiseptic, antispasmodic, carminative, cholagogic, diaphoretic, and vasodilator properties.

Mentha piperita. Peppermint. Typical minty fragrance with mentholic undertones. It has a clean, clearing, penetrating odor. Invigorating; ideal travel companion, calms the stomach. Used to bathe tired and sweaty feet. A good insect repellent. Has a cooling effect on the body. Stimulating, used for headaches and nausea, very cooling. Breath freshener.

Mentha pulegium. Pennyroyal, European. It is a reputed abortifacient, antiseptic, blood purifier, carminative, cholagogue, diaphoretic, digestive, emmenagogue, expectorant, insect repellant, natural flea repellant on pets, pectoral, refrigerant, spasmolytic, stimulant, sudorific, uterine stimulant, uterine vasodilator.

Mentha rotundifolia. Apple mint. Traditional use: The leaf is used in strong decoction to heal chapped hands. Add to bath water for an invigorating bath. Use in potpourri and herb bags. Medicinally, inhale drops of essential oil, or sprinkle on a handkerchief, for relief from heavy colds. Macerate the leaves in oil; then massage affected areas for migraines, facial neuralgia and rheumatic and muscular aches, especially in winter.

Mentha spicata. Spearmint. The aroma of spearmint is not as sharp and intense or vital as peppermint, as it contains no menthol. It is often described as minty with a slightly fruity aroma. Used for colic, indigestion, flatulence, intestinal cramps, fevers, nausea, antidepressant, relieves mental strain and helps to soothe headaches. The chief constituent of spearmint oil (essential oil of Mentha spicata) is L-carvone (more than 60%). Spearmint is used for culinary purposes and in the aroma and flavor industry. It has antifungal, antimicrobial and antiviral activities. The essential oil in the leaves is antiseptic, though it is toxic in large doses. An essential oil is obtained from the whole plant, the yield is about 4kg of oil from 1 tonne of leaves and is used as a food flavoring and in oral hygiene preparations.

Mentha viridis. Spearmint. See ***M. spicata*** [Syn. *M. viridis*].

Menthol. Menthol is a secondary alcohol obtained from peppermint oil or other mint oil, or it is prepared synthetically from the hydrogenation of thymol. It is quite effective in insect bite preparations because it is antipruritic and mild anesthetic properties. The cooling effect makes it a useful additive in shaving products. Applied to the skin, menthol produces vasodilation followed by a feeling of numbness, coolness and mild local anesthesia. Menthol is used traditionally to relieve headaches, toothache and neuralgia. Menthol has a beneficial effect on the irritated and inflamed mucous membranes of the upper respiratory tract, where it shows anti-germ properties, stimulates secretion and has a mild anesthetic effect on the mucosa. It has a soothing effect on the nose and throat at the first stages of a cold.

Michelia alba. Michelia Alba Flower Oil is the volatile oil obtained from the flowers of *Michelia alba*, Magnoliaceae. A flowering tree of the magnolia family, originally found in India, also called the "Joy Perfume tree" as it was one of the main floral ingredients in that perfume. Traditionally used in Indian incense as well.

Michelia champaca. Michelia Champaca Flower Oil is the volatile oil obtained from the flowers of the Champac, *Michelia Champaca* L., Magnoliaceae. The Michelia champaca , Common Name is "Champaka". The flower is considered as very precious among the Indian flowers. The smell of champaca flower is unique and attracting. Sometimes, it is called Golden Champaca. The Major components are as follows: Linalool (76%), 2- phenethyl alcohol (6.38%) etc. Champaka oil is non-toxic, non-irritant and generally non-sensitizing, although some people do have an allergic reaction to the oil. Due to its emmenagogue properties it should not be used in pregnancy. Using too much of this oil it is considered a deeply relaxing oil. The oil of *M. champaca* and the chemical constituent, parthenolide contained in this plant were found to give positive patch test reactions on individuals who were contact-sensitive to sesquiterpene lactones. The finding suggests that such lactones may be the allergens of Michelia species.

Myristica fragrans. Nutmeg. The inhaled aroma is stimulating and effective in fighting mental fatigue. High concentrations of this potent oil, however, can produce sedative effects. Warning: Nutmeg oil is toxic if used in large quantities, and can be stupefying. Use with caution on the skin.

Myrocarpus fastigiatus. Myrocarpus Fastigiatus Oil (cabreuva wood oil) is the volatile oil obtained from *Myrocarpus fastigiatus*, Leguminosae Syn. Cabreuva oil. Cabreuva essential oil has a delicate, sweet, woody-floral scent. It has a pale yellow color. The oil is steam distilled from from the hard wood. Cabreuva essential oil has a delicate, sweet, woody-floral scent. It has a pale yellow color. The oil is steam distilled from from the hard wood. It is described as antiseptic, balsamic, cicatrisant. Cabreuva oil has been used for centuries to treat wounds and scars. It is also sometimes used as an aphrodisiac for sexual debility because it has a direct action on the hypothalamus. It has a delicate woody aroma. It is nontoxic but sometimes slightly irritating to the skin.

Myrocarpus frondosus Wood Oil. Cabreuva is a tree found in Paraguay. It provides an essential oil that is sometimes used as a natural cold remedy. It has been used as a salve to help treat wounds, burns, and scars. The plant is steam distilled to extract the essential oil. Cabreuva oil is an antiseptic and is used on scars to help heal the skin and reduce the scar area.

Myroxylon balsamum (balsam tolu) oil. It has a sweetish scent, reminiscent of vanilla and green olivesand is used in perfumes as a source for Balsam. Balsam of Peru is used as a flavoring and fragrance in many products but can cause allergic reactions. Peru balsam has uses in medicine, pharmaceutical, in the food industry and in perfumery. It has been used as a cough supressant, in the treatment of dry socket in dentistry, in suppositories for hemorrhoids, the plants have been reported to inhibit *Mycobacterium tuberculosis* as well as the common ulcer-causing bacteria, *H. pylori* in test-tube studies, so it used topically as a treatment of wounds and ulcers, as an antiseptic and used as an anal muscle relaxant. Peru Balsam can be found in nappy rash creams, hair tonics, antidandruff preparations, and feminine hygiene sprays and as a natural fragrance in soaps, detergents, creams, lotions, and perfumes. Allergic reactions reported are generally skin rashes and dermatitis when the balsam comes into contact with the skin - even in small amounts found in soaps, perfumes, and other common body care products. These allergic reactions are attributed to the gum`s benzoic acids. Balsam contains 50% to 64% volatile oil and 20% to 28% resin. The volatile oil contains benzoic and cinnamic acid esters. The benzoic and cinnamic acids are believed to be the main active constituents of the resin. The oil contains about 60% cinnamein, a volatile oil that is extracted by steam distillation and used commercially in the perfume, cosmetic, and soap industries. Typical composition: α-bourbonene, α-cadinene, α-calacorene, α-copaene, α-curcumene, α-muurolene, α-pinene, benzaldehyde, benzoic, benzoic-acids, benzyl-alcohol, benzyl-benzoate, benzyl-cinnamate benzyl-ferulate, benzyl-isoferulate, β-bourbonene, β-elemene, cadalene, calamenene, caryophyllene, cinnamaldehyde, cinnamein, cinnamic-acids, cinnamyl-benzoate, cinnamyl-cinnamate, cis-ocimene, coumarin, δ-cadinene, dammaradienone, delta-cadinene, dihydromandelic-acid, eugenol, farnesol, ferulic-acid, gamma-muurolene, hydroxyhopanone, l-cadinol, methyl-cinnamate, nerolidol, oleanolic-acid, p-cymene, peruresinotannol, peruviol, resin, styrene, sumaresinolic-acid, tannin, toluresinotannol-cinnamate, vanillin, and wax.

Myroxylon pereirae. Peru Balsam. Peru balsam is extracted from the cortex of the tree *M. balsamum* var. *pereiae* that grows in Central America, e.g. along the coast of El Salvador. Peru balsam is used as a fragrance and flavor additive in foods. Peru balsam is weakly antiseptic and traditionally has been applied in pharmaceuticals for the treatment of wounds, eczema and pruritis (itching). It has been used for the care and relief of hemorrhoids.

Myrtus communis. Myrtle oil is obtained from *Myrtus communis* contains alpha-pinene, 1,8-cineole and limonene as its major constituents. α-pinene (14.2%); limonene (2.7%); 1,8- cineole (16.4%); linalool (20.2%); terpineol (3.2%); linalyl acetate (5.8%); geranyl acetate (2.4%); spathulenol (8%); sativen (3.6%). The essential oil calms the nerves, promotes a good sleep and aids respiration. A mild oil, it is suited for children. Makes a fresh and sweet perfume in its own right. Tonifies and clears skin. There are no indications of any special precautions when using this oil.

N

Nardostachys grandiflora. Jatamansi. Also known as Indian Nard [Syn. *N. jatamansi*]. Cooling, and used to reduce fevers. Soothing for the skin, excellent for skin irritations and allergies. Said to be good for wounds that do not heal. Helps to promote restful sleep, reduces tress and aids relaxation. Can assist in reducing inflammation. It is a member of the Valerian family. An essential oil is obtained from the root and young stems. It is harvested before the leaves unfurl. It is used in perfumery and as a hair tonic where it is said to make the hair grow faster and also to turn it black. The dried leaves are used as an incense. Essential oil is hypotensive in dogs, dosages for hypotensive humans, less than lab animal dosage. Preliminary clinical trials of jatamansone exhibited reduced aggressiveness, restlessness, stubbornness, as well as less insomnia.

Nardostachys grandiflora. Spikenard. Cooling, and used to reduce fevers. Soothing for the skin, excellent for skin irritations and allergies. Said to be good for wounds that do not heal. Helps to promote restful sleep, reduces tress and aids relaxation. Can assist in reducing inflammation. It is a member of the Valerian family.

Nardostachys jatamansi. Jatamansi. See ***N. grandiflora*** Oil [Syn: *N. jatamansi*].

Nardostachys jatamansi. Spikenard. See ***N. grandiflora*** Oil [Syn: *N. jatamansi*].

Nepeta cataria. Catnip. The tops and leaves are medicinal; they have a strong, characteristic odor and a peculiar, bitterish taste. It is very much liked by cats. Its sedative action on the nerves adds to its generally relaxing properties. Tea made from the leaves and flowers of this herbaceous perennial has traditionally been sipped to relieve coughs. The leaves and shoots can be used as ingredients in sauces and soups. The oil extracted from catnip plants is used in natural mosquito repellents.

O

Ocimum basilicum. Basil, French. It is called *tulsi* in India and used for Ayurvedic medicine. Sacred to Krishna and Vishnu. the leaves are heart-shaped and are considered a love symbol in Italy. The oil clears the head and is uplifting. It is useful in nervous conditions: good for anxiety, depression, hysteria, indecision and nervous debility. It is good for earache, colds, sinus, migraine, muscular spasm. It is good for sluggish and congested skin and is an insect repellant.

Ocimum basilicum. Basil, Sweet. See ***O. basilicum***.

Ocimum basilicum. Basil, Tropical. See ***O. basilicum***.

Origanum majorana. Marjoram, Sweet. A slightly sweet, but definitely herbaceous, woody and campherous, if not slightly medicinal, odor. It is used for aching muscles, sprains, cramps, rheumatism and other conditions where its relaxing effects are needed.

Origanum vulgare. Oregano, Green. A warming oil, it assists improvement of circulation, digestion, mental clarity and alertness. It is used to relieve muscle aches and pains and is said to assist in increasing physical endurance and energy. It may assist in reducing cellulite. In China it is used additionally to treat itchy skin conditions. Origanum oil was investigated for its antifungal properties. Origanum oil at 0.25mg/mL was found to completely inhibit the growth of *Candida albicans*. In addition, both the germination and the mycelial growth of *C. albicans* were found to be inhibited by Origanum oil in a dose-dependent manner.

Ormenis multicaulis. Ormenis Multicaulis Flower Oil is an essential oil obtained from the flowers of the Moroccan Chamomile, *Ormenis multicaulis*, Compositae. Chamomile oil, wild Moroccan. Odor: Agrestic family: minty, terpenic, fruity. *Ormenis multicaulis* is one of three varieties of chamomile used in perfumery and cosmetics. Significantly different from *Matricaria recutita* (blue chamomile) or *Anthemis nobilis* (Roman chamomile), Wild Chamomile is native to northwestern Africa. It grows abundantly in the wild in northern Morocco and has been distilled in that country since the nineteen-seventies.

Osmanthus fragrans. Osmanthus Fragrans Flower Extract is an extract of the flowers of the osmanthus, *Osmanthus fragrans*, Oleaceae. The odor is described as being in the fruity family, mainly apricot, floral, sweet and somewhat animalistic.

P

Pelargonium capitatum. There are three main essential oils of geranium, namely *P. fragrans*, which has a pine-nutmeg aroma, *P. odoratissimum*, which has an apple scent and *P. capitatum*, which has the scent of roses. In aromatherapy, it is considered both sedative and uplifting, so is much favoured in the treatment of anxiety states. In skin care it is considered gently antiseptic, cleansing, refreshing and a gentle tonic. It is ideal for sluggish, congested or oily skin and is said to improve the circulation of poor complexions. It has also been used for eczema, dermatitis, oedema and dry skin conditions in massage preparations.

Pelargonium graveolens. Geranium, Bourbon. It was used by the ancients as a remedy for wounds and tumors.It makes a very refreshing and relaxing bath oil with a delightful light rose perfume and a fresh green note. Said to balance sebum levels. Also good for sluggish, congested and oily skins and is a good skin cleanser.

Perilla frutescens. Perilla. Beefsteak plant or Zi Su Zi. It is found in Southern China, Taiwan, Indochina and India. The seeds are used. The perillaldehyde present in the plant inhibits fungi and both Gram-positive and Gram-negative bacteria. Perilla oil was especially effective against *Propionibacterium acnes*, *Staphylococcus aureus*, both of which can cause acne; and against the fungi *Mucor mucedo* and *Penicillium chrysogenum*, both food molds. Other components in the oil: Limonene was very active against both fungi and *P. acnes*, ß-caryophyllene very active against *P. acnes*.

Petroselinum crispum. Parsley. The medicinal action is due largely to the essential oil which gives Parsley strong diuretic, stomachic, carminative, irritant and emmenagogue properties. Parsley is a source of apiole (dimethoxysafrole), a volatile oil (ether) that is used in perfumery. Apiol is an antipyretic and, like myristicin, is a uterine stimulant, so that strong doses can be toxic, cause hemorrhaging, nervous disorders (polyneuritis) and may cause abortion. Apiol was once available as capsules for use as an abortifacient. A Russian product called *Supetin* was used for stimulating uterine contractions during labor (85% parsley juice).

Peumus boldus. Boldo. Boldo volatile oil is stated to be one of the most toxic oils. The oil in five drop doses has been found useful in genitor-urinary inflammation and has been used as a remedy for gonorrhea and liver diseases. The reputed diuretic and mild urinary antiseptic properties of boldo are probably attributable to the irritant nature of this volatile oil. Although Commission E has approved the use of boldo for the treatment of dyspepsia as well as for stomach and intestinal cramps, it must be noted that the volatile oil in the leaves contains about 40% ascaridole, a rather toxic component that poses certain problems from a health point of view (although it was used clinically in the past as an anthelmintic agent). Since no chronic toxicity testing has been carried out, it has been recommended that prolonged use of the herb or any consumption by pregnant women be avoided. Boldo essential oil also contains termpinen-4-ol, the irritant and diuretic principle in juniper oil. Application of the undiluted oil to the hairless backs of mice has an irritant effect. The oil contains irritant terpines including terpinen-4-ol, the irritant principle in juniper oil. In view of the toxicity data and the irritant nature of the volatile oil, excessive use of boldo should be avoided.

Picea abies. Picea Abies Leaf Oil is the volatile oil expressed from the needles of the Norway Spruce, Picea abies (L.), Pinaceae. The leaves and branches, or the essential oils, can be used to brew spruce beer. The tips from the needles can be used to make spruce tip syrup. Native Americans in New England also used the sap to make a gum which was used for various reasons, and which was the basis of the first commercial production of chewing gum.

Picea mariana. Picea Mariana Leaf Oil is an essential oil obtained from the leaves of the spruce, *Picea mariana*, Pinaceae. Also known as black spruce oil. Contains pinenes, limonene, bornyl acetate, tricyclene, phellandrene, myrcrene, thujone, dipentene, cadinene. Spruce Black essential oil is one of Canada's best kept secrets. It is sweeter, and softer than most evergreen oils with a balsamic, resinous odour with green woody notes. Overall, the scent is very clean, fresh and pleasant. Black Spruce Needle Essential oil has a long history of use in saunas, steam baths, and as an additive to baths and massage products in spas. Black Spruce essential oil should be applied to the skin only in dilution as it has the potential to be a possible skin irritant.

Picea sitchensis, a therapeutic quality aromatherapy essential oil, steam distilled using wild-crafted plants and traditional methods, from Norway, obtained from the needles of the tree. Major components: β-myrcene, β-phellandrene, borneone, bornyl acetate, 1,8-cineol, borneol. Pine baths are easy to use and are used for nervous diseases as well as neuralgic and rheumatic conditions. It is important to specify the temperature and duration of the bath when prescribing pine baths for domestic use.

Pimenta acris. Bay oil was one of the ingredients used in Bay Rum, which was an excellent hair tonic. It is extracted from Pimento or Allspice and contains natural antioxidant and antimicrobial properties. The name Allspice came from the association of the delicious aroma with cinnamon, nutmeg and cloves and apart from the obvious culinary uses has been used in the past for 'hysterical paroxysms'! It was formerly listed in the British and U.S. Pharmacopoeia. It is frequently used for its natural heady exotic aroma.

Pimenta dioica. Pimento. Pimento comes from the Spanish *pimienta* meaning "pepper," after the similarity in shape to peppercorns. The spice was first imported into Britain in the early 17th century and variously called Pimienta de Chapa and Pimienta de Tabasco, before Ray in 1693 described it as Allspice because of its combination of flavors of cinnamon, nutmeg and cloves. The essential oil is useful for flatulent indigestion and for hysterical paroxysms. An extract made from the crushed berries by boiling them down to a thick liquor and spread onto linen makes a stimulating plaster for neuralgic or rheumatic pains. Any pharmacologic activity associated with the plant is most likely due to the presence of eugenol that also has local antiseptic and anaesthetic properties. Eugenol also has antioxidant properties and allspice may serve as a potential source of new natural antioxidants. Furthermore, allspice appears to have *in vitro* activity against yeasts and fungi.

Pimpinella anisum. Aniseed. Anise oil is useful in destroying body lice, head lice and itching insects where the oil can be used by itself, so is useful for pediculosis, the skin condition caused by lice. It may be used for scabies, where it may be used externally in an ointment base. It is used either in oil or ointment base as a stimulating liniment and against vermin.

Pinus nigra. Pine. The European Black Pine *P. nigra* is a variable species of pine, occurring across southern Europe from Spain to the Crimea, and also in Asia Minor, Cyprus, and locally in the Atlas Mountains of northwest Africa. The turpentine obtained from the resin of all pine trees is antiseptic, diuretic, rubefacient and vermifuge and is used as a rub and steam bath in the treatment of rheumatic affections. It is also very beneficial to the respiratory system and so is useful for coughs, colds, influenza and tuberculosis. Externally it is a very beneficial treatment for a variety of skin complaints, wounds, sores, burns, boils etc., and is used in the form of liniment plasters, poultices, herbal steam baths and inhalers.

Pinus roxburghii. Turpentine. Turpentine and its related products have a long history of medicinal use, where they have been employed as topical counterirritants for the treatment of rheumatic disorders and muscle pain. A gum derived from turpentine was used in a traditional Chinese medicine to relieve the pain of toothache.

Pinus sylvestris. Pine, Scotch. Scots pine. Distillation of the pine wood by steam pressure gives pine oil, which has a scent like that of juniper oil. The buds, the thick distilled resin of the tree (turpentine) are used and the essential oil obtained by the steam distillation of the needles. It contains essence of terpentine (pinene, camphene, terpenes, etc.), mallol, essential oil, pinene, sylvestrine, bornyl acetate, cadinene, pumilone. The buds contain more than 200g of resin or rosin per Kg. It is used to impart its refreshing scent to bath essences. Pine baths are used for nervous diseases as well as neuralgic and rheumatic conditions. Oil of terpentine is used medicinally in veterinary practice as a rubefacient and vesicant, and is valuable as an antiseptic, for rheumatic swellings and for sprains and bruises, and to kill parasites. Rosin is used by violinists, gymnasts and for various paper and other uses. Residual tar is used as an antiseptic, stimulant, diuretic and diaphoretic action.

Piper nigrum. Black Pepper. It is anti-inflammatory, anticatarrhal, expectorant, supportive to the digestive glands, and traditionally used for rheumatoid arthritis. Considered an aphrodisiac and sexual stimulant.

Pistacia lentiscus. Mastic tree. The resin that oozes when the tree is tapped is known as mastic and smells like turpentine. It is widely employed in dentistry as a cement for filling decayed teeth and is also used in varnishes. In the East it is used as a breath sweetener.

Plumeria rubra. Frangipani flowers retain their scent after drying and are used in scent bags and potpourris. Folklore: It is also known as the Pagoda Tree, Temple Tree and is found in Malaya and has immense clusters of waxy flowers with the most delicious scent. It is used for freshening the clothes (like lavender is used in England).

Pogostemon cablin. Patchouli. Strong and a characteristic oriental and musty odor. It helps reduce skin oiliness, soothes skin problems and burns, reduces inflammation and is mildly antiseptic. It is a nerve sedative and antidepressant, is very beneficial for the skin and may help prevent wrinkles or chapped skin. It is also said to regenerate tissue and helps relieve itching from hives and other pruritic conditions.

Polianthes tuberosa (tuberose) oil**.** Polianthes Tuberosa Flower Oil is an essential oil obtained by high vacuum distillation of the extract of the flowers of the Tuberose, *Polianthes tuberosa* L., Amaryllidaceae. The tuberose is also used traditionally in Hawaii to create leis and was considered a funeral flower in Victorian times. Its scent is described as a complex, exotic, sweet, floral. In Taiwan, the tuberose has become a trendy food ingredient and can be found in many five-star hotels.

Pseudotsuga menziesii. Pine, Douglas. Distilled from the needles. Clearing and deodorizing. It is an excellent air freshener and has antiseptic properties that give it good disinfectant properties. It is stimulating, good for circulation and useful for colds, flu, and bronchitis conditions, especially decongestant baths. High quality oils do not smell as basic as toilet cleaners and have a lighter, more delicate aroma.

R

Ravensara aromatica. Ravensara. Sometime called Clove Nutmeg and is a member of the Lauraceae (laurel) family. The smell is slightly medicinal with a camphoraceous or eucalyptuslike note with a very slightly sweet back note of fruitiness. In Madagascar it is known as "the oil that heals" because of its antiseptic activity and for its useful properties in respiratory problems. It has anti-infectious, antiviral and antibacterial properties. It has shown to help with insomnia and muscle fatigue. Five leaf essential oils of *R. aromatica* Sonn. from Madagascar have been analyzed by GC and GC/MS were found to contain mainly methyl chavicol (79.7%), methyl eugenol (8.5%) and limonene (3.1%). Another study analyzed 28 samples of formally identified *R. aromatica* leaf oils and divided the leaf oils into four types: a <90% methyl chavicol type, a 74—-92% methyl eugenol type, a terpinene (25–28%) and limonene (15–22%) type, and a sabi-nene (25–34%), linalool (7–21%) and terpinen-4-ol (6–12%) type. Ravintsara Oil (*Cinnamomum camphora*) should not be confused with Ravensara Oil.

Rhododendron anthopogon. Anthopogon. Alpine Rosebay, Sunpati. The stems and leaves are used in Tibetan herbalism. They have a sweet, bitter and astringent taste and they promote heat. They are antitussive, diaphoretic and digestive and are used to treat lack of appetite, coughing and various skin disorders. In Nepal, the leaves are boiled and the vapour inhaled to treat coughs and colds. It a sweet herbal, faint balsamic essence and is used in perfumery.

Rhododendron anthopogon. Rhododendron. Alpine Rosebay, Sunpati. The stems and leaves are used in Tibetan herbalism. They have a sweet, bitter and astringent taste and they promote heat. They are antitussive, diaphoretic and digestive and are used to treat lack of appetite, coughing and various skin disorders. In Nepal, the leaves are boiled and the vapor inhaled to treat coughs and colds. It has a sweet herbal, faint balsamic essence and is used in perfumery.

Rosa centifolia flower oil**.** Though the perfumery industry can now produce excellent synthetic copies of rose oil, there is still no substitute for the real thing. In aromatherapy, this oil has been shown scientifically to increase the brain wave activity, in the areas of the brain responsible for concentration and alertness, while decreasing the heart rate. This makes it ideal for the treatment of stress and anger. Avicenna (Abu Ali Sina, who lived from 980 to 1037 AD) stated "Rose oil increases the might of the brain and quickness of the mind".

Rosa damascena. Rose, Damask. *R. damascena* is more commonly known as the Damask rose or simply as "Damask," or sometimes as the Rose of Castile, is a rose hybrid, derived from *R. gallica* and *R. moschata*. DNA analysis has shown that a third species, *R. fedtschenkoana*, is associated with the Damask rose. The Damask rose is commonly used to flavor food and to make rose water. Rose extract and rose oil have a host of beneficial effects on the skin and are great for promoting a youthful complexion with good tone, elasticity and an even-colored complexion.

Rosmarinus officinalis. Rosemary. The typical fragrance is sweetly herbal and slightly medicinal with a hint of camphor. Rosemary is forthright and strong, helping to improve mental clarity aiding concentration and used to enhance meditation. It may be beneficial for problem skin conditions and dandruff. It is anticatarrhal, anti-infectious, antispasmodic and is a useful component in decongestant baths. It helps overcome mental fatigue. The Commission E approved the internal use of rosemary leaf for dyspeptic complaints and external use as supportive therapy for rheumatic diseases and circulatory problems. No reported drug interactions, side effects and recommended use is 4–6 g of cut leaf for infusions, powder, dry extracts and other galenical preparations for internal and external use. AHPA-BSH has classified rosemary as a Class 2b/emmenagogue herb where it can stimulate uterine contraction and induce miscarriage, establishing counter-indication of use during pregnancy.

S

Salvia lavandulaefolia. Sage, Spanish. *S. lavandulaefolia* (Spanish sage) enhances memory in healthy young volunteers.

Salvia officinalis. Sage. It has been used in Europe for skin conditions such as eczema, acne, dandruff and hair loss. It has been recognized for its benefits in relieving mental fatigue. The aromatherapy benefits are said to be uplifting and relaxing. There are many different sages and each have a unique aroma. Oral-Mice LD50: 800mg/kg. The oil of sage, can induce hypoglycemia, tachycardia, convulsions, muscle cramps and respiratory disorders. The Commission E has approved the internal use of sage leaf for dyspeptic symptoms and excessive perspiration, and external use for inflammations of the mucous membranes of nose and throat with recommended dry leaf intake, 1–3 g, three times daily or fluid extract 1,1 (g/mL), 1–3 mL, three times daily. Sage is none the less classified as a AHPA-BSH Class 2b herb, not advised for long term use or during pregnancy, and not to exceed the recommended dose of 4–6 g daily.

Salvia sclarea. Clary Sage. Antispasmodic oil, good for menstrual difficulty, high blood pressure, muscle cramps, respiratory problems, emotional and physical tensions. The seeds become mucilaginous in water and may be used to extract foreign bodies from the eye. The essential oil, which is known as sage clary or Muscatel oil, is obtained by steam distillation. An ointment made with clary leaves will help draw out inflammation and bring boils and spots to a head.

Salvia stenophylla. Mountain sage, mainly from South Africa. The most abundant compounds in the Blue Mountain sage oil are α-bisabolol (46.5%), limonene (38.1%), δ-3-carene (24.9%), γ-terpinene (20.3%), *p*-cymene (18.4%) and (E)-nerolidol (53.6%). The dominant compounds in another source declare *S. stenophylla* oils to include: cis-lanceol (20.9%), δ-3-carene (18.8%), β-bisabolene (11.3%), 1,8-cineole (10.9%), and α-bisabolol (10.4%). It has been used traditionally as a disinfectant by burning it in huts after sickness, and it is also mixed with tobacco for smoking. The essential oil from Blue Mountain sage is straw-like in color. The oil is used to relieve stress and promote relaxation. It is used as a massage oil and an agent of aromatherapy. It may also be diluted, placed in the palms of the hands, and inhaled, a method used to clear the sinuses and bronchi.

Santalum album. Sandalwood. The chief uses of sandalwood are as a deodorant, in skin care, for cutaneous inflammation and as an antiseptic. It is claimed to be a skin softener and stimulates peripheral blood circulation in the skin. It protects against skin diseases and allergic conditions, is hemostatic or styptic and removes skin blemishes.

Santalum austrocaledonicum. Santalum Austrocaledonicum Wood Oil is the volatile oil obtained from the wood of *Santalum austrocaledonicum*, Santalaceae. *Santalum austrocaledonicum* grows in New Caledonia and Vanatua. The odor is recognisable as sandalwood but is more terpenic and ambery than East Indian sandalwood, less radiant and less intensely woody, but with some of the creaminess. Australian sandalwood grows in the drier inland regions of Western and South Australia and has an odor described as soft, woody and extremely tenacious with a rather dry-bitter slightly resinous top note. Australian sandalwood essential is extracted by a combination of solvent extraction and co-distillation is adopted. It has a very similar composition to East Indian sandalwood essential oil and, despite the use of two different botanical names, the species are probably almost identical and so is likely to have similar therapeutic actions to East Indian sandalwood essential oil

Santalum spicata wood oil**.** The inner heartwood is used to prepare this essential oil, which is traditionally used to treat dry, inflamed or irritable skin. It is antiseptic and is used externally for the treatment of problem skin conditions. The aroma of sandalwood is a rich woody oriental note, which is considered by aromatherapists to be relaxing. The chief uses of sandalwood are as a deodorant, in skin care, for cutaneous inflammation and as an antiseptic. It is claimed to be a skin softener and stimulates peripheral blood circulation in the skin. It protects against skin diseases and allergic conditions, is haemostatic or styptic and removes skin blemishes. No adverse effects are expected from the topical application of this oil.

Santalum spicatum. Sandalwood. See ***S. album***.

Sassafras albidum. Sassafras. An oil is distilled from the root or the root bark of *Sassafras albidum*, or from the wood of certain species of *Ocotea*. It contains safrole. Neither sassafras nor the oil should be taken internally. Sassafras oil has rubefacient properties and was formerly used as a pediculocide.

Sassafras officinale root oil is the volatile oil obtained from the root of the Sassafras, Sassafras officinale, Lauraceae. Most of the data available refers to *Sassafras albidum*, a tree indigenous to the eastern United States of America. Sassafras yields about 3-9% of a volatile oil with safrole as the major constituent (80-90%), other cinnamic acid derived compounds, including sesamin, about 6% tannin, about 9% of a reddish-brown phlobaphene, resin and starch. Over the years the oil obtained from the roots wood have been used as a scent in perfumes and soaps. Specifically indicated for topical use: Oil in *pediculosis capitis*. Safrole, except for normal content in the natural essences used and provided that the concentration does not exceed: 100ppm in the finished cosmetic product, 50ppm in products for dental and oral hygiene, and provided that safrole is not present in toothpastes intended specifically for children.

Satureja montana. Savory, Mountain. Savory oil is sometimes used as a local application to carious teeth, for relieving toothache; and its tincture is a valuable carminative. It is an antiseptic oil with antibacterial and antifungal properties and so would be useful in problem acneic skin conditions.

***Schinus molle* oil.** A monograph published in 1976 on Brazilian peppertree's essential oil indicated no toxicity in animals and humans ingesting or applying the essential oil topically. Today, herbalists and natural health practitioners in both North and South America use Brazilian peppertree mostly for colds, flu, and other upper respiratory infections; as a remedy for hypertension and for irregular heartbeat; for fungal infections and Candida; and as a female balancing herb for numerous menstrual disorders, including menstrual cramps and excessive bleeding.

Sinapsis alba. Mustard. Sinapsis is a stimulating external application, the rubefacient action causing mild irritation to the skin, stimulating the circulation to that area and relieving muscular and skeletal pain, sciatica, neuralgia and various internal inflammations. Mustard oil can be mixed with rectified alcohol (1:40 oil to alcohol) and used as a lotion for gouty pains, lumbago and rheumatism. Mustard oils are absorbed through the skin and eliminated via the lung, so that the antibacterial action can take effect there.

Styrax benzoin. Styrax Benzoin (Benzoin) Extract. It is used as an antiseptic, astringent and expectorant, and the tincture is widely used as a skin protectant and as an antiseptic and styptic on small cuts. It is under the domination of the Sun or Jupiter. It has a sweet, vanillalike smell. It is a scented gum that has been used in cosmetics for hundreds of years. The ancient civilizations thought it a grand remedy for driving away evil spirits and was often used in fumigations and incense. It is an ingredient in Friar's Balsam. It was often referred to in old herbals as "gum benzoin," "balsam" or "gum benjamin." Virgin Milk, an old fashioned toilet water, included benzoin as well as lavender and ethanol. It was supposed to make the skin "clear and brilliant." Nowadays used as a fixative in perfumes.

Syzygium aromaticum. Clove. Bacteriacidal and insecticidal. Oil of cloves is used as a disinfectant and dental analgesic, and in alcohol as a stimulant. Applied externally, it can produce local anesthesia. Externally used on sores and infected areas and also used as a mosquito and moth repellant. Essential oils are obtained from the buds, stems, and leaves. Clove buds yield approximately 15–20% of a volatile oil that is responsible for the characteristic smell and flavor. The stems yield about 5% of the oil, and the leaves yield about 2%. Clove has been used for its antiseptic and analgesic effects, and has been studied for use in platelet aggregation inhibition, antithrombotic activity, and chemoprotective and antipyretic effects.

T

Tagetes minuta. Tagetes. Externally, it is used to treat hemorrhoids and skin infections. It is an insect repellant and branches are placed in among blankets and winter clothing to repel moths and other insects. If the dogs' baskets and kennels are lined with the plant, the fleas will soon leave, and a few leaves rubbed into the pet's coat is a good flea repellant.

Tanecetum vulgare. Tansy. A weak anthelmintic and mild irritant, tansy can be poisonous even when applied externally; therefore it is little used. American Indians used the plant to cause abortion. It is used as an an ointment for pruritis. In medieval times, the leaves were placed in beds and strewn over floors as their camphor smell kept away flies and fleas. If large amounts of any preparation containing tansy are taken, toxicosis with epileptic convulsions may result. The essence is an insecticide. As this is an emmenagogic plant (one that restores the menstrual flow), it should on no account be taken in any form by pregnant women.

Tarchonanthus Camphoratus. Tarchonanthus Camphoratus Oil is the volatile oil obtained from *Tarchonanthus camphoratus*

Thuja occidentalis. Thuja. It has been used, among other things, for psoriasis, rheumatism, and topically for warts. It is useful in the treatment of psoriasis and rheumatism. Thuja oil is a very poisonous essential oil and should only be used under strict medical supervision.

Thymus capitatus. Thymus Capitatus Herb Oil is an essential oil obtained from the herbs of the Spanish Origanum, Thymus capitatus, Lamiaceae.

Thymus mastichina flower oil. Aromatic scent: Spicy, warm, herbaceous, much more camphoraceous and medicinal smelling than the Sweet Marjoram. History: This Marjoram is actually wild growing Thyme. However, it is commonly referred to as Wild Marjoram. The results showed that linalool (26 -30 %) was the most important compound in leaves and flowers of plants from Sesimbra. In Algarve plants, 1,8-cineol or eucalyptol was the most significant volatile compound of which concentrations ranged from (23 – 33 %). In general these results are in accordance with that already studied by hydrodistillation and GC-MS. Used up to 6% in fragrances. The leaves and especially the essential oil contained in them are strongly antispetic, deodorant and disinfectant.

Thymus vulgaris. Thyme. The essential oil is found to be useful for overcoming fatigue and physical weakness after illness. It is antimicrobial, antibacterial, antifungal, antiviral and is strongly germicidal. It is used for joint pain, backache and sciatica in a hot bath. Inhaled, thyme oil uplifts the spirit, relieves depression and is an excellent decongestant and cold treatment.

Thymus zygis. Thyme, Thymol. See also ***T. vulgaris***. Oil of thymol has a powerful antiseptic action for which it is used in mouthwashes and toothpastes. Thyme is also effective against ascarids and hookworms. It can be used externally for warts or to encourage the flow of blood to the surface. Thyme baths are said to be helpful for fatigue, neurasthenia, rheumatic problems, paralysis, bruises, swellings, and sprains. It is also used for the skin and vaginitis. A salve made with thyme oil is antiseptic and can be used for shingles. It is used as a paint in ringworm, in eczema, psoriasis, broken chilblains, parasitic skin infections and burns.

V

Valeriana officinalis. Valerian. Very soothing and relaxing, especially during difficult times. Encourages sleep and restfulness, calms the nerves during sleep. The oil has a very characteristic musty odor.

Vanilla planifolia. Vanilla. Vanilla is comforting, calming and soothing. The fragrance relaxes and softens anger, frustration and irritability when used in massage oils. It is also considered to be an aphrodisiac by many.

Verbena officinalis leaf rectified oil. Verbena Officinalis Leaf Rectified Oil is a rectified essential oil obtained from the leaves of the plant Verbena officinalis L., Verbenaceae. Vervain sometimes called Verbena is an essential oil that will strengthen the nervous system whilst relaxing any tension and stress. It may be used as a mouthwash against caries and gum disease. It should not be confused with Lemon verbena which is Lippia citriadora and is banned. The pro-apoptotic activity of Verbena officinalis essential oil and its main component citral have been verified. The major compounds being 3-hexen-1-ol (7.28%), 1-octen-3-ol (32.76%), linalool (4.66%), verbenone (20.49%) and geranial (7.22%). [CAS: 84961-67-1; EINECS: 284-657-0]. Function: Perfuming. The actual or estimated LD_{50} value: 5,000 mg/kg body weight. AICS status (NICNAS Australia): AICS compliant.

Vetiveria zizanioides. Vetiver. A scented grass with a woody, sultry almost smoky aroma. Deeply relaxing, the essential oil is sedating and ideal in massage oil or bath. Traditionally used in men's toiletries.

Viola odorata leaf extract. Viola Odorata Extract is an extract of the leaves of the Sweet Violet, Viola odorata L., Violaceae. Viola odorata is a species of the genus Viola native to Europe and Asia, but has also been introduced to North America and Australasia. It is commonly known as Sweet Violet, English Violet, Common Violet, or Garden Violet. Viola Odorata Extract is an extract of the flowers and leaves of the Sweet Violet. This plant is anti-inflammatory and emollient. An essential oil from the flowers and leaves is used in perfumery, and it can also be used in aromatherapy.

Z

Zingiber officinale. Ginger. Externally, it is a rubefacient and used for rheumatic pains and stimulant of peripheral circulation in bad circulation, e.g. chillblains and cramps. It is diaphoretic (promotes sweating). Ginger baths decrease muscle soreness and muscle stiffness. Used also in morning sickness and travel sickness. Ginger has been used for centuries as a cooking spice and medicinally demonstrates a diverse range of applications having biological properties such as the ability to modulate platelet aggregation, serve as an analgesic, anti-inflammatory, hypoglycemic, antimicrobial, antiparasitic activity and antioxidant. Also commonly used for nausea. Ginger and its constituents induce apoptosis in human cancer cell lines and display anticancer properties against spontaneous tumors in animal models. In humans, the administration of ginger up to 6 g/day is relatively safe, yielding few side effects with the exception of a few subjects who experience nausea and drowsiness. The Commission E approved the internal use of ginger for dyspepsia and prevention of motion sickness. Powdered rhizome, 0.25–1.0 g, three times daily.

AHPA-BSH recommends not exceeding the recommended 2–4 g/day, also warning against long-term use or using during pregnancy.

The Toxicology of Essential Oils

It has always been a challenge to find out the toxicological profile of essential oils and despite some excellent reference works on the topic, there has never been a single reference that supplied all the required data.

Misconceptions

Sadly, many of the books published on the topic of aromatherapy have not been written by scientists or toxicologists, and this has meant that there is a great deal of poor quality information circulating that has propagated and found its way on to the Internet.

Not tested on animals

The use of the phrase "not tested on animals" is not only illegal but is also totally without truth, since in nearly all cases these essential oils have been tested on animals at one time or another in the past.

The LD_{50} value (lethal dose 50%) using animals (usually based on rats or mice) gives a value for the death of half the animals tested, so that an LD_{50} of 2g/kg refers to the total weight of material tested/body mass of test animal that caused death in half the number tested. Thus if a human weighed 70kg, then it might be expected that 140g could be consumed before it was fatal.

Natural and safe

The idea that because a material is natural and so must be safe could never be further from reality. There are essential oils that are so toxic that they should never be consumed or applied to the skin without extreme caution. The risks of sensitization, irritation, photo-toxicity or being an abortifacient are fact. Essential oils contain a rich blend of highly functional molecules, some of which are beneficial and others that are not.

General considerations

Finding the results of the oral and dermal studies is very time consuming, and a review of the oils and their effects in a spreadsheet have been compiled. (See **Appendix I, Essential Oils Safety**.) This compilation will allow a quick review as to the suitability of an oil in a given product.

Quick summary of unsuitable essential oils

It is useful to check to see if there is a risk associated with the essential oil before looking at the spreadsheet.

Essential oils that may be dangerous in aromatherapy

Almond oil bitter (*Prunus amygdalus*) contains hydrocyanic acid; Armoise oil (*Artemisia herba-alba*) contains thujones; Boldo leaf (*Peumus boldus*) contains ascaridole; Calamus oil (*Acorus calamus*) contains β-asarone type compound; Chenopodium (Wormseed) oil (*Chenopodium ambrosioides*) contains ascaridole; Croton oils (*Croton tiglium*, *C. oblongifolius*); Horseradish oil (*Amoracia rusticana*) contains allyl isocyanate, phenylethyl isocyanate; Lanyana oil (*Artemisia afra*) contains thujones; Mustard oil (*Brassica nigra, B. juncea*) contains allyl isocyanate; Parsley herb oil (*Petroselinum crispum*) contains dill apiole; Pennyroyal oil (*Mentha pulegium*) contains pulegone; Perilla oil (*Perilla frutescens*) contains perilla ketone; Savin oil (*Juniperus sabina*) contains sabinyl acetate; Sassafras oil (*Sassafras albidum*) contains safrole; Summer Savory oil (*Satureja hortensis*); Tansy oil (*Tanacetum vulgare*) contains thujones; Wintergreen oil (*Gaultheria procumbens*) contains methyl salicylate; and Wormwood oil (*Artemisia absinthium*) contains thujones.

Oils that may cause irritation

Bay oil West Indian (*Pimenta racemosa*), Clove oils (stem, leaf, bud) *Syzygium aromaticum*, and Coriander oil (*Coriandrum sativum*) high linalool content; Ho oil (*Cinnamomum camphora* var. *linaloolifera* and *Cinnamomum camphora* var. *glavescens*) high linalool content; Kuromoji oil (*Linda umbellata*) high linalool content; May Chang oil (*Litsea cubeba*), Melissa oil (*Melissa officinalis*), Origanum oil (*Origanum vulgare*), Pimento berry and leaf oils (*Pimenta officinalis*); Rosewood oil (*Aniba rosaedora*) high linalool content; Summer Savory oil (*Satureja hortensis*), Winter Savory oil (*Satureja montana*), Tagetes oil (*Tagetes minuta*), Tea Tree oil (*Melaleuca alternifolia*), Thyme oil (*Thymus vulgaris*) and Turpentine oil (*Pinus sylvestris*).

Oils that are known to cause sensitization in massage

Cassia oil (*Cinnamomum cassia*) contains cinnamic aldehyde, coumarin; Cinnamon bark oil (*Cinnamomum zeylanicum*) contains cinnamic aldehyde; Costus oil, abs, concrete (*Saussurea lappa*) contains sesquiterpene lactones; Elecampane oil (*Inula helenium*) contains sesquiterpene lactones; Fig leaf absolute (*Ficus carica*); Massoia bark oil (*Cryptocarya massoia*)contains massoia lactone; Melissa oil (*Melissa officinalis*) contains citral; Oakmoss (*Evernia prunastri*); Treemoss (*Evernia Furfuracea*); Opoponax (*Commiphora erythrea*); Peru balsam and oil (*Myroxylon pereirae*); Styrax (*Liquidamber* spp); Verbena absolute and oil (*Lippia citriodora*); Tea absolute (*Camellia sinensis*); Turpentine oil (*Pinus* spp.); Lemon Myrtle oil (*Backhousia citriodora*) contains high citral/citronellal content; and Inula (*Inula graveolens*) contains sesquiterpene lactones.

Oils that are phototoxic

Amni visnaga oil (*Amni visnaga*); Angelica root oil (*Angelica archangelica*); Bergamot oil expressed (*Citrus aurantium bergamia*); Cumin oil (*Cuminum cyminum*); Fig leaf absolute (*Ficus carica*); Grapefruit oil expressed (*Citrus paradisi*); Lemon oil cold pressed (*Citrus medica limonum*); Lime oil expressed (*Citrus aurantifolia*); Mandarin oil cold-pressed (*Citrus reticulata*); Opoponax oil, absolute, resinoid (*Commiphora erythrea*); Orange oil bitter (*Citrus aurantium amara*); Parsley leaf oil (*Petroselinum crispum*); Petitgrain Mandarin oil (*Citrus reticulata* var. *mandarin*); Rue oil (*Ruta graveolens*); Tagetes oil and absolute (*Tagetes minuta*); Tangerine oil cold-pressed (*Citrus reticulata*); and Verbena oil (*Lippia citriodora*).

Pregnancy and effects on the reproductive system (abortifacient)

Chaste Tree Oil (*Vitex agnus-castus*); Plectranthus Oil (*Plectranthus fruticosa*); Parsley leaf oil (*Petroselinum crispum*); Spanish sage oil (*Salvia lavandulaefolia*); and Savin oil (*Juniperus sabina*).

Pregnancy and effects on the reproductive system (avoid during pregnancy)

Balsamite contains camphor; Camphor, White (*Cinnamomum camphora*) contains camphor; Ho leaf oil (*Cinnamomum camphora* var. *linaloolifera* and *Cinnamomum camphora* var. *glavescens*) contains safrole and camphor; Hyssop oil (*Hyssopus officinalis*) contains pinocamphone; Dill seed oil (*Anethum graveolens*) contains apiol; Juniper oil (*Juniperus pfitzeriana*); Parsley leaf and seed oils (*Petroselinum crispum*) contains apiol; Plectranthus Oil (*Plectranthus fruticosa*) contains sabinyl acetate; Spanish sage oil (*Salvia lavandulaefolia*) contains sabinyl acetate; and Savin oil (*Juniperus sabina*) contains sabinyl acetate.

Pregnancy and effects on the reproductive system (use with caution during pregnancy)

Wormwood oil (*Artemisia absinthium*) contains thujones; Brazilian Cangerana oil (*Cabralea cangerana* Sald.) [Syn. *C. glaberrima*] contains safrole; French Lavender (*Lavandula stoechas*) contains camphor; Perilla oil (*Perilla frutescens*) contains perilla ketone; Rue (*Ruta graveolens*) contains chalepensin [3-(α-,α-dimethylallyl) psoralen,13%]; Tree Moss (*Evernia furfuracea*) contains atranorin, physodic acid, furfuracinic acid and chloro-atranorin.

Pregnancy and effects on the reproductive system (not used orally, rectally or vaginally during pregnancy)

Anise (*Pimpinella anisum*) contains anethole; Fennel (*Foeniculum vulgare*) contains anethole; Lavandin (*Lavandula hybrida*) contains camphor, French Lavender (*Lavandula stoechas*) contains camphor; Mace (*Myristica fragrans*) contains safrol and myristicin (which are both contraindicated in pregnancy); Nutmeg (*Myristica fragrans*) contains safrole and myristicin; Rosemary (*Rosmarinus officinalis*) contains camphor; Spike Lavender (*Lavandula latifolia*) contains higher camphor levels than other lavender spp.; Star Anise (*Illicium verum*) contains anethole; and Yarrow (*Achillea millefolium*) contains camphor.

Oils that are carcinogenic

Birch tar oil crude (*Betula panda*) contains polynuclear hydrocarbons; Cade oil crude (*Juniperus oxycedrus*) contains polynuclear hydrocarbons; Calamus oil (*Acorus calamus*) contians β-asarone type; *Cinnamomum porrectum* oil Croton oils (*Croton tiglium*, *C. oblongifolius*); Ocotea cymbarum oil; and Sassafras oils (*Sassafras albidum*) contain safrole.

Oils that are carcinogenic to rodents (methyl eugenol)

Bay oil, West Indian (*Pimenta racemosa*) to 12.6% (methyl eugenol); Basil oils (*Ocimum* spp.) some chemotypes to 65% (methyl eugenol); Melaleuca oils (e.g. *Melaleuca bracteata*) to 50% (methyl eugenol); Nutmeg oil (*Myristica fragrans*) to 1.2% (methyl eugenol); Pimento oils (*Pimenta officinalis*) to 15% (methyl eugenol); Rose oils (*Rosa* spp.) to 3.0% (methyl eugenol).

Legal requirements

There are 24 potential allergens that have to be calculated out not only for the single oil, but that have to be accumulatively accounted for in a blend of essential oils. These must be declared at 0.01% in rinse-off products such as shower gels and bath foams and at 0.001% in the case of leave-on products such as lotions and creams.

The list is as follows: amyl cinnamyl alcohol, amyl cinnamal, anise alcohol, benzyl alcohol, benzyl benzoate, benzyl cinnamate, benzyl salicylate, cinnamyl alcohol, cinnamal, citral, citronellol, coumarin, eugenol, farnesol, geraniol, hexyl cinnamal, hydroxycitronellal, isoeugenol, butylphenyl methylpropional, limonene, linalool, hydroxyisohexyl-3-cyclohexene carboxaldehyde, methyl 2-octynoate, α-isomethyl ionone, *Evernia prunastri* (Oakmoss) Extract, *E. Furfuracea* (Treemoss) Extract.

Other requirements

Essential oils are not single components but complex blends. For example, the typical composition of ylang-ylang oil (from the European Flavour & Fragrance Association list of potential allergens in essential oils) contains the materials shown on the next page.

Potential allergen	***Cananga odorata***
Benzyl Alcohol	0.000
Benzyl Salicylate	3.000
Cinnamyl Alcohol	*
Cinnamal	*
Citral	*
Coumarin	*
Eugenol	0.700
Geraniol	1.500
Isoeugenol	*
Anise Alcohol	*
Benzyl Benzoate	5.000
Benzyl Cinnamate	*
Citronellol	*
Farnesol	2.000
Limonene	*
Linalool	3.000

Each of these materials have the potential to cause a skin reaction and their safety must be determined the safety in a particular product. In order to assess the safety, one has to look at the current technique.

The International Fragrance Association (IFRA) has introduced the Quantitative Risk Assessment (QRA) approach for fragrance ingredients. It is highly recommended that readers acquaint themselves with its Web site.

The first useful table is the Sensitization Assessment Factor (SAF), where the safety assessor may determine the Category Consumer Exposure Level (mg/cm2/day) that is driven by the product type in that category with the combined highest consumer exposure level and highest Sensitization Assessment Factor (SAF).

The Category is determined from the IFRA Categories for Dermal Sensitization, QRA Approach, arranged by Category.

The safety is then determined by the category and the method of delivery for each of the potential allergens. This is a laborious process, but the Research Institute for Fragrance Materials (RIFM) and IFRA have published a number of useful papers that have made those calculations already.

If 1% of *Cananga odorata* essential oil was used in a lipstick, this would equate to 0.007% eugenol in the example. At this point, consult the table in **Appendix I**.

Clearly this is well within the safety limits for eugenol for ylang-ylang oil when used at 1%. The process is now repeated for all of the other allergens and any other potential active that may cause problems.

Conclusions

The use of essential oils can be safe provided that the percentage use, product application, target consumer and all of the toxicology data have been carefully evaluated and considered. It is never wise to use an essential oil without first diluting it in carrier oil. The responsibility for product safety should not rely on the information provided in this chapter.

Chapter 10

Natural Actives

Introduction

There are many occasions when formulators want to be sure that the products they are making contain a discernible level of active materials. There are many extracts that are standardized against a particular active component, but at the level used in the formula may be insufficient to satisfy its functional purpose. In these cases the addition of active material may be a solution.

The industry has been slowly adding to the portfolio of these materials and it was thought appropriate to conduct a review of single active materials sold at >90% active. Sources of supply have been provided, but it is possible that this is not a comprehensive list and that not all the actives have been discovered.

All of the actives described below have been found available commercially.

Allantoin

This material is also known as glyoxydiureide or 5-ureidohydantoin and its history is very interesting. During World War I it was observed that the wounds of soldiers in trenches appeared to heal more quickly than those kept under sanitary conditions. Upon investigation it was found that the trenches were infested with maggots, which were crawling on their wounds. Further investigation revealed that the substance secreted by the maggots, which was contributing to the improved healing, was allantoin. It also occurs in tobacco seeds, sugar beet, wheat sprouts and comfrey (0.8%) as well as in rice polishings and in the bark of horse chestnut. Allantoin is a topical vulnerary and is a skin ulcer therapy.

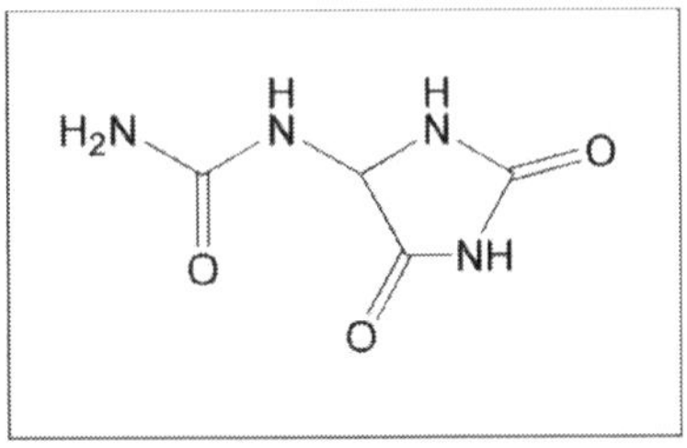

Allantoin

In the veterinary field it has been used topically to stimulate the healing of suppurating wounds and resistant ulcers. In the 5th edition 1941, it reported that it was used externally to stimulate cell proliferation in 0.3-0.5% aq. solutions. It was for ulcers, non-healing wounds, fistulas etc.

Supplier: Clariant. INCI name: Allantoin

Arbutin

Arbutin is traditionally obtained from the aerial part of Bearberry (*Arctostaphylos uva ursi*) by hydroalcoholic extraction, followed by purification steps by fractionated recrystallization and filtrations on active charcoal.

Arbutin may be considered as a 'pro-hydroquinone' that avoids the damaging effects of hydroquinone but keeps the efficacy as an antiseptic and a melanogenesis inhibitor by inhibiting competitively tyrosinase. The material is useful not only for cultural skin whitening but also for the skin fairing of freckles and liver spots (age spots).

The inhibitory action of arbutin, kojic acid and ascorbic acid on tyrosinase activity in human melanocytes has been compared. All of these materials are active whitening cosmetic components and hydroquinone was used as a positive control. The depigmenting effect of linoleic acid, which has been reported to inhibit melanin synthesis, was also compared with those of arbutin, kojic acid and ascorbic acid. Arbutin dose-dependently reduced tyrosinase activity at final concentrations between 0.01 mM and 1.0 mM, at which no change in cell viability could be seen. This action was about 1/100 that of hydroquinone, and was stronger than that of kojic acid and ascorbic acid. Linoleic acid did not reduce tyrosinase activity at non-cytotoxic ranges. Furthermore, at concentrations of 0.5 mM, the amount of melanin was reduced significantly by arbutin.

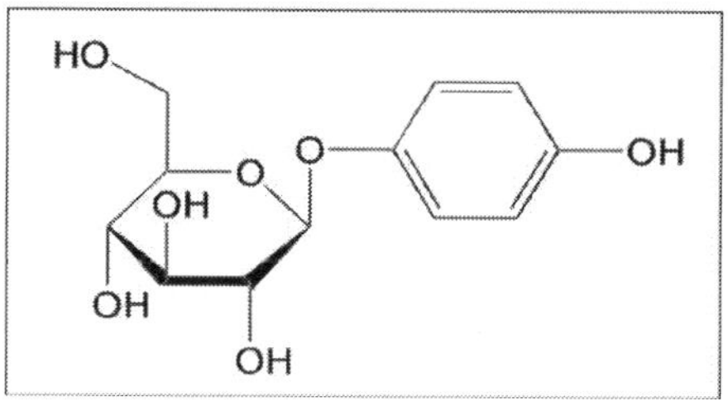

HO O O OH OH OH OH

Arbutin

These results suggested that arbutin inhibits melanin production by reducing tyrosinase activity and that the depigmenting action of this agent is stronger than that of kojic acid or ascorbic acid.

Supplier: Bioland, A&E Connock. INCI name: Arbutin. Supplier: Pentapharm. INCI name: Alpha-Arbutin.

Asiatic Acid, Asiaticoside and Madecassic Acid

These materials are found and standardized in Gotu Kola, Hydrocotyle, Indian Pennywort, or Talepetrako (*Centella asiatica*) with typical values asiaticoside approx. 36-44%, asiatic acid and madecassic acid approx. 44-66%.

Gotu kola extracts have been found to promote wound healing. Cell culture experiments have shown that the total triterpenoid fraction of the extract, at a concentration of 25 mcg/mL does not affect cell proliferation, total cell protein synthesis or the biosynthesis of proteoglycans in human skin fibroblasts. The fraction does, however, significantly increase the collagen content of cell layer fibronectin, which may explain the action in wound healing. The glycoside madecassoside has anti-inflammatory properties, while asiaticoside appears to stimulate the wound healing process.

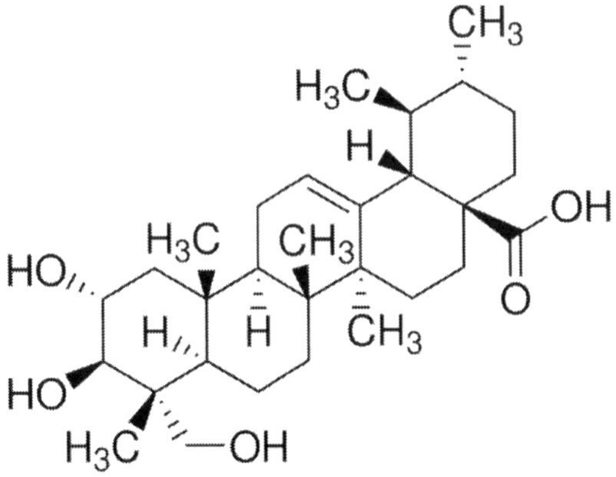

Asiatic acid

Asiaticoside was found active against leprosy (by dissolving the waxy coating of *Mycobacterium leprae*) and an oxidized form, oxyasiaticoside, inhibited growth HOof tubercle bacillus in vitro (0.015 mg/mL) and *in vivo*. It also helps heal ugly skin lesions common to systemic lupus erythematosus and herpes simplex by promoting a rapid thickening of the skin and an increased blood supply to the connective tissue. This constituent is also responsible for accelerated growth of hair and nails as well. A leaf extract (standardized to asiaticoside) was evaluated in clinical patients with soiled wounds and chronic atony, resistant to treatment; results showed complete healing in 64% and improvement in 16% of 20 patients.

Supplier: Indena, Bayer, Cognis. INCI name: Asiaticoside, Asiatic Acid, Madecassic Acid

Astaxanthin

Astaxanthin is a fat-soluble carotenoid with a unique molecular structure that makes it an extremely effective antioxidant and a red natural color. The micro-alga *Haematococcus pluvialis* is the richest natural source for astaxanthin, capable of accumulating up to 5% of its dry biomass. It is the most abundant carotenoid pigment found in aquatic animals and the marine world. It is found in sea foods such as salmon, trout, red sea bream, shrimp, lobster and caviar. In fish, astaxanthin has been shown to have a synergistic effect on other important antioxidants such as Vitamins C and E.

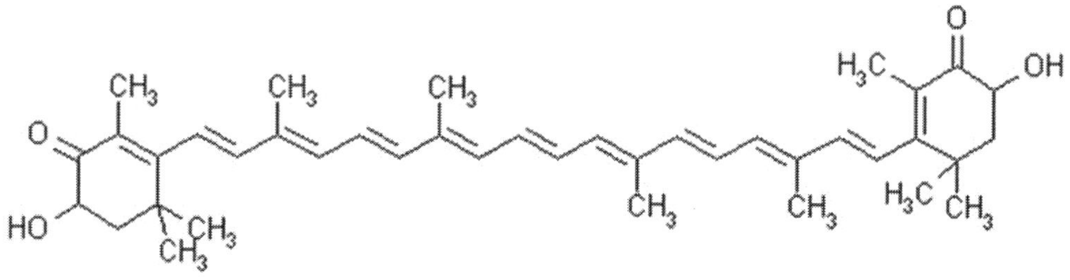

Astaxanthin

Supplier: U.S. Nutra (through Chesham Speciality Ingredients). INCI name: Astaxanthin

Avenanthramide

Avenanthramide is produced by Super Critical Fluid extraction using CO_2 from the whole oat grain (*Avena sativa*). This group of phenolic amides, which are unique to oats, are known as avenanthramides and are the principle components responsible for reducing skin irritations and redness. These compounds have only been found in oats.

Avenanthramides show redness reduction against UV irritation and the efficacy starts at 0.4 ppm. The material also shows superior anti-irritant and antioxidant properties. Avenanthramides are responsible for anti-inflammatory, anti-itch properties and show antihistamine activity.

Supplier: Symrise. INCI name: Hydroxyphenyl Propamidobenzoic Acid

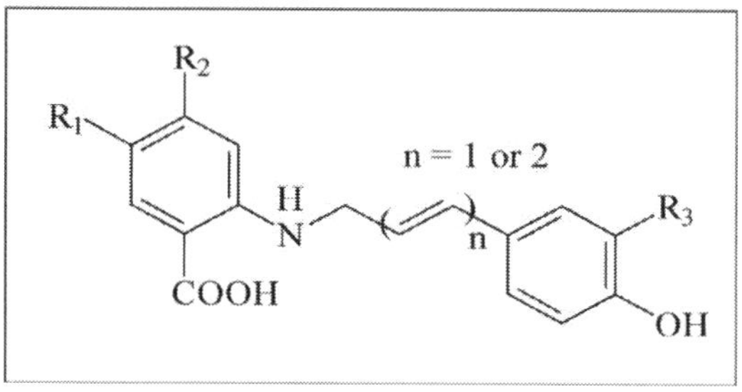

Avenanthramide

Azulene

Azulene is a main component of chamomile extract that is used as a coloring agent (*azul* means "blue" in Spanish) in cosmetics and ointments, and used as a folk medicine to treat migraine, arthritis, rheumatic diseases, and allergy.

It has antioxidant and antiinflammatory activities. Examples of common substituted azulenes obtained naturally are vetivazulene (4,8-dimethyl-2-isopropylazulene) from vetiver oil, and guaiazulene (1,4-dimethyl-7-isopropylazulene) from constituent of guaiac wood oil.

Chamazulene

Guaiazulene has been used as an anti-ulcer drug. Various biological activities such as antiphlogistic, anti-allergic, bacteriostatic from azulene class compounds are expected. Sadly this wonderful blue (blue-black) active no longer seems to be available commercially. Azulene is a prime component of the essential oil of chamomile flowers, *Matricaria recutita* L. and related plant species. Products containing azulene generally also contain the other characteristic components of chamomile's essential oil. Azulene extracts are used in skin creams for reducing skin puffiness and wrinkles and is also known for its anti-irritant and vulnerary properties.

It has never been clear whether the material was extracted from one of the chamomiles or from *Myoporum crassifolium* Forst or *Vanillosmopsis erythropappa* Schultz-Bip in which this material is also found. Maybe the source has become too scarce for there to be a reliable commercial supply.

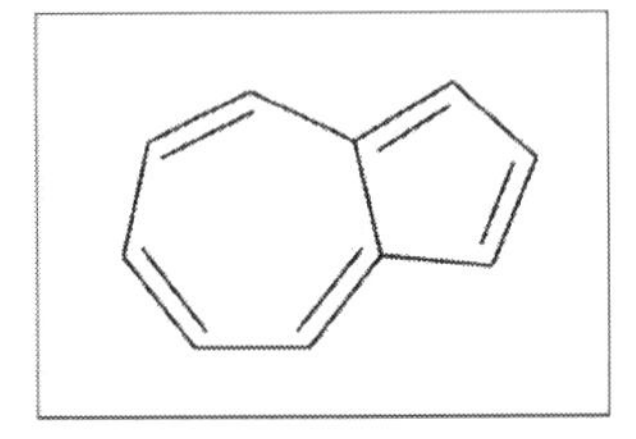

Azulene

Supplier: Symrise. INCI name: Azulene.

Betulinic acid

Betulinic acid (3b-hydroxy-lup-20(29)-en-28-oic acid, BA) is a plant product with a pentacyclic triterpene skeleton of the lupane type. It may be described as a steroidal triterpene. The name is derived from the corresponding alcohol betulin, which is contained in the white cork of birch tree trunks.

It is also found in the bark of several species of plants, including the Ber tree (*Ziziphus mauritiana*), the white birch (*Betula pubescens*), and many other plants such as the tropical carnivorous plants *Triphyophyllum peltatum* and *Ancistrocladus heyneanus*; , a member of the persimmon family; *Tetracera boiviniana*; the jambul (*Syzygium formosanum*); flowering quince (*Chaenomeles sinensis*); and *Pulsatilla chinensis*.

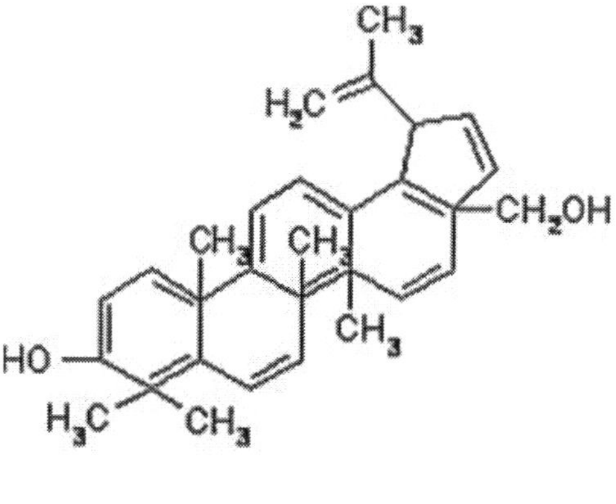

Betulin 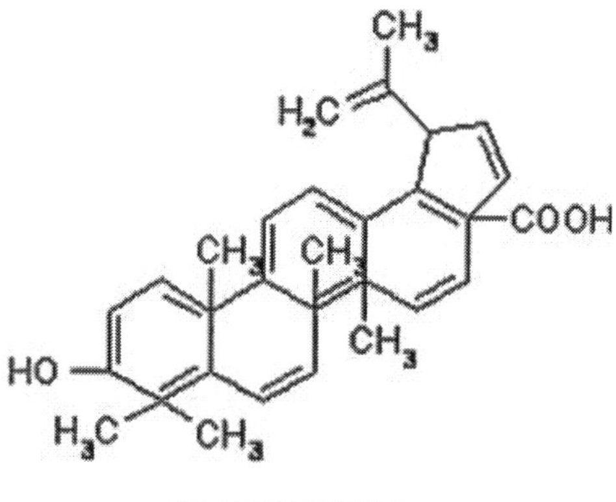 **Betulinic acid**

It is traditionally used in Brazilian folk medicine for its antibacterial properties and for asthma treatment. The commercial source is extracted from the birch (*Betula alba*) bark.

Betulinic acid exhibits strong, glucocorticoid-like activity by promoting the synthesis of proteins acting as modulators of the inflammation process. It was demonstrated that betulinic acid inhibits TPA-induced inflammation and has the ability to inhibit tumor formation. Betulinic acid is a selective inhibitor of human melanoma that functions by induction of apoptosis. It is also studied as an inhibitor of type 1 HIV.

Supplier: Soliance. INCI name: Birch (Betula alba) Bark Extract

Bisabolol

Chemically it is 3-cyclohexene-1-methanol-a,4-dimethyl-a-(4-methyl-3-pentenyl)-a,4-dimethyl-a-(4-methyl-3-pentenyl)-3-cyclohexene-1-methanol. It is an antiphlogistic active principle derived from nature. It is an optically active, monocyclic, unsaturated sesquiterpene alcohol. It is obtained from natural raw materials. It has been identified as a natural component of various plants. For example, it has been found in the essential oils of *Matricaria chamomilla* (German Chamomile), *Myoporum crassifolium* Forst and *Vanillosmopsis erythropappa* Schultz-Bip.

It is used for sensitive skins, sun burn, rubbing, chafing, mechanical scraping (e.g. shaving), soreness, nappy rash and after using chemicals on the skin (e.g. depilation). Tests have showed that a 1% solution of bisabolol in propylene glycol reduced the irritation reaction of the skin after the application of ammonia, if applied before the irritant.

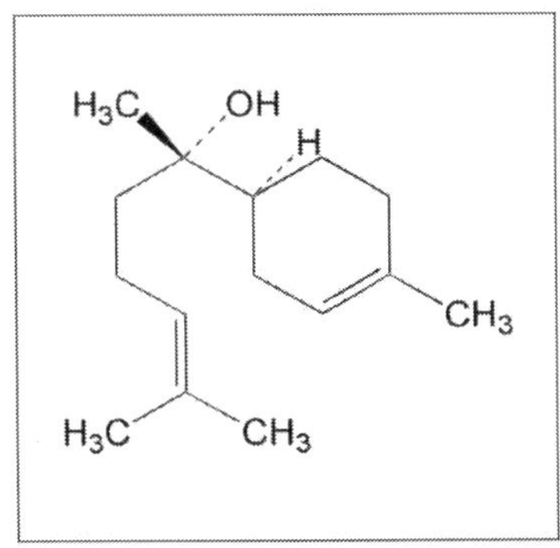

Bisabolol

A summary of the results from three tests showed that (+/-)-alphabisabolol rac. is at least 50% more effective than (-)-alpha-bisabolol nat. During the treatment of UV-induced erythema on human skin (-)-alpha-bisabolol nat. was also found to exert an antiphlogistic effect. Bisabolol also possesses an unequivocal anti-inflammatory activity [Jakovlev and Schlichtegroll; Duke], in which the natural product has consistently a higher activity than the synthetic product.

Supplier: BASF, Symrose. INCI name: Bisabolol.

Boswellin or Boswellic Acid

Boswellic acids are a series of pentacyclic triterpene molecules that are produced by plants in the genus Boswellia. Like many other terpenes, boswellic acids appear in the resin of the plant that exudes them; it is estimated that they make up 30% of the resin of *Boswellia serrata.*

Boswellic acids have been found to inhibit leukotriene synthesis via inhibition of 5-lipoxygenase, but did not affect the cyclooxygenase activities and therefore synthesis of prostaglandins. This data suggests that boswellic acids are specific inhibitors of leukotriene synthesis acting either by interlocking with the 5-lipoxygenase enzyme or by blocking its mobility. The mechanism of boswellic acids and their derivatives were compared with that of nordihydroguaiaretic acid. Both compounds inhibited enzyme 5-lipoxygenase at comparable dose levels. Unlike nordihydroguaiaretic acid, however, boswellic acids did not impair the other two enzymes metabolizing arachidonic acid, cyclooxygenase and 12-lipoxygenase. Inhibition of the 5-lipoxygenase pathway by various isomers and derivatives constituting the mixture of boswellic acids of *Boswellia serrata* have been studied. It has been found that the acetyl-11-keto-ß-boswellic acid provides the best inhibitory action.

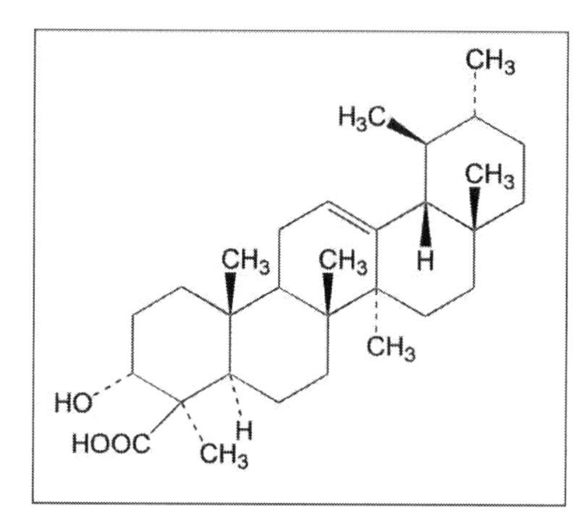

Boswellic Acid

Boswellic acid is an excellent anti-inflammatory.

Supplier: Sabinsa. INCI name: Boswellia Serrata Resin Extract

Caffeine

Caffeine is a bitter, white crystalline xanthine alkaloid that is a psychoactive stimulant drug. Caffeine is found in varying quantities in the beans, leaves and fruit of some plants, where it acts as a natural pesticide that paralyzes and kills certain insects feeding on the plants.

It is most commonly consumed by humans in infusions extracted from the cherries of the coffee plant and the leaves of the tea bush, as well as from various foods and drinks containing products derived from the kola nut. Other sources include Yerba Mate (*Ilex paraguensis*), Guarana berries (*Paullinia cupana*), and the Yaupon Holly (*Ilex cassine*).

Caffeine is a central nervous system (CNS) stimulant, having the effect of temporarily warding off drowsiness and restoring alertness. It is also used topically as a skin stimulant and often employed in anticellulite products. The U.S. Food and Drug Administration list caffeine as a "multiple purpose generally recognized as safe food substance."

Caffeine

Caffeine inhibits phosphodiesterase and increases the c-AMP level in cells that leads to the activation of protein kinase. Once activated, this enzyme phosphorylates triacylglycerol lipase that may then lead to fat degradation. In more detail, caffeine inhibits the degradation of cAMP (cyclic-AMP) to AMP that leads to an accumulation of cAMP. It is this increased amount of cAMP that stimulates the protein kinase to phosphorylate triacylglycerol lipase and so as the lipase is activated so it cleaves triacylglycerol (or fat) into fatty acids and glycerol.

Androgens, especially testosterone and dihydrotestosterone can have a disadvantage in male skin when hair becomes thinner in the scalp and there is reduced epidermal permeability barrier repair capacity. Dihydrotestosterone acts as an adenyl cyclase inhibitor. Caffeine can act as a phosphodiesterase inhibitor and is able to counteract the testosterone-induced effects on barrier function and so beneficial particularly for male skin.

There are a multitude of products on the market that contain a blend of active materials among which caffeine is a component, but there are fewer suppliers of the pure material.

Supplier: Merck, A&E Connock. INCI name: Caffeine.

Carotene

The carotenoids form one of the most important groups of natural pigments and more than 400 naturally occurring carotenoids have been identified.

Carotenes are fat-soluble and often synthetically produced natural pigments. The colors range from yellow to red. The pigment is sensitive to oxidation because of the conjugated double bonds and the molecule may be isomerized if involved in a heating processing. There are a number of related chemicals

This is a group of yellow/orange colors extracted from such diverse sources as algae, carrots and palm oil. It is also available as a "nature identical" product. Crystalline β-carotene is sensitive to air and light. Vegetable fat and oil solutions and suspensions are quite stable in products. The most commonly encountered carotenoids are fucoxanthin, lutein, violaxanthin, ß-carotene, lycopene and apo-carotenoids.

The carotenoids—apart from the chlorophylls—are the largest group of oil-soluble pigments found in nature. They consist of molecules with long chains made up of carbon, hydrogen and mostly oxygen (ß-carotene consists of only carbon and hydrogen). One of the carotenoid's characteristics is that the color varies (according to the type of carotenoid) from yellow to red-orange. Carotenoids, like chlorophylls, exist in green plants. They are responsible for the yellow color of flowers and the pigments of many fruits and vegetables like carrots, paprika and tomato.

The first discovery of the color in carrots was the reason for the generic name of carotenoids. It is converted by mammals into vitamin A and as a result is called provitamin A.

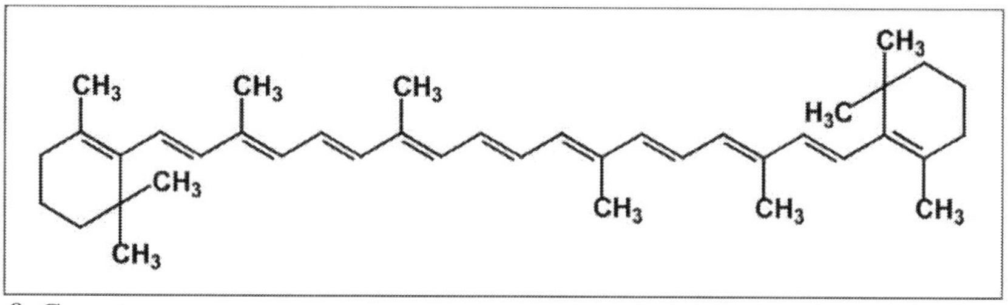

ß-Carotene

ß-carotene is one of the major yellow colors used in the food industry and the largest use is by the dairy industry (butter, cheese and ice cream). The use of ß-carotene has almost entirely replaced E102 Tartrazine yellow, which for various reasons has received bad press in recent years. In addition to its color, ß-carotene is one of the popular free radical scavengers and antioxidants. There are many active solutions of carrot oil available (e.g. CLR).

Suppliers: Overseal Foods, Univar, Unifect, DSM. INCI name: Beta-Carotene.

Chlorophyll

The name of chlorella is derived from the Greek word *chlor*, meaning "green," and the Latin word *ella*, meaning "tiny." The Greek word was also the origin or the word chlorophyll, the green pigment found in plants and which is a by-product of photosynthesis. Approximately 2-3% of chlorella is chlorophyll.

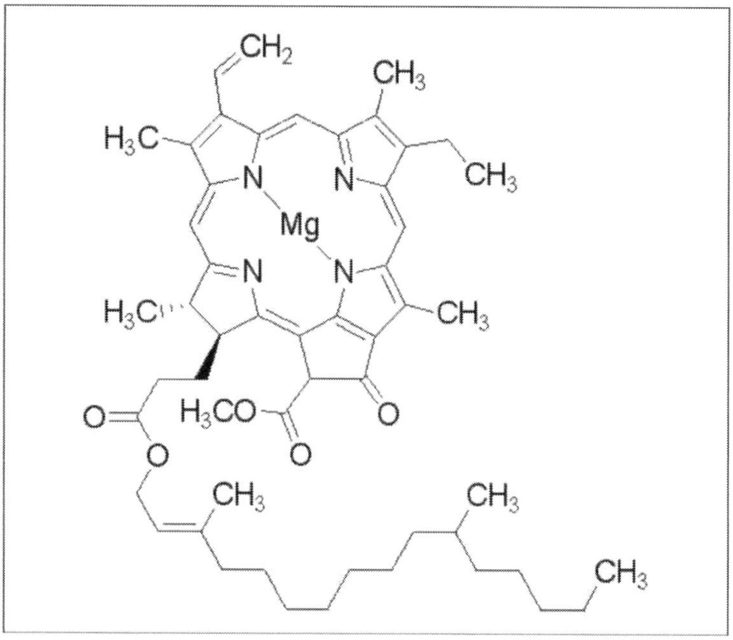

Chlorophyll

Extracted from grass and alfalfa, this is present in all green plants and has always been a part of man's diet. Gives a moss green color. Naturally oil-soluble. It is also found in green vegetables such as spinach or *Spinacia oleracea* and the common stinging nettle or *Urtica dioica.*

It is used to color soaps, oils, fats, waxes, confectionery, preserves, liquors, cosmetics, perfumes. It is used as a deodorant, and in veterinary medicine has been used orally to reduce odors, and topically to promote healing of skin lesions.

A study on the problem of foot odor or pedal bromidrosis was conducted. One hundred patients who had noticeable pedal bromidrosis were chosen in a series of ward surveys. One half of these men were issued placebo capsules and the others were issued chlorophyll capsules. Only the pharmacist knew which patients had which capsules. The patients were categorized according to the intensity of foot odor. Odour impressions were made weekly. The results of the study indicate that the attention given to the feet was the important factor in the improvement that most had, not the medication itself .

Copper chlorophyll

Derived from plants (as in Chlorophyll) but gives a brighter more intense green color due to the replacement of the naturally occurring magnesium in the chlorophyll by copper. It is naturally oil-soluble. This is produced as the copper chlorophyll, but a saponification process renders this form water-soluble. The color is a bright green to green/blue.

Suppliers: Overseal Foods, Unifect, Univar. INCI name: not allowed in the USA.

Curcumin

Curcumin is extracted from the dried root of the turmeric rhizome *Curcuma longa.* The process of extraction requires the raw material to be ground into powder, and washed with a suitable solvent that selectively extracts coloring matter. This process after distillation of the solvent yields an oleoresin with coloring matter content in the region of 25–35% along with volatile oils and other resinous extractives. The oleoresin so obtained is subjected to further washes using selective solvents that can extract the curcumin pigment from the oleoresin. This process yields a powdered, purified food color, known as curcumin powder, with more than 90% coloring matter content and very little volatile oil and other dry matter of natural origin. The characteristic yellow color of turmeric is due to the curcuminoids, first isolated by Vogel in 1842.

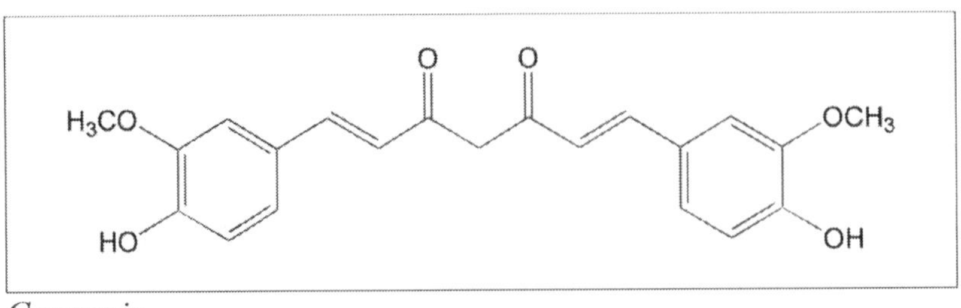

H3CO O O OCH3 HO OH

Curcumin

Curcumin provides a water-soluble orange-yellow color. It is a natural extract obtained by solvent extraction from the dried rhizomes of turmeric (used in Indian cuisine as a flavoring agent). Curcumin may be used to compensate for fading of natural coloring in pre-packed foods. Recognized as an anticarcinogenic agent during laboratory tests.

It is widely used in traditional Indian medicine to cure biliary disorders, anorexia, cough, diabetic wounds, hepatic disorders, rheumatism and sinusitis. Turmeric paste in slaked lime is a popular home remedy for the treatment of inflammation and wounds.

Curcumin has antioxidant, anti-inflammatory, antiviral and antifungal actions. Studies have shown that curcumin is not toxic to humans. Curcumin exerts anti-inflammatory activity by inhibition of a number of different molecules that play an important role in inflammation. Turmeric is effective in reducing postsurgical inflammation. It is also useful as an anti-inflammatory agent for arthritis and other inflammatory conditions such as dysmenorrhoea, asthma, infections, eczema, psoriasis and injuries. In the case of arthritis and injuries, a combination with more analgesic herbs may be necessary to provide adequate symptom relief.

Suppliers: Overseal Foods, Sabinsa, Unifect, Univar. INCI name: Curcumin.

Darutoside

Darutoside has been obtained from the plant *Siegesbeckia orientalis* L. by ethanolic extraction, followed by fractionated recrystallisation. Other varieties of *Siegesbeckia* such as *S. pubescens* and *S. glabrescens* contain darutoside as well as darutigenol, aglycon (without the sugar moiety) and isodarutoside. It is also found in *Palafoxia arida. Siegesbeckia orientalis* is, however, the primary source of darutosides. The plant is native to Ethiopia, but mainly found in Madagascar; also grows in Japan, Australia and the Dekkan peninsula. In Reunion and Mauritius islands it is known under several names referring to its healing properties: Divine grass, Collecolle (meaning "glue-glue") or Guerit-vite (meaning "heals rapidly"). Popular uses of the plant are externally for wound healing and for soothing inflammation. It has been examined for firming treatments, anti stretch mark treatment and antiaging preparations.

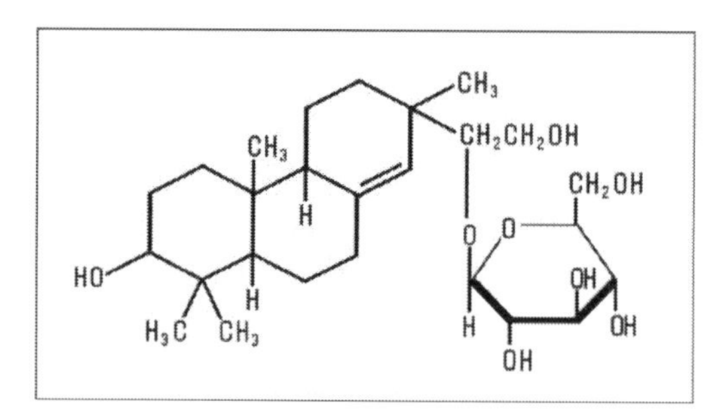

Darutoside

Supplier: Sederma. INCI name: Darutoside or *Siegesbeckia orientalis* Extract.

Diosgenin

The pharmaceutical use that led to the source of progesterone (one of the first female contraceptives) was originally sourced from Wild Yam (*Dioscorea villosa*) in the form of diosgenin. Eventually the stocks of Wild Yam fell to such low levels that industry was finding it hard to support the demand and so the much more widely available material Fenugreek (*Trigonella foenum-graecum*) was used as an alternative source of diosgenin. However the diosgenin by this time was being converted to another hormonal steroid, namely estrone.

Diosgenin improves skin elasticity, increases skin hydration, reduces skin roughness, is wrinkle-smoothing and shows an improvement in the skin lipid content.

There are many useful blends that contain diosgenin like Unisteron Y-50 (Induchem), Viazest Yam (Aldivia), Trogenine (Solabia), Wild Yam (Greentech), Actiphyte of Mexican Wild Yam, Actigen Y (Active Organics), Titramis Wild Yam (Alban Muller).

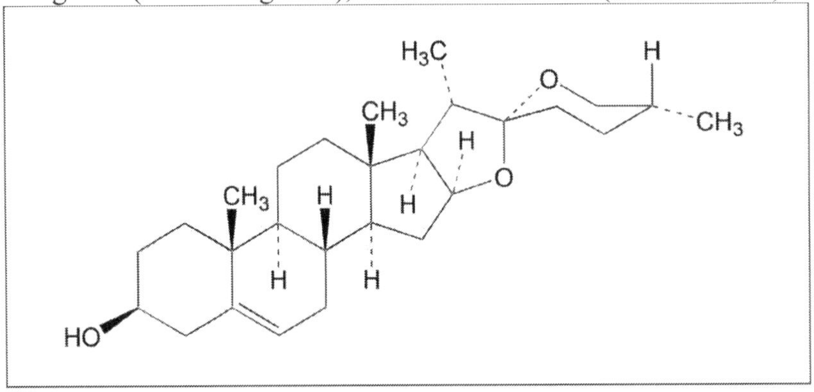

Diosgenin

Supplier: K,-W. Pfannenschmidt GmbH, Biospectrum, Sabinsa. INCI name: Diosgenin

Diosmin or Diosmine

It is a naturally occurring flavonic glycoside; rhamnoglycoside of diosmetin, which can be isolated from various plant sources. It is isolated from lemon peel (*Citrus limon* L., Rutaceae), from *Xanthoxylum avicennae*, Rutaceae) and from the flowers of *Sophora microphylla* Ait, Leguminosae.

It has been studied for its efficacy in animal models in high protein edema. It has also been evaluated for inflammatory granuloma and in a clinical study in post-phlebitic ulcers. It is used in acute hemorrhoids and has been the subject of a clinical trial in venous insufficiency. It may also be considered to be an excellent capillary protectant where it has been used with success in varicose veins.

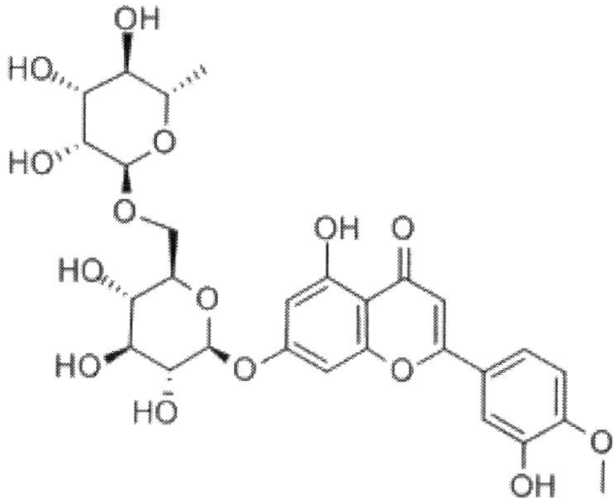

Diosmin

Supplier: Princeps, Innothera

Esculoside or Esculin

Esculoside is obtained from the bark of horse chestnut (*Aesculus hippocastanum* L.) by extraction with hot water, followed by precipitation of the tannins and fractionate recrystallization with active carbon filtration steps. It can also be obtained by partial synthesis from esculetol. Esculoside has also been found in an Australian Pittosporaceae, *Bursaria spinosa* and in *Fraxinus japonica.*

It has been shown to have anti-inflammatory activity.

It has been used in the treatment of red skin blotches, of heavy legs and edema [Rehder; Tarayre et Lauresser] as well as increasing the capillary resistance, decreasing capillary permeability and having anti-inflammatory activity.

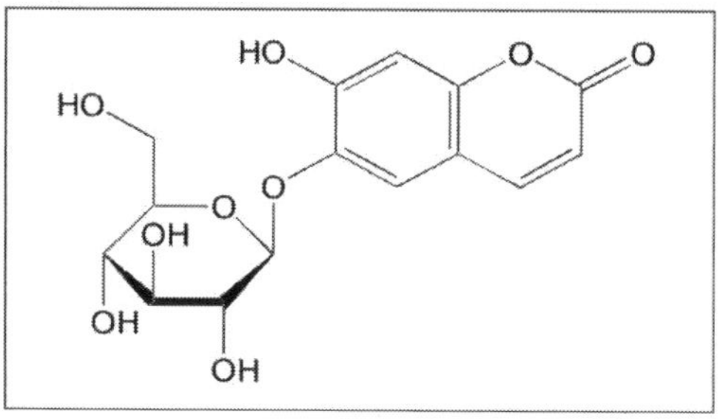

Esculoside

Supplier: Sederma, Euromed, Indena. INCI name: Esculin.

Farnesol

Farnesol is a natural organic compound that is an acyclic sesquiterpene alcohol found as a colorless liquid, insoluble in water, but miscible with oils. It is present in many essential oils such as citronella, neroli, cyclamen, lemon grass, tuberose, rose, musk, balsam and tolu and is used in perfumery to emphasize the odors of sweet floral perfumes. Farnesol is the only stereoisomer present in many essential oils but occurs mixed with *cis,trans* farnesol in petitgrain oil and several other oils.

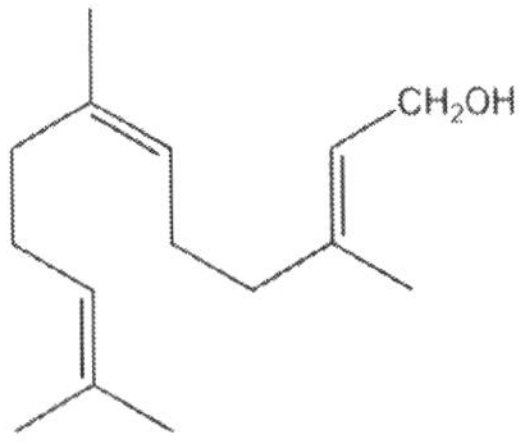

CH_2OH

Farnesol

In perfumery, it is used to emphasize the odour of sweet floral perfumes, such as lilac and cyclamen.

One study showed that HGQ, a mixture composed of three products occurring in plant or animal species, farnesol, glyceryl monolaurate and phenoxyethanol, killed corynebacteria and that it could be recommended for use as an effective deodorant. It is one of the 26 potential allergens that need to be declared in cosmetic products (0.01% in rinse-off products and 0.001 in leave-on).

Supplier: Symrise, Lipo. INCI name: Farnesol

Ferulic acid

Ferulic acid (4-hydroxy-3-meth-oxycinnamic acid) is a natural UV-absorber (maximum absorbance between 300 and 320 nm) derived and extracted from rice bran. Ferulic acid is available as an odorless and pale yellow crystalline powder. It is also a powerful antioxidant. Ferulic acid, unlike *p*-cinnamic acid and caffeic acid, scavenges superoxide anion radicals and also inhibits lipid peroxidation induced by trans-*Ferulic Acid*

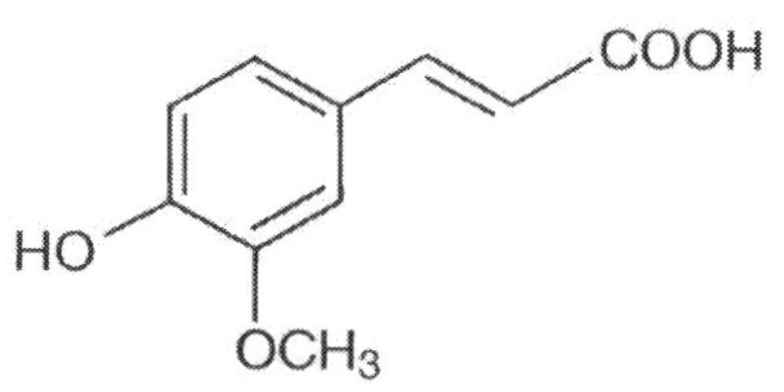

COOH
HO
OCH_3

trans - Ferulic Acid

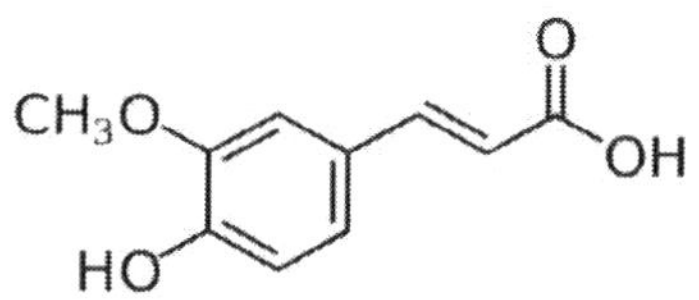

O
CH_3O
OH
HO

In 1886 Hlasiwetz and Barth in Innsbruck, Austria, isolated a dibasic acid from *Ferula foetida* and named this compound ferulic acid. Five years later, it was isolated again from *Pinus laricio* Poir by Bamberger. It is now often extracted from rice bran (*Oryza sativa*).

Supplier: Tsuno Rice, Ichimaru Pharcos, A&E Connock. INCI name: Ferulic Acid

Gamma-oryzanol

The labelling index and arrested mitosis values were significantly elevated by the topical application of *gamma*-oryzanol. Elevation of these values was paralleled by an increase in sebum production (reported in a previous paper) showing that sebaceous gland cells are stimulated by *gamma*-oryzanol. The paper further stated that other than sex hormones, *gamma*-oryzanol is the only substance known to stimulate sebaceous glands by topical application.

The antioxidant effects displayed by *gamma*-oryzanol are widely known, and its antioxidative power exhibited in the oxidation of oils and fats has been reported to be superior in animal oils and fats to vegetable ones.

Gamma-oryzanol also exhibits UV-absorbing properties and has absorption maxima at about 290 and 315 nm. It has tyrosinase-inhibiting action. In addition it inhibits erythema formation and has anti-inflammatory action. This study was carried out on mice, where it was noted that the clipped naked skin of mice exposed to a UV lamp developed erythema, whereas the mice treated with oils containing *gamma*-oryzanol led to the prevention of swelling and blisters. Lesions that were already inflamed were not relieved.

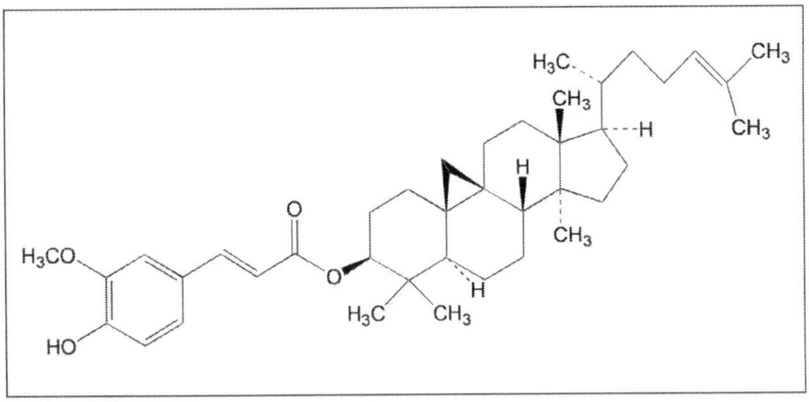

gamma-*Oryzanol*

Supplier: Tsuno Rice, Ichimaru Pharcos, A&E Connock, K,-W. Pfannenschmidt GmbH, Oryza Oil. INCI name: Oryzanol.

Glycyrrhetic acid, Glycyrrhetinic Acid, Enoxolone

Glycyrrhetic acid (3-ß-hydroxy-11-oxo-12-en (18-ß)-30-oic acid), synonym: 18-ß-glycyrrhetinic acid (enoxolone). Glycyrrhetinic acid is a pentacyclic triterpene obtained through the hydrolysis of glycyrrhyzic acid, a glucoside that comes from the liquorice root

Glycyrrhyiza glabra L., was isolated at the beginning of the 20th century. This chemical structure was defined after research work conducted by Ruzicka (1936–1943) and Beaton Spring around 1955.

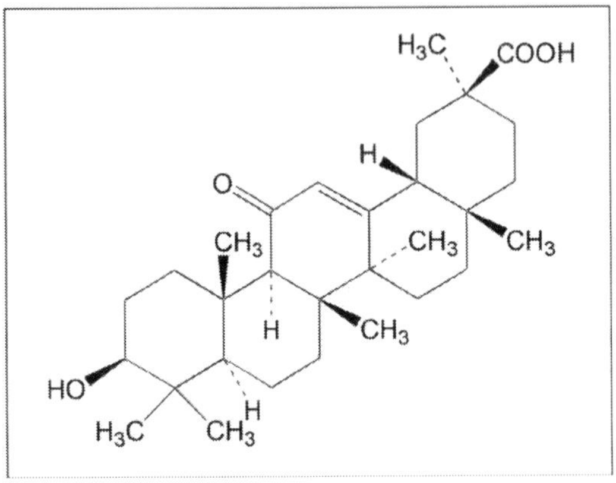

Glycyrrhetinic Acid

It is an anti-inflammatory and can be used in cases of acne, eczema, sun erythema and skin blotch. It is often used in combination with other actives for healing (synergy with asiaticoside (see **Asiatic Acid, Asiaticoside and Madecassic Acid** listing - from the aerial parts of *Centella asiatica*), and for alopecia (with Minoxidil). It is used in products specific for anti-allergy.

There are a number of useful salts. Potassium glycyrrhizinate has lower pharmacological activity, whereas it is a good surfactant and it has emulsifying properties and is soluble in water. Its uses are (in toothpaste) to soothe irritation, to prevent dental caries and to increase foaming power. Glycyrrhizic acid is anti-inflammatory, used in dental decay toothpaste, herpes simplex, conjunctivitis, osteoporosis (with calcitonine), surfactant (emulsifying).

Monoammonium glycyrrhizinate is a taste enhancer and is 60 times sweeter than sugar. It is almost insoluble in alcohol, soluble in dilute alcohol, soluble in hot water. Dipotassium glycyrrhizinate is an anti-inflammatory and used against dental decay in toothpaste.

Supplier: Amsar Private Ltd., Soliance, Lipo, Maruzen, Burgundy, Cognis. INCI name: Glycyrrhetinic Acid.

Hexylresorcinol

It is present in significant amounts in the bran fractions of wheat, rye and other cereals. This material is most normally used in mouthwashes and throat lozenges because it has antiseptic and anesthetic properties. It has been used topically on small skin infections. It is a strong inhibitor of tyrosinase and peroxidase/H_2O_2, it is also an effective skin lightener and even-toner for normal skin and effective skin lightener for hyper-pigmented skin. The European Union (EU) Council of Ministers has approved hexylresorcinol as a new additive and safe to use in EU foodstuffs (June 2006; Source: Keith Nuthall)

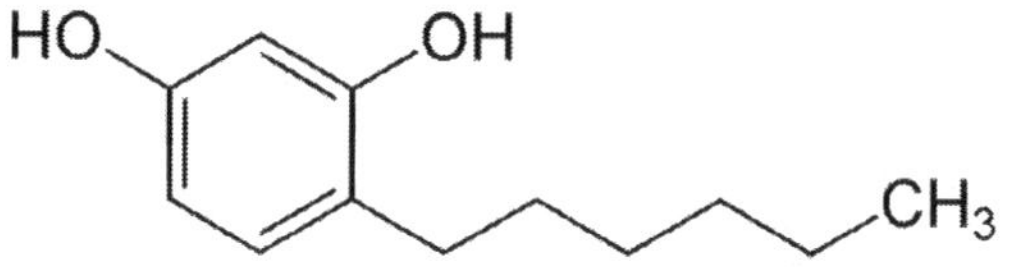

HO OH CH$_3$

Hexylresorcinol

Supplier: Kumar Organic Products, Sytheon Ltd. INCI name; Hexylresorcinol

Hinokitiol

Hinokitiol is a white crystalline acidic substance first isolated from the essential oil of Formosan Hinoki (*Chamaecyparis taiwanensis*) [Nozoe]. This substance was also found in the essential oil of Aomori Hiba tree (*Thujopsis dolabrata*) at a later date. Both of these species of trees are known for their high degree of resistance against wood decay. Subsequent research led to a proposed structure of a seven-membered carbocycle containing two double bonds (which is very rare in natural products), as well as carbonyl and enol radicals [Nozoe, T., Katsura]. After a further examination and revision, it was determined it to be 2-hydroxy-4-isopropyl 2,4,6-cycloheptatrieni-one (4-isopropyltropo-lone) with 3 double.

It has been shown that hinokitiol shows antimicrobial properties at low dosages and has a wide antimicrobial spectrum against general bacteria and fungi. Hinokitiol is effective against MRSA (Methicillin-Resistant *Staphylococcus aureus*) and studies have been carried out between antibiotics and hinokitiol, studying the occurrence of resistant bacterial strains like *Escherichia coli*, *Staphylococcus* sp and others. The results suggest that hinokitiol, unlike classical antibiotics, may be used to treat infections without the undesirable side effect of generating mutant resistant strains.

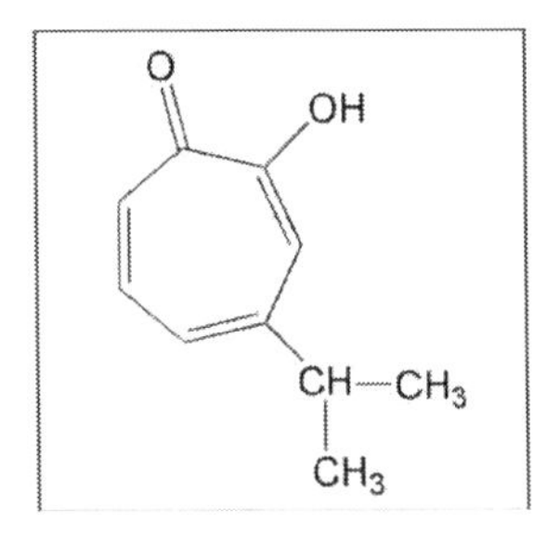

Hinokitiol

Supplier: A&E Connock, S. Black. INCI name: Hinokitiol

Kojic acid

Kojic acid is a chelating agent produced by several species of fungi, especially *Aspergillus oryzae*, which has the Japanese common name of *koji*. It is a by-product in the fermentation process of malting rice, when producing sake (Japanese rice wine). It is used in food and cosmetics to help preserve against color changes. Kojic acid also has antibacterial and antifungal properties.

Kojic acid markedly inactivated isolated tyrosinase by chelation. In cultured human melanocytes, tyrosinase activity per well was slightly reduced at the concentration range between 0.1 mM and 0.5 mM but was rapidly dose-dependently reduced at higher concentration. The inhibitory effect of kojic acid on tyrosinase activity in the cell culture system is smaller than that of arbutin at concentrations that do not affect cell viability, even though marked inactivation was observed in isolated tyrosinase. There are conflicting reports on the effectiveness in kojic acid.

Kojic acid may take two to three months to show efficacy and thus may seem rather slow to be effective. However, kojic acid does not have any side effects during the 10-to 20-month period during normal application on the skin.

Kojic Acid

Supplier: A&E Connock, Sino Lion, Ichimaru Pharcos, Creations Couleurs, Tri-K Industries. INCI name: Kojic Acid

Lutein

Lutein from the Latin *luteus* meaning "yellow" is a xanthophyll (E161a) that provides a yellow-red color. It is related to carotene (E160a) and is normally found in green leaves so is available as a natural plant extract. Lutein is found in plants such as dandelion (*Taraxacum officinalis*) along with violaxanthin, St. John's Wort (*Hypericum perforatum*), marigold (*Calendula officinalis*) and orange peel (*Citrus aurantium amara*). Interestingly, lutein is often found in combination with violaxanthin. It has also been found in Mexican tarragon (*Tagetes lucida* Cav.) and in green leafy vegetables such as spinach and kale. Lutein is employed by organisms as an antioxidant and for blue light absorption. Lutein is a lipophilic molecule and is generally insoluble in water.

With aging there is an increased expression or activity of matrix metalloproteinases that degrade and remodel the structural extracellular matrix. In addition, exposure of skin to ultraviolet radiation leads to loss of cell viability, membrane damage and deposition of excessive elastotic material. Lutein has antioxidant, anti-inflammatory and photoprotective properties. The mechanism to lutein's antiaging and anticarcinogenic effects include the inhibition of MMP to TIMP ratio in dermal fibroblasts and melanoma cells, and the inhibition of cell loss, membrane damage and elastin expression in ultraviolet radiation exposed fibroblasts.

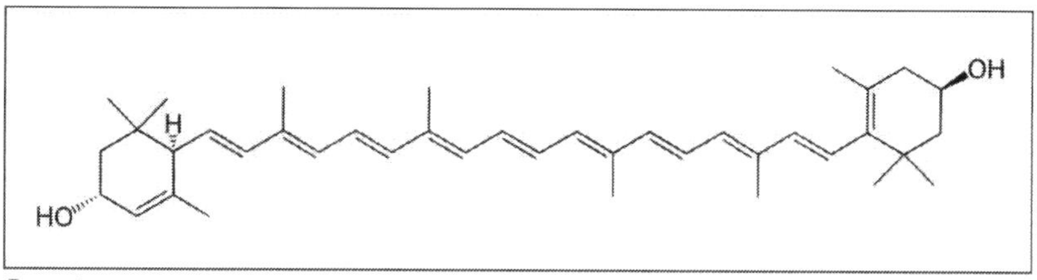

Lutein

One study investigated, with the use of a mouse endotoxin-induced uveitis (EIU) model, the neuroprotective effects of lutein against retinal neural damage caused by inflammation. The data revealed that the antioxidant lutein was neuroprotective during EIU, suggesting it might suppress retinal neural damage during inflammation.

Supplier: D.D. Williamson, Overseal, Vevy, Kemin. INCI name: Xanthophyll

Lycopene

Lycopene is a bright red carotenoid pigment and phytochemical found in tomatoes and other red fruits and vegetables, such as red carrots and papayas (*Carica papaya*) and is also reported in other plant sources such as *Rosa rubiginosa* (rose hip), *Taxus baccata* (yew), *Calendula officinalis* (marigold) and *Citrullus lanatus* (watermelon), though clearly at much lower levels. It is insoluble in water.

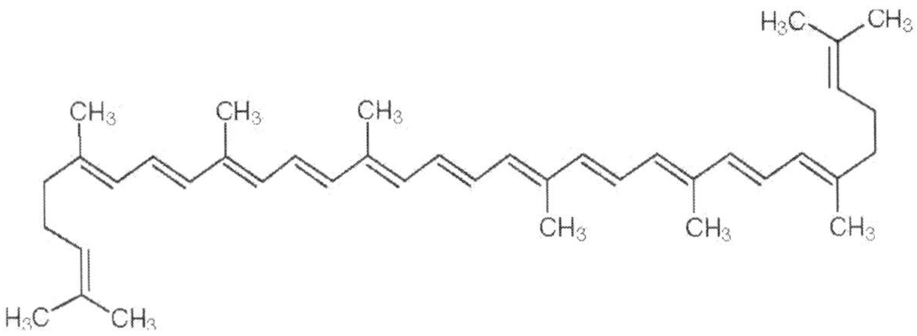

Lycopene

Because preliminary research has shown an inverse correlation between consumption of tomatoes and cancer risk, lycopene has been considered a potential agent for prevention of some types of cancers, particularly prostate cancer.

It has a similar structure to the other carotenes and is a powerful antioxidant and free radical scavenger. Due to its strong color and nontoxicity, lycopene is a useful food coloring.

Supplier: Sensient, Unifect, S. Black, Overseal, Univar. INCI name: Lycopene

Madecassoside

Centella asiatica extracts are renowned for their remarkable activities on the skin. The biological activity is because of the four active molecules from the triterpene series: asiatic acid (see **Asiatic Acid, Asiaticoside and Madecassic Acid** listing), madecassic acid, asiaticoside and madecassoside. Besides their contribution to the natural defense of the plant against environmental aggressions, these molecules have been proved to play a major role in skin care and skin diseases (wound care, dermis restoration and anti-inflammatory activity).

The activity of madecassoside on keratinocytes was investigated using an ex vivo study over five days with daily applications of an oil-in-water emulsion containing 3% madecassoside. The histologic observations and the corresponding images analysis demonstrated that madecassoside significantly induced Aquaporin-3 expression, especially in the basal keratinocytes, and in the filaggrin synthesis. The subsequent applications involved the cellular homeostasis, the hydric flow regulation and the stratum corneum organization.

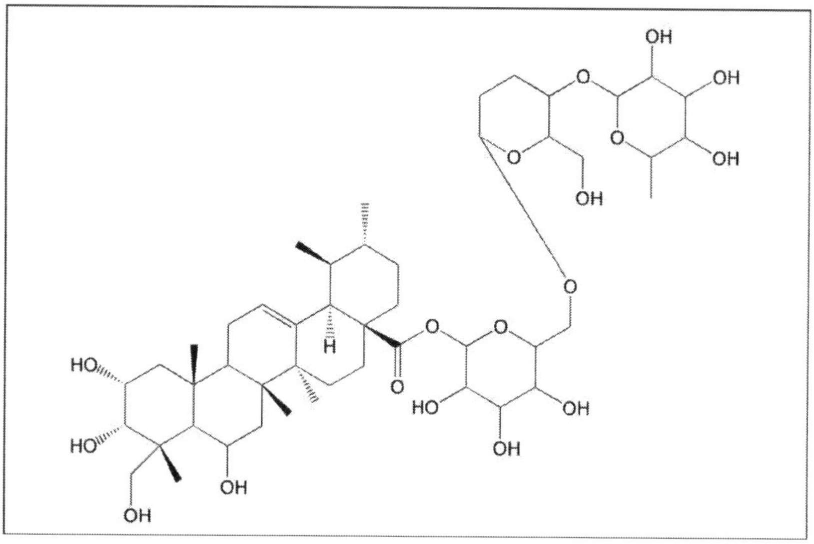

Madecassoside

Madecassoside modulates inflammation and has a modulating effect that works in two ways. It decreases inflammatory ligand release and decreases keratinocyte sensitivity to inflammation that so prevents and progressively reduces the self-maintained disorders of the inflammation and their consequences on the hyperproliferation of keratinocytes. Madecassoside is an ingredient that restores the epider-mis/dermis homeostasis and preserves essential cell functions. In the restructuring of the extracellular matrix madecassoside maintains the structure of the extracellular matrix and contributes to the improvement of the dialogue between the epidermis and dermis by assuring an increased expression of type I and III collagens. An excellent material for wound healing.

Supplier: Bayer Health Care, Unifect,

Menthol

Menthol is a secondary alcohol obtained from peppermint oil or other mint oil, or it is prepared synthetically from the hydrogenation of thymol. It is quite effective in insect bite preparations because it is antipruritic and mild anaesthetic properties. The cooling effect makes it a useful additive in shaving products. Applied to the skin, menthol produces vasodilation followed by a feeling of numbness, coolness and mild local anaesthesia. Menthol is used traditionally to relieve headaches, toothache and neuralgia.

Menthol has a beneficial effect on the irritated and inflamed mucous membranes of the upper respiratory tract, where it shows anti-germ properties, stimulates secretion and has a mild anaesthetic effect on the mucosa. It has a soothing effect on the nose and throat at the first stages of a cold [Weiss].

Supplier: A&E Connock, Jeen Intl., Symrise. INCI name: Menthol

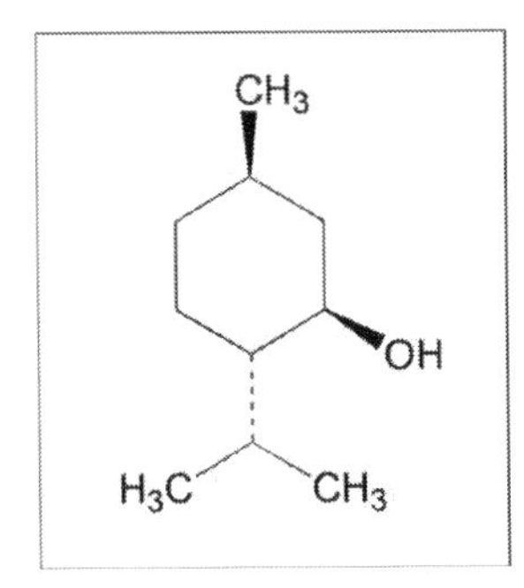

Menthol

Menthyl lactate

Menthyl Lactate is a menthol-derived cooling agent that possesses a very mild inherent odor and long tasting cooling effect. It can be easily mixed with perfume oils or glycols (as solvents) before added to shampoos, shower gels and foam bath, etc.

Supplier: Givaudan, A&E Connock, Prod'Hyg, Sino Lion, Symrise, LCW-Sensient, Corum.

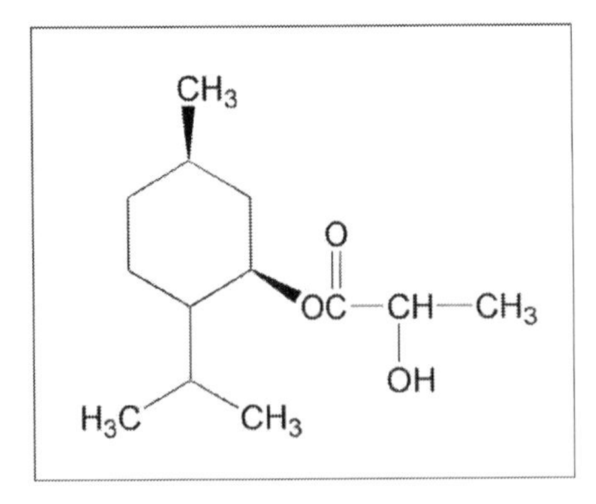

Menthyl Lactate

INCI name: Menthyl Lactate

Perillaldehyde

Perillaldehyde, or perilla aldehyde, is a natural organic compound found most abundantly in the annual herb Perilla or Beefsteak Plant (*Perilla frutescens*), but also in a wide variety of other plants and essential oils. It is a monoterpenoid containing an aldehyde functional group.

Perillaldehyde, or volatile oils from perilla that are rich in perillaldehyde, are used as food additives for flavoring and in perfumery to add spiciness. Perillaldehyde can be readily converted to perilla alcohol, which is also used in perfumery.

It was discovered that *Perilla frutescens* (Shiso oil) which contained perillaldehyde, (74%) and limonene (12.8%) had antimicrobial activity, which was mainly attributed to the perillaldehyde. It was found that perillaldehyde inhibited fungi and both Gram-positive and Gram-negative bacteria and that perilla oil was especially effective against *Acnes propionibacterium* and *Staphylococcus aureus* (both of which can cause acne).

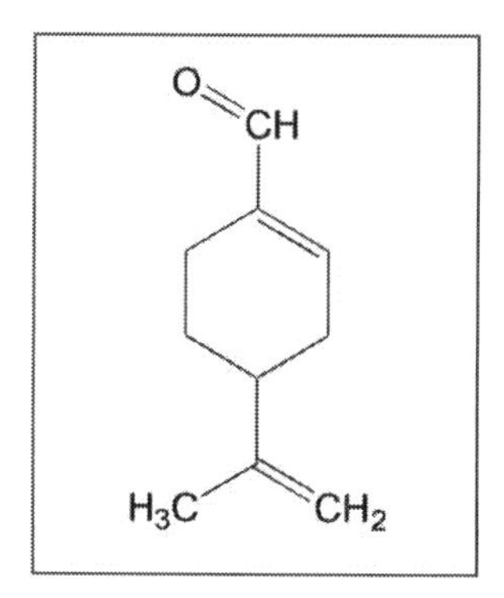

Perillaldehyde

Supplier: Dr André Rieks. INCI name: Perillaldehyde

Phytic acid

Phytic acid (known as inositol hexaphosphate (IP6), or phytate) is the principal storage form of phosphorus in many plant tissues, especially in the grass family (wheat, rice, rye, barley, etc.) and beans. Phosphorus in this form is generally not bioavailable to humans because humans lack the digestive enzyme, phytase, required to separate phosphorus from the phytate molecule.

Phytic acid protects skin from external aggressions and pollution as well as being a good chelating agent that protects products from color changes due to metal ions. The product is often sold as a 50% solution.

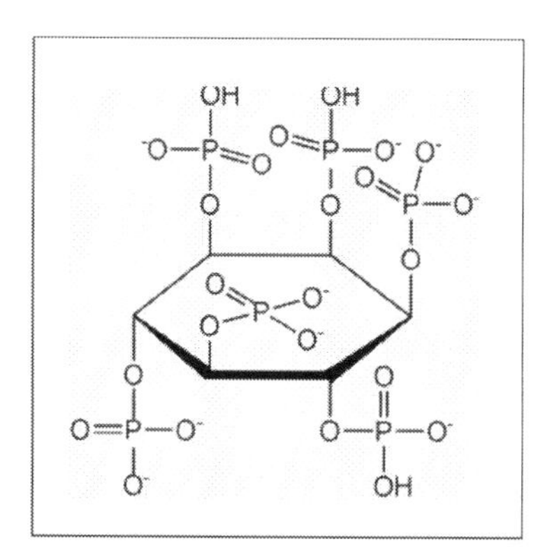

Phytic Acid

Supplier: Tsuno, Dr. Straetmans Chemische Produkte GmbH. INCI name: Phytic Acid.

Pongamol

Pongamol comes from the Karanja Tree (*Pongamia spp*) that grows extensively across India. It is extracted from the seed oil and purified. The oil is commonly used in Indian Ayurvedic medicine for the treatment of skin conditions, for skin protection and rheumatic pain. It is taken internally for bronchitis and whooping cough. Oil is applied to skin diseases, in scabies, sores and herpes, and cases of eczema have benefited by applying a mixture of the oil and zinc oxide.

Some plants have the ability to absorb a large amount of UV irradiation and it is likely that the compound found in *Pongamia* species protects the plant against UV irradiation.

Pongamol helps to protect skin against UV aggressions, limit the signs of photo-aging and reduce skin damage caused by the environment and UV rays. The material is not listed in Annex VII (permitted sunscreens) of the cosmetic regulations.

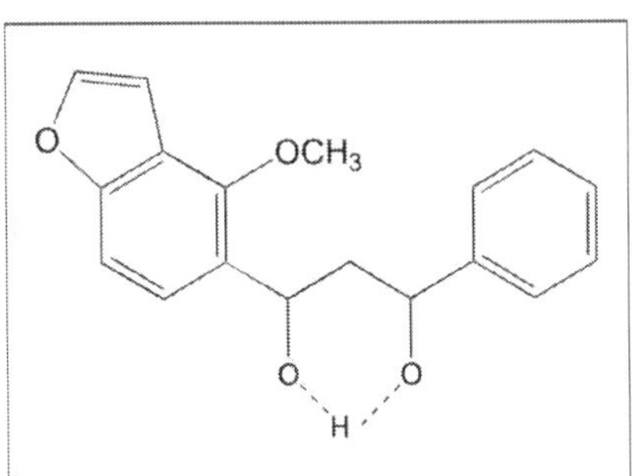

O
OCH_3
O O
H

Pongamol

Greentech supply an emulsion of the active plant. Supplier: Givaudan (formerly Quest). INCI name: Pongamol.

Resveratrol

Resveratrol is a phytoalexin (*trans*-3,5,4'-trihydroxystilbene), an antioxidant polyphenol from red wine and recent evidence has supported the assumption that it is largely responsible for red wine's protective effects on blood vessels, by inhibiting lipid peroxidation of low-density lipoprotein (LDL). It is found in the skins of red grapes, and is synthesized by the plant in response to attack by pathogens such as bacteria or fungi (especially by the *Botrytis* fungus).

It has been the subject of intense interest in recent years due to a range of unique antiaging properties. It was reported that a resveratrol-based skin care formulation had a 17 times greater antioxidant activity than idebenone. The role of resveratrol in prevention of photoaging was reviewed and compared with other antioxidants used in skin care products.

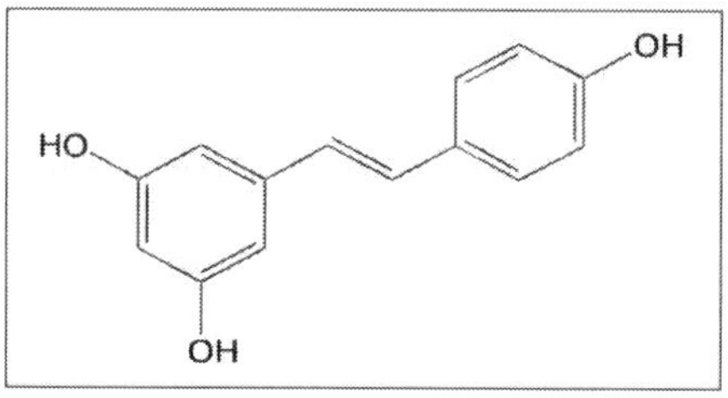

Resveratrol

It has been clinically proven to have a two-step antiwrinkle activity and also has antiaging activity based on Sirtuin-1 activation.

Supplier: Breko, Soliance, BioSpectrum, Kaden Biochemicals (Symrise), Pfannenschmidt GmbH. INCI name: Resveratrol

Rosmarinic acid

Rosmarinic acid is a natural polyphenol antioxidant carboxylic acid found in many Lamiaceae herbs used commonly as culinary herbs such as lemon balm (*Melissa officinalis*), rosemary (*Rosmarinus officinalis*), oregano (*Origanum* sp), sage (*Salvia vulgaris*), thyme (*Thmnus vulgaris*) and peppermint (*Mentha piperita*). Other species include self-heal (*Prunella vulgaris*), Chinese basil (*Perilla frutescens*), wild mint (*Hyptis verticillata*) and painted nettle (*Coleus blumei*), in the Apiaceae family (e.g. *Sanicula europaea*), and Boraginaceae (e.g. *Anchusa officinalis* and *Symphytum officinale*).

Chemically, rosmarinic acid is an ester of caffeic acid with 3,4-dihydroxyphenyl lactic acid. It is a red-orange powder that is slightly soluble in water, but well-soluble in most organic solvents.[2]

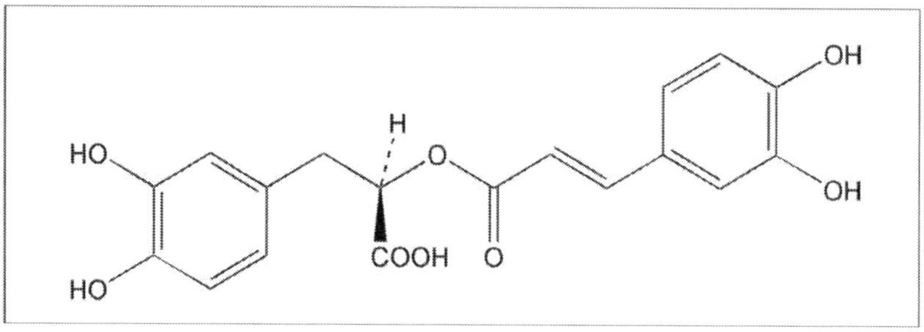

Rosmarinic Acid

Rosmarinic acid has significant antioxidant, anti-inflammatory and antimicrobial activities. It has been used topically as a non-steroidal anti-inflammatory drug. Rosmarinic acid has good percutaneous absorption, tissue distribution, and bioavailability that makes it a good ingredient suitable for transdermal administration [Ritschel].

Rosmarinic acid is effective in reducing both gingival inflammation and plaque accumulation. It also has an antibacterial activity and is proven to reduce inflammation, which makes it ideal for topical skin infections of the epidermis and oral mucosa. It is often used as part of a blend of materials in order to produce a natural preservative.

Supplier: Sabinsa . INCI name: Rosmarinic Acid

Salicylic acid

The name salicylic acid (2-Hydroxybenzoic acid) is derived from the Latin *salix* meaning "willow," from the bark of which it can be obtained. It is a beta hydroxy acid (BHA) and is probably best known for its use in anti-acne treatments. Salicylic acid was also isolated from the herb meadowsweet (*Filipendula ulmaria*, Syn. *Spiraea ulmaria*).

Salicylic acid is known for its ability to ease aches and pains and reduce fevers. These medicinal properties, particularly fever relief, have been known since ancient times, and it was used as an antiinflammatory drug.

Salicylic acid is a skin exfoliant and has been used in conjunction with AHAs (alpha hydroxy acids) or alone for this purpose. It is a key ingredient for the treatment of acne, psoriasis, calluses, corns and warts. It works as a keratolytic by increasing the rate of skin cell turnover causing the cells of the epidermis to shed more readily, this helps prevent pores from clogging and allows for new cell growth. It has also been used in shampoos for the treatment of dandruff and has some skin-whitening properties.

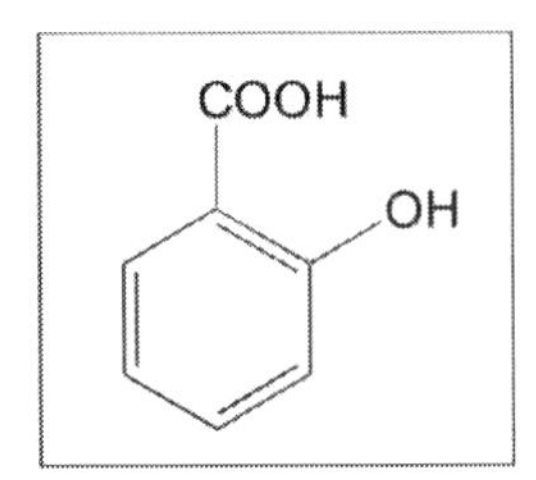

Salicylic acid

The U.S. Food and Drug Administration recommends the use of sun protection when using skin care products containing salicylic acid (or any other BHA) on sun-exposed skin.

Salicylic acid is a preservative listed in Annex VI, but it is fairly poor for yeast, mold, bacteria when used alone. Maximum permitted level is 0.5% (as the acid).

Sodium β-Sitosterol Sulfate

Sterols, also known as phytosterols, are fats present in all plants, including fruits and vegetables. β-sitosterol is the major phytosterol in higher plants and possesses anti-inflammatory (similar to cortisone), antipyretic, antineoplastic, and immune-modulating properties. Phytosterols also reduced experimentally induced edema.

β-sitosterol significantly increases keratin expression in human skin after 24 and 48 hours.

It protects against chemical insults and treated skin exhibits a preserved keratin layer with minor edema between epidermal cells following an insult with strong detergent. It significantly improves skin hydration and reduces loss of skin hydration and restores the skin barrier function.

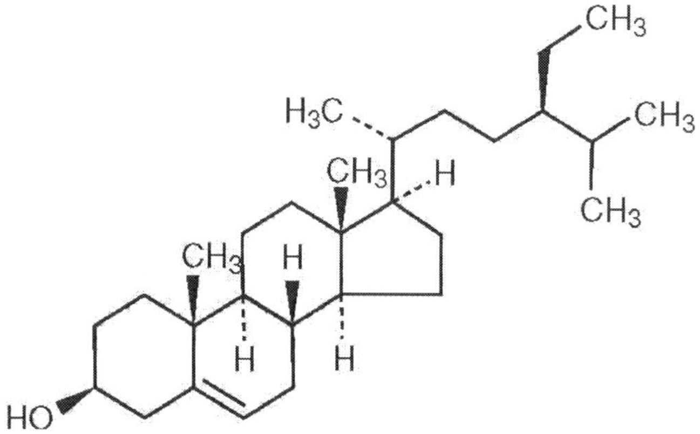

β-Sitosterol

Studies have shown that skin roughness is reduced and lines and wrinkles are greatly reduced.

There are numerous commercial preparations that contain standardized amounts of mixed phytosterols. There is a mixture of naturally occurring phytosterols derived from a by-product of soybean oils. It is rich in β-sitosterol, campesterol and stigmasterol, which also contains brassicasterol, β-sitostanol, stigmastanol and campestanol (ex Fenchem through Cornelius).

Supplier ISP-Vincience. INCI name: Sodium Beta Sitosterol Sulfate.

Ursolic acid

Ursolic acid, also known as urson, prunol, micromerol and malol, is a pentacyclic triterpenoid compound that naturally occurs in many plants, including apples (*Prunus malus*), basil (*Ocimum basilicum*), bilberries (*Vaccinium myrtilus*), cranberries (*Vaccinium macrocarpon*), elder flower (*Sambucus nigra*), peppermint (*Mentha piperita*), rosemary (*Rosmarinus officinalis*), lavender (*Lavandula angustifolia*), oregano (*Oreganum vulgare*), thyme (*Thymus vulgaris*), hawthorn (*Crataegus monogyna*) and prunes (*Prunus domestica*). Apple peels contain a high quantity of ursolic acid and related compounds.

Ursolic acid and its alkali salts (potassium or sodium ursolates) were formerly used as emulsifying agents in pharmaceutical, cosmetic and food preparations. In recent times ursolic acid has been found to be medicinally active both topically and internally and its anti-inflammatory (oral and topical), antitumor (skin cancer), and antimicrobial properties make it useful in cosmetic applications.

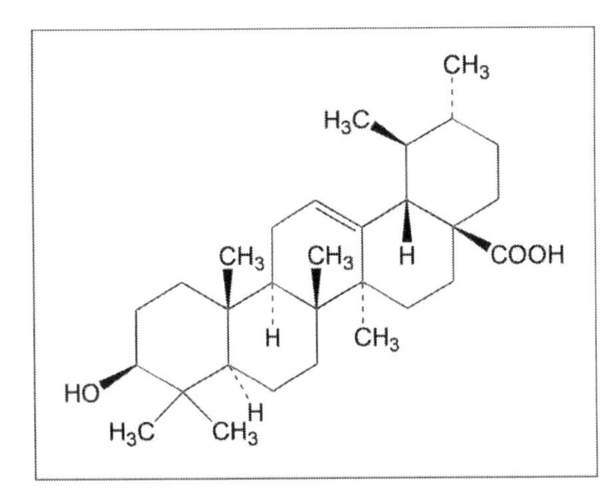

Ursolic Acid

Oleanolic acid is an isomer of ursolic acid and the compound is available naturally in its free form in *Olea europaea* (olive) leaves.

There is also some evidence to show that ursolic acid may be of relevance in hair tonics to encourage hair growth. Hair tonics containing oleanolic acid are also patented in Japan.

Supplier: BioSpectrum, Sabinsa. INCI name: Ursolic Acid.

Chapter 11

Isoflavones, Phytohormones and Phytosterols

INTRODUCTION

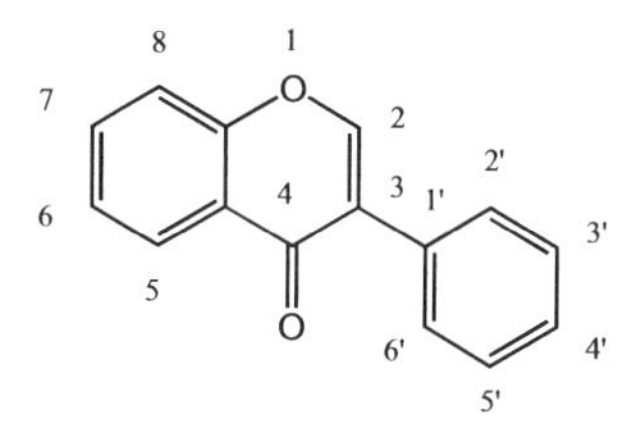

Isoflavones have the phenyl group attached to the 3-position, whereas in flavones the phenyl group is attached to the 2-position. The isoflavones are mainly found to occur within the Leguminosae (specifically in the sub-family Papilionoideae), although the literature shows many other species that contain these chemical. Isoflavones are are also found in other botanical families such as the Compositae, the Iridaceae, the Myristicaceae, and the Rosaceae.

These isoflavones can act as steroidal mimics by filling the stereochemical space that could be occupied by oestrogenic compounds. It is this spacial chemistry that helps explain the effects of many nutritional herbal supplements and topical preparations.

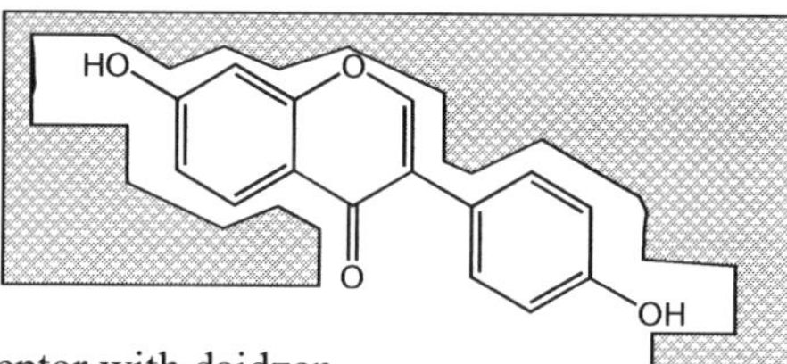

Fig.1 Estrogen receptor with daidzen

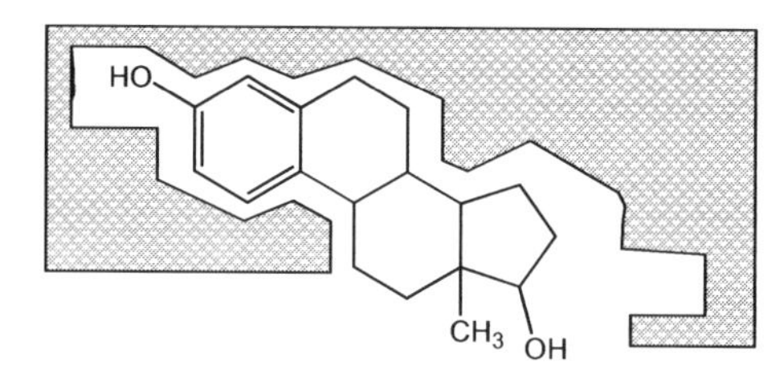

Fig. 2 Estrogen receptor with 17β-estradiol

Daidzein is a phyto-estrogen, but is also called a phenolic estrogen, to distinguish it from a steroidal estrogen like 17β-estradiol. The activity of phytoestrogen is much weaker than the steroidal estrogen, varying from 0.005-2% [Brand]. The estrogenic properties are insufficient in strength to replace steroidal estrogens, but they do have significant value when it comes to reducing the effects of ageing and improving the quality of the skin.

Phyto-oestrogens may also be viewed in relation to the phytochemical division of terpenoids, which comprise the largest group of natural plant products. All terpenoids are derived biogenetically from isoprene. The largest group of terpenoids are the triterpenoids, which *include, amongst other divisions, the triterpenoid and steroid saponins, and, the phytosterols. The phyto-oestrogens fall into these three categor*ies.

In addition, nature has a rich portfolio of phytosterols. It is easy to understand why sterols like stigmasterol (Fig.3) and β-sitosterol (Fig.4) have an effect that is anti-inflammatory and capable of reducing swelling and erythema, when their structure is compared to corticosterone (Fig.5) and hydrocortisone (Fig.6).

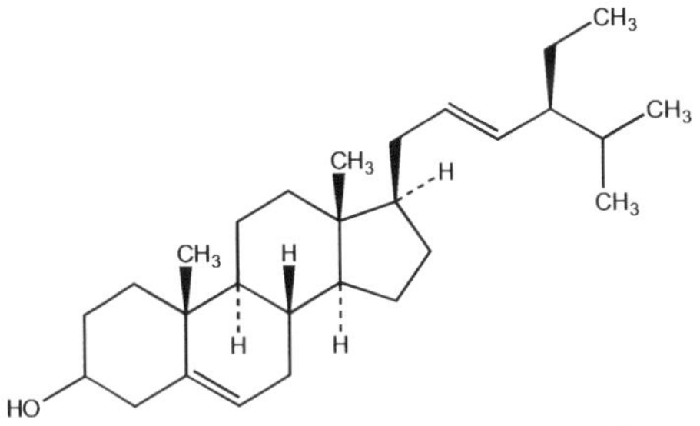

CH_3 CH_3 CH_3 H CH_3 CH_3 H H H HO

Fig.3 Stigmasterol

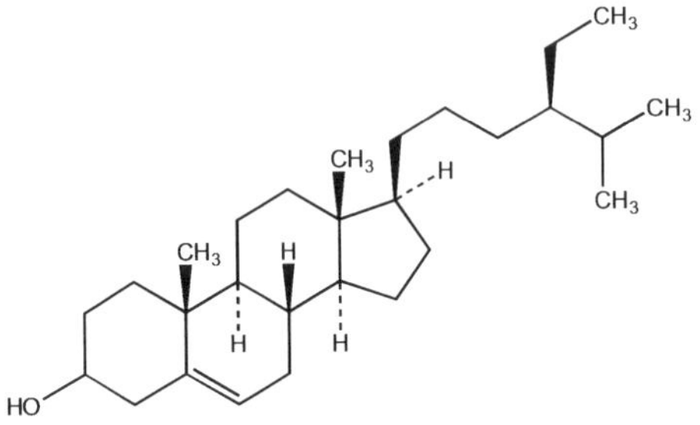

CH_3 CH_3 CH_3 H CH_3 CH_3 H H H HO

Fig.4 β-sitosterol

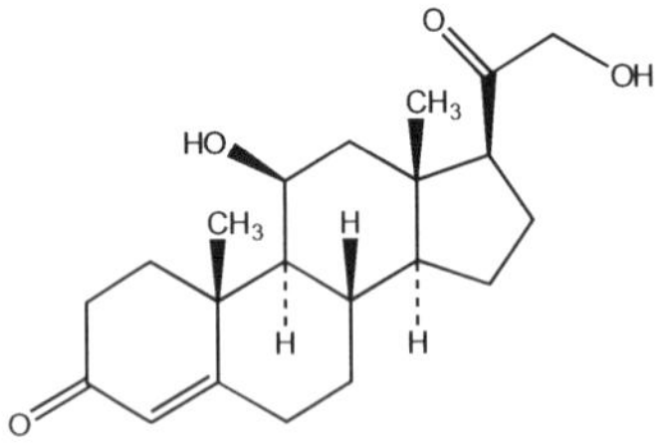

O OH CH_3 HO CH_3 H H H O

Fig.5 Corticosterone

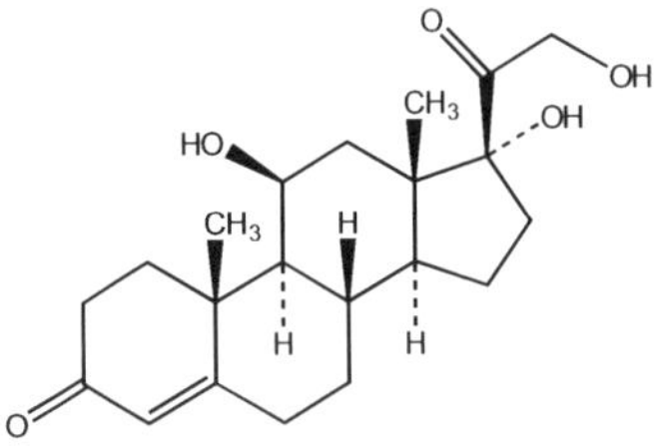

O OH CH_3 OH HO CH_3 H H H O

Fig. 6 Hydrocortisone

ISOFLAVONES

The most commonly occurring isoflavones are

Biochanin-A	5,7-dihydroxy-4'-methoxyisoflavone
Daidzein	4',7-dihydroxyisoflavone
(+/-)-Equol	4',7-isoflavandiol
Formonometin	7-hydroxy-4'-methoxyisoflavone
Glycitein	4',7-dihydroxy-6-methoxyisoflavone
Genistein	4',5,7-trihydroxyisoflavone
Genistein-4',7-dimethylether	5-hydroxy-4',7-dimethoxyisoflavone
Prunetin	4',5-dihydroxy-7-methoxyisoflavone

with the associated glucosides

Genistin	glucosyl-7-genistein
Glycitin	4',7-dihydroxy-6-methoxyisoflavone-7-d-glucoside
Ononin	formononetin-7-O-glucoside
Sissotrin	biochanin A-7-glucoside

The comparison of effects and functions of plants containing the same isoflavones shows remarkable similarity.

Daidzein as an example of an isoflavone

Daidzein is a solid substance that is virtually insoluble in water. Its molecular formula is $C_{15}H_{10}O_4$, and its molecular weight is 254.24 daltons. Daidzein is also known as 7-hydroxy-3-(4-hydroxyphenyl)-4*H* -1-benzopyran-4-one and 4', 7-dihydroxyisoflavone. Daidzin, which has greater water solubility than daidzein, is the 7-beta glucoside of daidzein.

Daidzein is an isoflavone. It is also classified as a phytoestrogen since it is a plant-derived nonsteroidal compound that has estrogen-like biological activity. Daidzein is the aglycone (sometimes called the aglucon) of daidzin (see Fig.1). The isoflavone is found naturally as the glycoside daidzin and as the glycosides 6"-O-malonyldaidzin (Fig.9) and 6"-O-acetyldaidzin (Fig.10). Daidzein and its glycosides are mainly found in the Leguminosae family that includes soy beans and chickpeas.

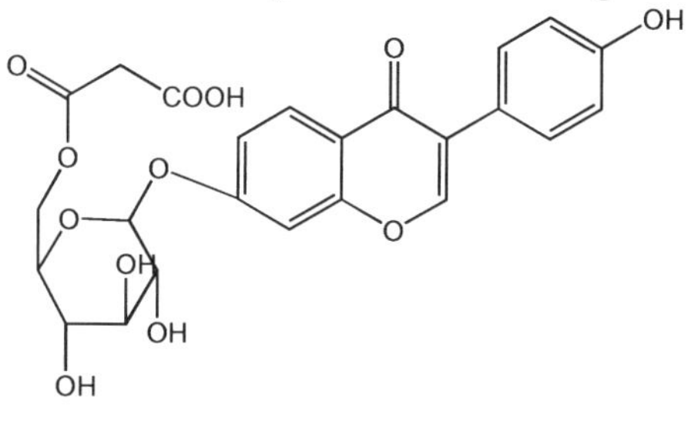

Fig.9 Malonyldaidzin

Soybeans and soy foods are the major dietary sources of these substances. Daidzein glycosides are the second most abundant isoflavones in soybeans and soy foods; genistein glycosides are the most abundant.

Nonfermented soy foods, such as tofu, contain daidzein, principally in its glycoside forms. Fermented soy foods, such as tempeh and miso, contain significant levels of the aglycone.

Fig.10 Acetyldiadzin

Kudzu Vine (*Pueraria labata*)

The roots of *Pueraria labata* is an herbal medicine commonly known as the kudzu vine. It has been used for centuries in traditional Chinese medicine for the treatment of alcohol abuse and thought to be effective because of the daidzein and daidzin found in the herb. A study on Syrian Golden Hamsters suppressed the alcohol choice.

White Kwao Krua (*Pueraria mirifica*)

In addition to genistein, daidzein (see above), daidzin and genistin, the plant contains a some unique isoflavones, kwakhurin, kwakhurin hydrate (Fig.11) and puerarin (Fig.12) to name but a few.

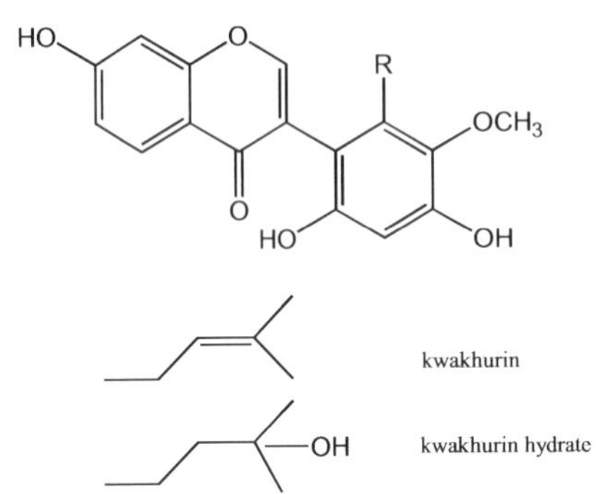

Fig.11 kwakhurin, kwakhurin hydrate

The roots also contain mirificoumestan (Fig.13), deoxymiroestrol (Fig.14) and coumestrol (Fig.15). The traditional use of the plant is clearly for the hormonal properties, since in Thailand it is used for breast development.

When *Pueraria mirifica* is taken as a dietary supplement, its phytoestrogen constituents will naturally alleviate symptoms occurring as a result of the aging process and a deficiency in estrogen levels, e.g. sagging breasts, wrinkled skin, bone loss, grey hair, etc. These aging signs and symptoms will, to a certain extent, be reversed.

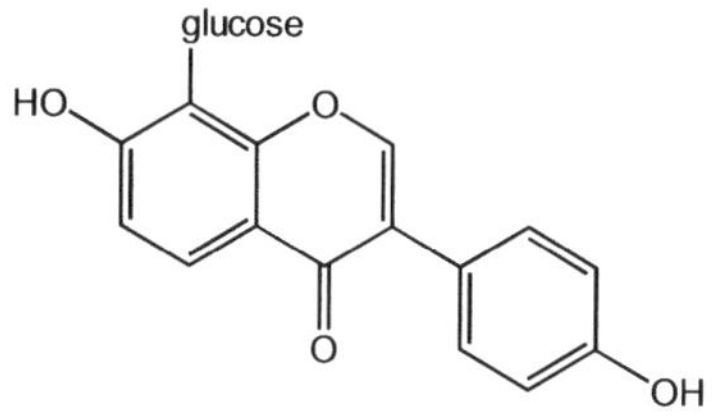

Fig.12 puerarin

The rich source of sterols and phyto-hormones also indicates the plant for the topical treatment of wrinkles and aging skin conditions.

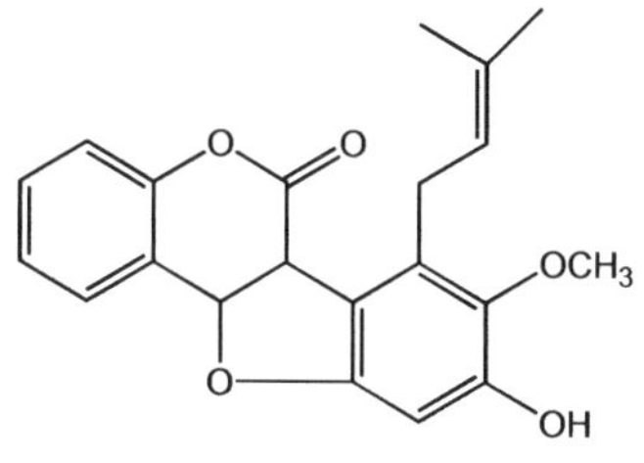

Fig. 13 mirificoumestan

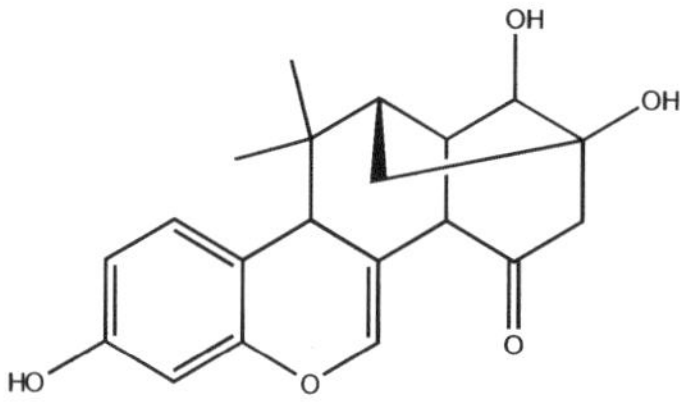

Fig.14 deoxymiroestrol

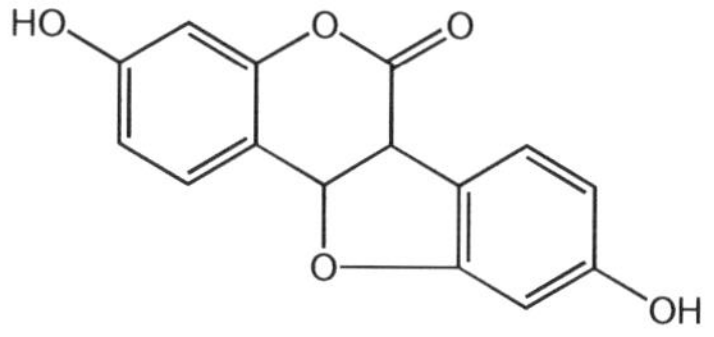

Fig. 15 coumestrol

Plants that contain coumestrol: *Brassica oleracea* var. *gemmifera* var. *gemmifera* DC [Brassicaceae] Shoot 400ppm, *Pisum sativum* L. [Fabaceae] Fruit 300ppm, *Medicago sativa* subsp. *sativa* [Fabaceae] Shoot 190ppm, *Pisum sativum* L. [Fabaceae] Seed 0.6ppm, *Glycine max* (L.) MERR. [Fabaceae] Seed, Shoot, Leaf, Plant, Root, *Medicago sativa* subsp. *sativa* [Fabaceae] Leaf, Plant, Sprout Seedling, Root, Leaf, Root, Seed, *Phaseolus vulgaris* subsp. var. vulgaris [Fabaceae] Fruit, Leaf, Sprout Seedling, *Psoralea corylifolia* L. [Fabaceae] Root, *Pueraria montana* subsp. var. *lobata* (WILLD.) Maesen & S. M. Almeida [Fabaceae] Stem, Root, *Spinacia oleracea* L. [Chenopodiaceae] Leaf, *Taraxacum officinale* Weber ex F. H. Wigg. [Asteraceae] Plant, *Trifolium pratense* L. [Fabaceae] Flower, Leaf, Shoot, *Vicia faba* L. [Fabaceae] Seed, *Vigna radiata* (L.) WILCZEK [Fabaceae] Seed, Sprout Seedling, *Vigna unguiculata* subsp. *unguiculata* [Fabaceae] Seed, Stem.

Red Clover (*Trifolium pratense* L) (Leguminosae).

The flowerheads are used and they contain the isoflavones; biochanin A, daidzein, formononetin, genistein, pratensein, and trifoside. The plant has alterative, antispasmodic, expectorant properties and is a sedative dermatological agent. Its main use is an alterative and for skin complaints such as psoriasis and eczema, as well as an expectorant use in coughs and bronchial conditions.

Biochanin A (Fig.16) and formononetin (Fig.17)are two isoflavones from red clover and are just like genistein and daidzein, except that they have methyl groups replacing the hydroxyl groups.

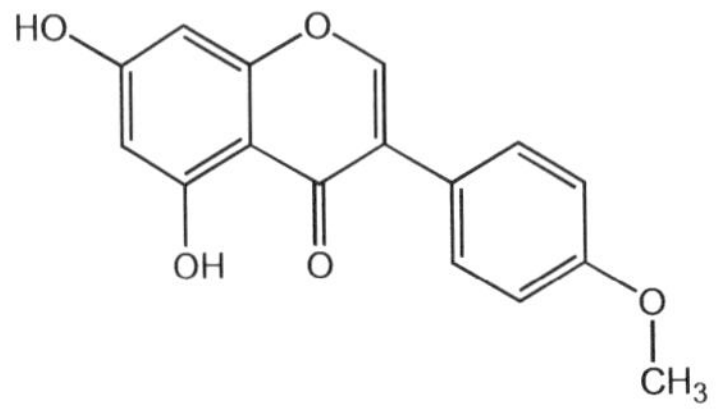

Fig.16 Biochanin A

Other sources of biochanin A are *Baptisia tinctoria* (Wild Indigo), *Medicago sativa* (Alfalfa), *Sophora japonica* (Japanese Pagoda Tree) and *Vigna radiata* (Mungbean).

These two isoflavones are considerably less estrogenic in their original forms, because the stereochemistry of the methoxy groups means they are not able to bind to the estrogen receptors as efficiently.

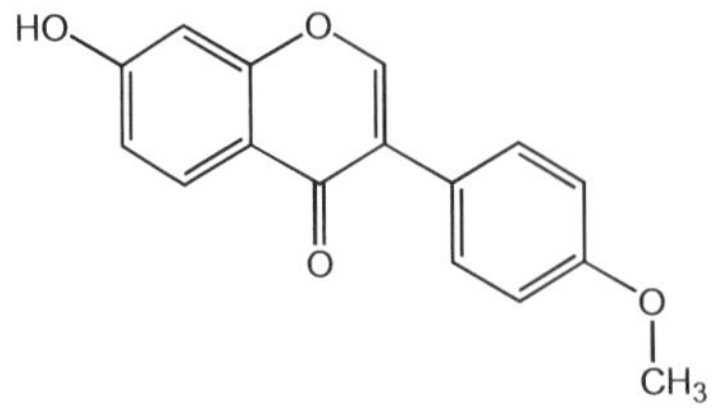

Fig.17 Formononetin

However, once these molecules are ingested, bacteria in the colon are able to remove the methyl groups - biochanin A becomes genistein (Fig.18) and formononetin becomes daidzein (Fig.1 see above). Daidzein can be further metabolized to equol (Fig.19).

Other sources of formononetin are *Astragalus membranaceus* (Astragalus), *Cimicifuga racemosa* (Black Cohosh), *Glycyrrhiza glabra* (Licorice root), *Medicago sativa* (Alfalfa), *Pueraria* spp. (Kudzu; Pueraria) and *Vigna radiata* (Mungbean). Internally, biochanin A and formononetin are then able to be a source of considerable estrogenic activity.

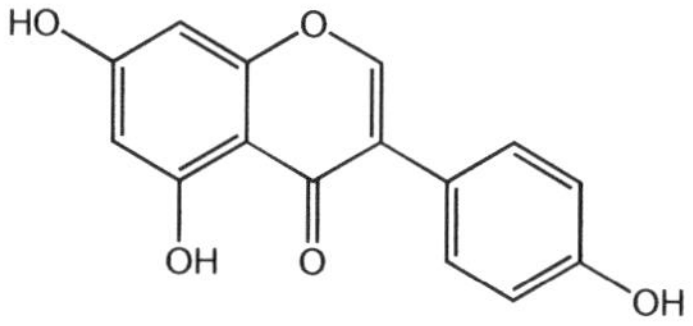

Fig.18 Genistein

Other sources of genistein are *Baptisia tinctoria* (Wild Indigo), *Cytisus scoparius* (Scotch Broom), *Glycine max* (Soybean), *Glycyrrhiza glabra* (Licorice root), *Medicago sativa* (Alfalfa), *Pueraria* spp. (Kudzu; Pueraria), *Sophora japonica* (Japanese Pagoda Tree) and *Vigna radiata* (Mungbean)

It may well be that these mechanisms give red clover its reputation as an alterative remedy, cleansing the system yet mild enough for many children's skin problems, even eczema. A lotion of red clover can be used externally to give relief from itching in skin disorders. Specific for acne, boils and similar eruptions, including eczema and skin problems especially where irritation is a factor.

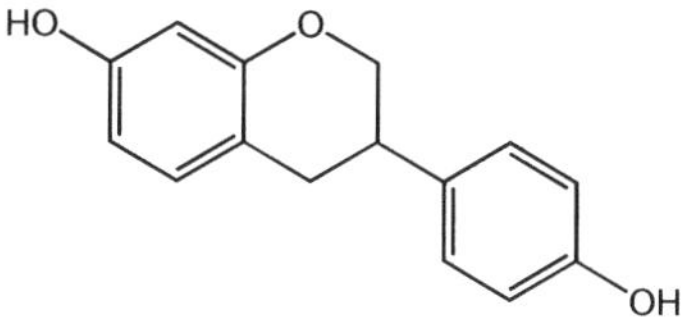

Fig.19 Equol

Historically, the flower tea has been used as an antispasmodic, expectorant and mild sedative. It is also recommended for athlete's foot, sores, burns, and ulcers. [Leung & Foster] and has been used in the herbal treatment of cancer, especially of the breast or ovaries.

Red Clover is also a very popular remedy as the alternative for hormone replacement therapy and is sold extensively for this purpose.

Sweet Yellow Melilot (*Melilotus officinalis*)

Melilot is soothing, lenitive, astringent, refreshing and anti-irritant and has similar properties to the red clover described above. It is also described as possibly having the additional properties of being anti-inflammatory, anti-oedema, a veinous astringent (haemorrhoids) and anaesthetic [Council of Europe].

However, it is perhaps not the isoflavones at force here, but maybe the β-sitosterol or coumarin the roots contain.

Melilotus officinalis L. extract, containing 0.25% coumarin (Fig 20) was studied on acute inflammation induced with oil of turpentine in male rabbits. M. officinalis had anti-inflammatory effects because it reduced the activation of circulating phagocytes and lowered citrulline production.

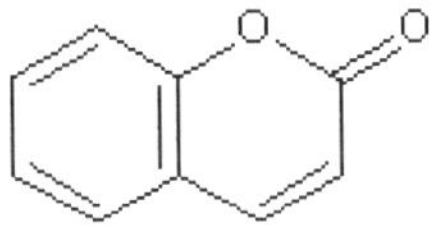

Fig. 20 coumarin

These properties were similar to those of hydrocortisone sodium hemisuccinate and coumarin.

PHYTOSTEROLS AND RELATED COMPOUNDS

The benefits of these phytosterols may be seen in the common herbal materials indicated for arthritis, such as Frankincense (*Boswellia serrata*). The boswellic acid present inhibits two inflammatory enzymes, 5-lipoxygenase (which produces leukotrienes) and human leukocyte elastase HLE (which degades elastase).

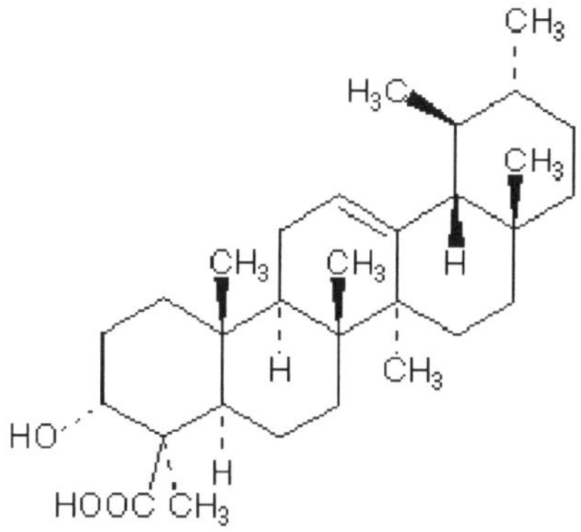

Fig. 21 Boswellic acid

Committee on Toxicity of chemicals in food, consumer products and the environment. Working group on phytocstrogens cellular and molecular mechanisms of phytoestrogen activity

The Department of Biochemical Pharmacology, Imperial College School of Medicine prepared a paper for discussion: "Assessment of the estrogenic potency of phyto-compounds". This reviewed the available information on cellular and molecular mechanisms and phytoestrogen estrogenic potencies. Out of the 28 points (statements for comment really) the following stood out:

Taking all estrogen receptor binding assays into account the review proposed the following rank order of phytoestrogen potency: estradiol >> coumestrol > 8-prenylnaringenin > equol >= genistein > biochanin A > daidzein > genistein glucuronide* > daidzein glucuronide* > formononetin (the activity of those compounds marked * may be due to the presence of activating enzymes present in the receptor preparation).

Phytoestrogens stimulated *in vitro* cell proliferation at concentrations of 0.1 - 10mM (3 - 4-fold less than estradiol). They did not induce the maximal proliferative effect of estradiol as higher concentrations inhibited proliferation. The majority of endogenous estrogens (> 90%) were not freely available but bound to plasma proteins. Phytoestrogens bound at 1/100th to 1/1000th the affinity of estradiol. The availability of phytoestrogens in plasma relative to estradiol will be greater. Coumestrol, 8-prenylnaringenin and equol were > 1000-fold less potent than estradiol and the isoflavones > 10 000-fold less potent.

The Wild Yam (*Dioscorea villosa*)

The Wild Yam (*Dioscorea villosa*) was the source of diosgenin (Fig. 21), a steroidal saponin used as the starting point for the commercial source of pregnanolone (Fig. 22) and progesterone (Fig.23) used as the first birth control pills. The root of *Dioscorea* is used for numerous purposes, but the major use is for the suppression of menopausal symptoms like hot flushes.

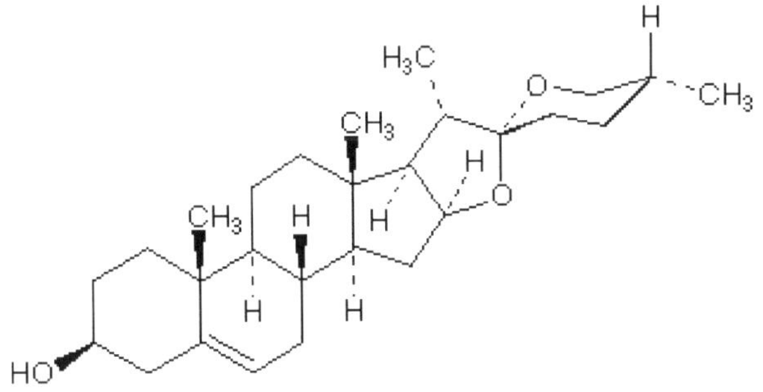

Fig. 22 diosgenin

There are many other sources of diosgenin

Trigonella foenum-graecum L. [Fabaceae] Seed 3300-19000ppm, *Solanum nigrum* L. [Solanaceae] Fruit 4000-12000ppm, *Daucus carota* L. [Apiaceae] Root 5400-6000ppm, *Dioscorea bulbifera* L. [Dioscoreaceae] Tuber 4500-4500ppm, *Medicago sativa* subsp. *sativa* [Fabaceae] Seed, *Agave sisalana* PERRINE [Agavaceae] Plant, *Aletris farinosa* L. [Liliaceae] Root, *Areca catechu* L. [Arecaceae] Seed, *Asparagus officinalis* L. [Liliaceae] Shoot, *Balanites aegyptiacus* (L.) DELILE [Balanitaceae] Fruit, *Chamaelirium luteum* (L.) A. GRAY [Liliaceae] Root, *Costus speciosus* (J. KONIG) SM. [Costaceae] Rhizome, *Dioscorea composita* HEMSL. [Dioscoreaceae] Plant, *Dioscorea* sp. [Dioscoreaceae] Root, *Dioscorea villosa* L. Plant, *Dioscorea villosa* L. [Dioscoreaceae] Tuber, *Jateorhiza palmata* MIERS [Menispermaceae] Root, *Lycium chinense* MILL. [Solanaceae] Flower, *Melilotus officinalis* LAM. [Fabaceae] Seed, *Momordica charantia* L. [Cucurbitaceae] Fruit, *Paris polyphylla* SM. [Liliaceae] Rhizome, *Smilax china* L. [Smilacaceae] Root, *Solanum dulcamara* L. [Solanaceae] Plant, *Tribulus terrestris* L. [Zygophyllaceae] Shoot.

During pregnancy, small frequent doses will help allay nausea. It is antispasmodic. It is valuable neuralgic affections, spasmodic hiccough and spasmodic asthma.

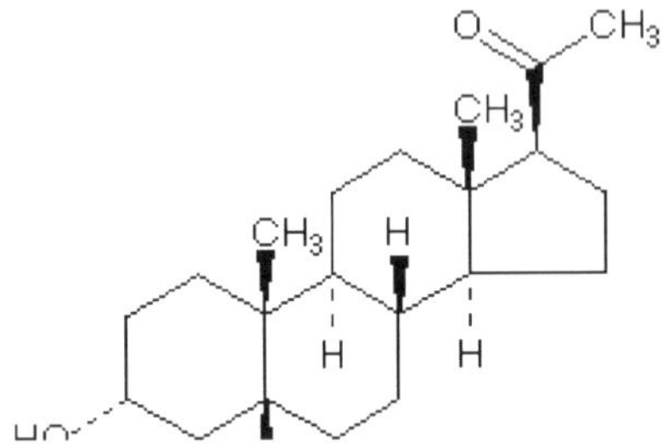

Fig. 23 pregnanolone

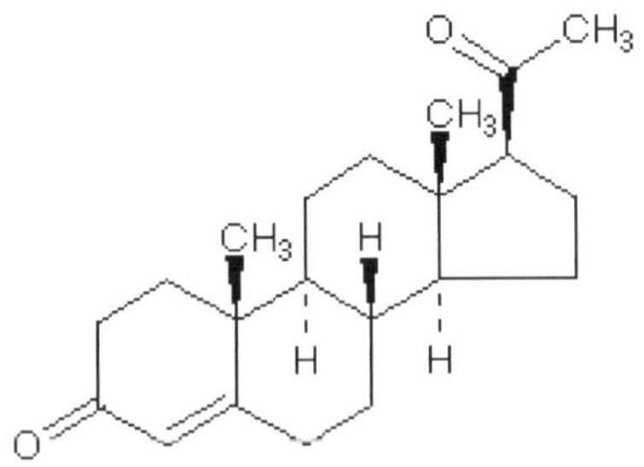

Fig. 24 progesterone

It is spasmolytic, a mild diaphoretic. It has potential in skin care and body care being anti-inflammatory and anti-rheumatic. It is also cited for dysmenorrhoea, ovarian and uterine pain, perhaps showing the power of this herbal root.

It is interesting to note that *Vitex agnus-castus* is a source of natural progesterone. Proprietary preparations containing this material have been available in Germany since the 1950s and many documented studies have investigated the use of these products to treat various gynaecological disorders [Newall]. The fruit of Vitex contains essential oils, iridoid glycosides, and flavonoids. Essential oils include limonene, 1,8 cineole, and sabinene.3 The primary flavonoids include castican, orientin, and isovitexin. The two iridoidglycosides isolated are agnuside and aucubin. Agnuside serves as a reference material for quality control in the manufacture of Vitex extracts. One other report demonstrated delta-3-ketosteroids in the flowers and leaves of Vitex that probably contained progesterone and 17-hydroxyprogesterone. The active constituents have been determined as 17-α-hydroxyprogesterone (leaf), 17-hydroxyprogesterone (leaf), androstenedione (leaf), δ-3-ketosteroids (leaf), epitestosterone (flower), progesterone (leaf), testosterone (flower and leaf).

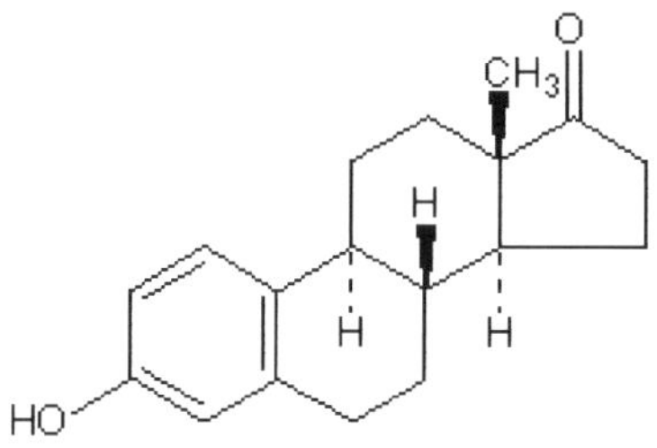

O
CH_3
H
H
H
HO

Fig. 25 estrone

It is highly unlikely that the diosgenin in the plant could ever be synthesised on the topical application to the skin to form a corticosteroid or hormonal derivative. However, it does seem likely that this material (being the precursor to these estrogenic molecules) will to some extent mimic the function of those pharmaceutical active materials and benefit the skin.

However, the production of wild yam was unable to sustain the demand for diosgenin as the starting precursor, for the production of birth control materials, which by this stage was dominated by estrone.

Fenugreek (*Trigonella foenum graecum*)

The world turned its attention to Fenugreek (*Trigonella foenum graecum*) for its source of diosgenin.

Fenugreek or Foenugreek seeds are emollient and accelerate the healing of suppurations and inflammations. Externally cooked with water into a porridge and used as hot compresses on boils and abscesses in a similar manner to the usage of linseed [Fluck].

Decoctions of whole plant are used as a bath for uterus infections. The seeds are tonic, restorative, aphrodisiac and galactagogue. Their emollient properties are useful for the itch. A cataplasm obtained by boiling the flour of the seeds with vinegar and saltpetre is used for swelling of the spleen. Extracts of the seeds are incorporated into several cosmetics claimed to have effect on premature hair loss, and as a skin cleanser [Iwu], and it is also reported in Java in hair tonics and to cure baldness. Many of the herbal materials found to have an effect on hair growth have a hormonal or hormonal-mimetic basis.

Likewise there are a number of references to fenugreek having galactagogue (increase milk in nursing mothers) activity, which again is indicative of an estrogen-like activity. The plant should be used with caution as Fenugreek is reputed to be oxytocic and *in vitro* uterine stimulant activity has been documented [Newall *et al*], so the use of fenugreek during pregnancy and lactation in doses greatly exceeding those normally encountered in foods is not advisable.

Pomegranate (*Punica granatum*)

Pomegranate is one of the many plants that contain substances with hormone-type action. The seeds of pomegranate, the ancient symbol of fertility, were found to contain an estrone identical with the genuine hormone. *Punica granatum* seeds are the best source of plant estrone to date.

The antioxidant and eicosanoid enzyme inhibition properties of pomegranate (*Punica granatum*) fermented juice and seed oil flavonoids were studied, which showed strong antioxidant activity (determined by measuring the coupled oxidation of carotene and linoleic acid) close to that of butylated hydroxyanisole (BHA) and green tea, and significantly greater than that of red wine. [Schubert *et al*].

This is clearly a fruit worthy of further exploration, especially as most of the information to date relates to the use of the bark, seeds and the roots as a taenicide (expelling worms). The rind is used as an astringent [Lust]. The leaf has antibacterial properties and is applied externally to sores [Stuart].

Other plants that contain estrone

Punica granatum L. Pomegranate (Seed) 17ppm, *Malus domestica* BORKH. Apple (Seed), *Zea mays* L. Corn (Seed Oil), *Humulus lupulus* L. Hops (Fruit), *Olea europaea* L. Olive (Seed), *Panax quinquefolius* L. American Ginseng (Plant), *Phaseolus vulgaris* L. Anasazi Bean (Flower), *Phoenix dactylifera* L. - Date Palm (Seed), *Prunus armeniaca* L. - Apricot (Seed).

Date Palm (*Phoenix dactylifera*)

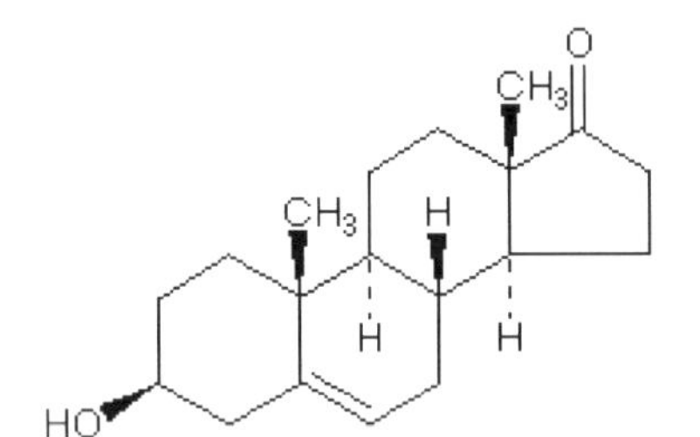

Fig.26 Prasterone or Dehydroepiandrosterone (DHEA).

Body hormones play a central role in skin appearance and are implicated in skin aging. Studies have shown that the decrease of these hormones plays an important role in skin endogenous aging, reduced skin thickness, and the disturbance of normal collagen turnover which, in turn, results in a decrease in collagen I and III synthesis. Date Palm has seven compounds with regenerative, anti-oxidizing, firming, and soothing properties, extracted from the kernel: phytosterols, phytosteroids, ursolic acid, isoflavons, policosonols, pro-vitamin A and vitamin E.

Some studies suggest that DHEA administration would have a beneficial effect against signs of aging. DHEA is known for its capacity to promote keratinization of the epidermis or to reinforce the barrier function of the skin.

The author compared on *ex vivo* skin, the effects of Date Palm kernel extract with those of DHEA in reference to DHEA as an anti-aging molecule. There was a decrease of wrinkles within only five weeks of Date Palm kernel extract application and also improved the skin structure in a way superior to that of DHEA.

The seed and the pollen have both been shown to contain estrone and this may further explain the reasons for this activity [Morton; Duke].

Plants with a future for topical application

In view of the benefits seen with those plants containing genistein and daidzein, we looked at other plants that might have potential as topical materials and also looked to see if they contained phytosterols and/or phyto-hormones.

The results were promising.

Calabar Bean (*Physostigma venenosum*)

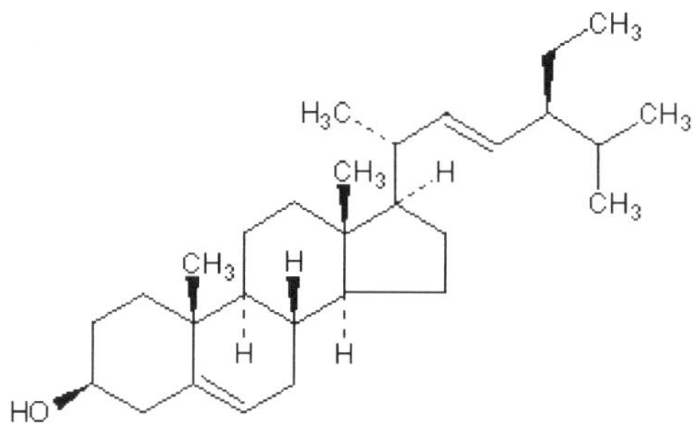

Fig. 27 Stigmasterol

Other plants that contains stigmasterol (the main contenders): *Annona cherimola* Mill. [Annonaceae] Seed 3080-4000ppm, *Panax quinquefolius* L. [Araliaceae] Plant 500ppm, *Vigna radiata* (L.) Wilczek [Fabaceae] Seed 230-230ppm, *Limonia acidissima* L. [Rutaceae] Fruit 150-150ppm, *Limonia acidissima* L. [Rutaceae] Leaf 120-120ppm, *Fagopyrum esculentum* Moench. [Polygonaceae] Seed 92ppm, *Centella asiatica* (L.) Urban [Apiaceae] Plant 40ppm, *Medicago sativa* subsp. *sativa* [Fabaceae] Fruit 40ppm, *Salvia officinalis* L. [Lamiaceae] Leaf 5ppm.

Suma or Brazilian Ginseng (*Pfaffia panniculata*)

Fig. 28 β-ecdysterone

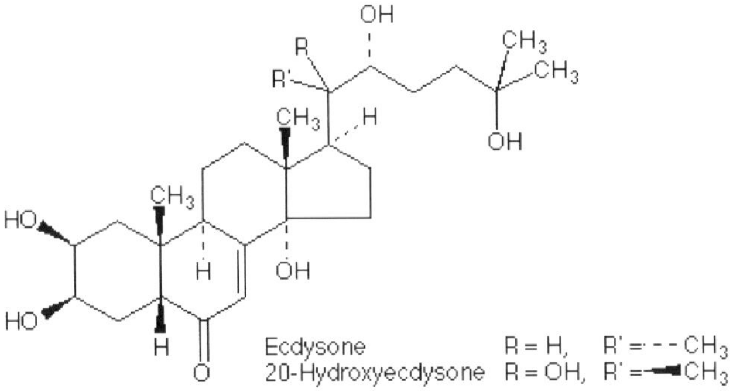

Other plants that contains β-ecdysterone: *Morus alba* L. [Moraceae] Leaf, *Chenopodium album* L. [Chenopodiaceae] Root. Ecdysterone: *Achyranthes aspera* BLUME [Amaranthaceae] Shoot, *Achyranthes bidentata* BLUME [Amaranthaceae] Root, *Morus alba* L. [Moraceae] Leaf, *Paris quadrifolia* L. [Liliaceae] Plant, *Polypodium aureum* L. [Polypodiaceae] Rhizome, *Taxus baccata* L. [Taxaceae] Leaf . [Phytochemical and Ethnobotanical Databases]

Cherimoya (*Annona cherimoya*)

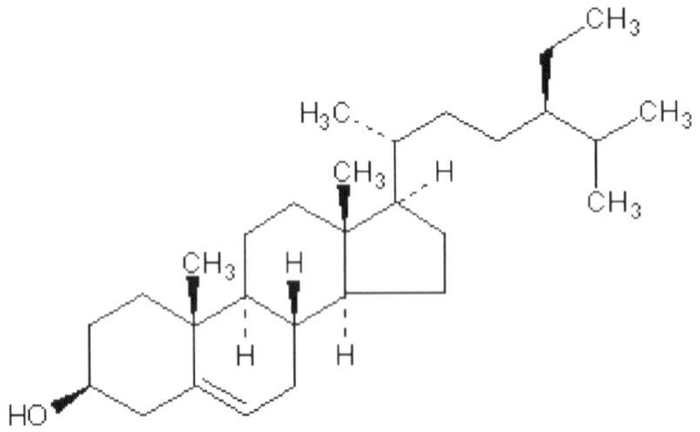

Fig. 29 β-sitosterol

Major plants that contain β-sitosterol : *Annona cherimola* Mill. [Annonaceae] Seed 10000-14000ppm, *Crataegus laevigata* (Poir.) DC [Rosaceae] Flower 6500-7800ppm, *Crataegus laevigata* (Poir.) DC [Rosaceae] Leaf 5100-6200ppm, *Nigella sativa* L. [Ranunculaceae] Seed 3218-3218ppm, *Oenothera biennis* L. [Onagraceae] Seed 1186-2528ppm, *Salvia officinalis* L. [Lamiaceae] Leaf 5-2450ppm, *Morus alba* L. [Moraceae] Leaf 2000ppm, *Senna obtusifolia* (L.) H. Irwin & Barneby [Fabaceae] Seed 1000-2000ppm, *Fagopyrum esculentum* Moench. [Polygonaceae] Seed 1880ppm, *Ocimum basilicum* L. [Lamiaceae] Leaf 896-1705ppm, *Zea mays* L. [Poaceae] Silk Stigma Style 1300ppm, *Salvia officinalis* L. [Lamiaceae] Stem 1214ppm, *Ocimum basilicum* L. [Lamiaceae] Flower 1051ppm, *Syzygium aromaticum* (L.) Merr. & L. M. Perry [Myrtaceae] Essential Oil 1000ppm, *Hippophae rhamnoides* L. [Elaeagnaceae] Seed 550-970ppm, *Glycine max* (L.) Merr. [Fabaceae] Seed 900ppm, *Nepeta cataria* L. [Lamiaceae] Shoot 900ppm, *Glycyrrhiza glabra* L. [Fabaceae] Root 500ppm, *Ocimum basilicum* L. [Lamiaceae]

Root 408ppm, *Viola odorata* L. [Violaceae] Plant 330ppm, *Cnicus benedictus* L. [Asteraceae] Seed 243ppm, *Ocimum basilicum* L. [Lamiaceae] Sprout Seedling 230ppm, *Withania somnifera* (L.) Dunal [Solanaceae] Root 200ppm, *Serenoa repens* (W. Bartram) Small [Arecaceae] Fruit 189ppm, *Turnera diffusa* Willd. ex Schult. [Turneraceae] Shoot 33ppm, *Agrimonia eupatoria* L. [Rosaceae] Shoot 25ppm, *Medicago sativa* subsp. *sativa* [Fabaceae] Fruit 5ppm. [Phytochemical and Ethnobotanical Databases]

Lima Bean or Butter Bean (*Phaseolus lunatus*)

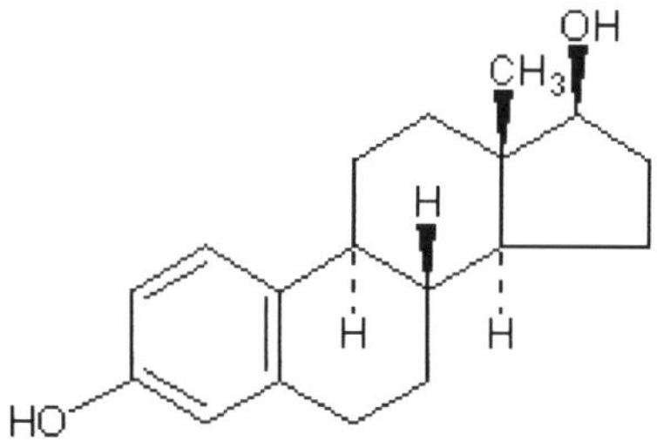

Fig. 30 Estradiol

Other plants that contains estradiol : *Humulus lupulus* L. [Cannabaceae] Fruit, *Panax ginseng* C. A. Meyer [Araliaceae] Root, *Panax quinquefolius* L. [Araliaceae] Plant, *Punica granatum* L. [Punicaceae] Seed. [Phytochemical and Ethnobotanical Databases]

Hops (*Humulus lupulus*)

The hop contains β-sitosterol, estradiol, stigmasterol and estrone. In addition it contains many other materials that are known for their sedative and relaxing attributes.

Regular doses of the herb can help regulate the menstrual cycle. It was the girls and women picking hops who first discovered that hops have an effect on genital organs. Before machines were introduced, hop pickers used to spend several weeks at this work, and it had always been known that menstrual periods would come early in young girls while they were doing this work. The reason is that hops contain plant hormones, particularly when very fresh, and these are similar to oestrogens. Considerable amounts have been found, 30,000 to 300,000 i.u. of oestrogen in 100g of hops. This also explains why hops will suppress sexual excitement in men. It has been shown that there are substances called anti-androgens that are able to cancel the effects of the male hormone (androgen).

It was found that hop extract not only recovered the proliferation of hair follicle derived keratinocyte (HFKs) suppressed by androgen but also stimulated the proliferation of HFKs. Furthermore, the effects of hop were evaluated using both animal tests and human volunteers *in vivo*. It was demonstrated that hop showed a potent acceleration on hair growth.

Sarsaparilla (*Smilax ornate*)

It is used in concoctions with other plants as a tonic or aphrodisiac. Sarsaparilla was formerly used in the treatment of syphilis, gonorrhoea, rheumatism and certain skin diseases. Used in soft drinks, the genins are also used in the partial synthesis of cortisone and other steroids. As part of a wider treatment for chronic rheumatism it should be considered as it is especially useful for rheumatoid arthritis. It has been shown that Sarsaparilla contains chemicals with properties that aid testosterone activity in the body.

Sarsaparilla contains saponins, sarsaponin and parallin, which yield isomeric sapogenins, sarsapogenin and smilogenin. It also contains sitosterol and stigmasterol in the free form and as glucosides. It is antirheumatic, antiseptic, antipruritic and is indicated for psoriasis, and other cutaneous conditions. Like other steroidal plants it is indicated for chronic rheumatism and rheumatoid arthritis. It is specifically used in cases of psoriasis especially where there is desquamation.

Chaper 12

Natural Anti-Irritant Plants

INTRODUCTION

The mechanism by which the skin becomes irritated and inflamed is both complex and dependent on numerous factors. No one pathway can be held to be entirely responsible for skin erythema or pruritic skin conditions.

Unlike modern allopathic drugs which are single active components that target one specific pathway, herbal medicines work in a way that depends on an orchestral approach. A plant contains a multitude of different molecules that act synergistically on targeted elements of the complex cellular pathway. Individually these elements may work quite effectively; however, time and again it has been proven that the overall effect is far superior when the whole plant is used.

The mechanism of inflammation shown below is one of the possible pathways.

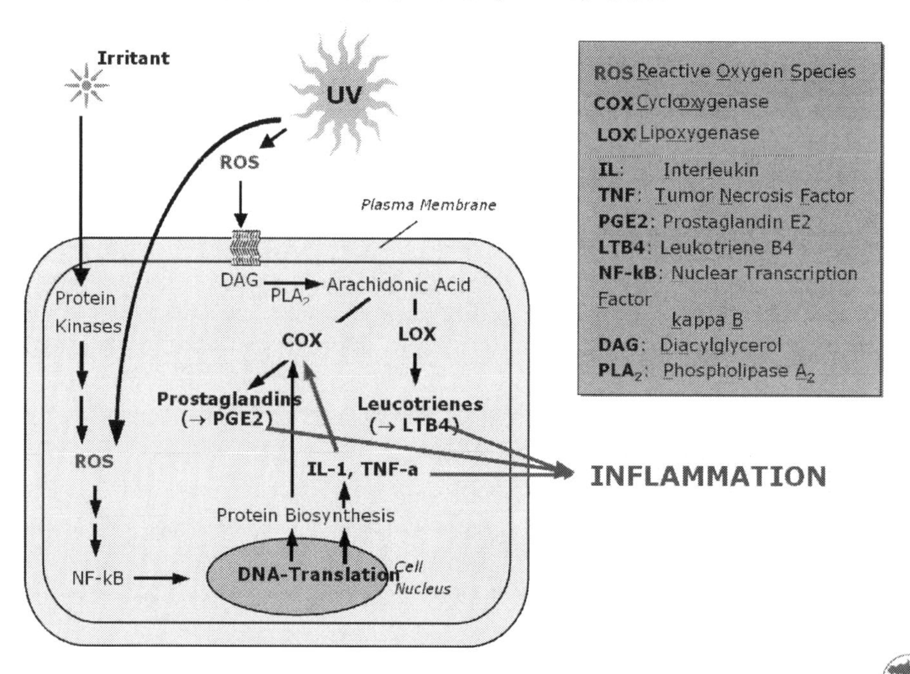

OVERVIEW

Some of the most commonly known plants to treat inflammation and erythema are German Chamomile (*Matricaria recutita*) and Roman Camomile (*Anthemis nobilis*) which contain a wide range of actives of which bisabolol, azulene derivatives and various flavonoids (particularly apigenin) are found to be the most functional components.

Other herbal materials renowned for their anti-inflammatory activity depend on entirely different chemical moieties.

Comfrey (*Symphitum officinale*) has always been considered in herbal folklore to be an ideal solution for sore, red, inflamed and damaged skin. [The scare over comfrey was as a result of an over-zealous ingestion of the herb that caused renal failure, but this event has no relevance in topical administration for which it is still permitted]. A typical analysis of the plant illustrates the complexity of the chemical compositions and also serves to demonstrate that different parts of the plant contain different actives which can be separated out into fundamental components.

Overall composition of the plant

Ash root 140,000 ppm, carbohydrates root 759,000 ppm, fat root 17,000 ppm, fiber root 72,000 ppm, gum root 50,000 - 100,000 ppm, protein root 94,000 ppm, resin root, water root 862,000 ppm

The support ingredients

The skin is dependant on a blend of minerals, vitamins and sugars to function and metabolise normally. The presence of minerals, vitamins and sugar related molecules (much loved by the skin) are essential for normal function.

Minerals

Aluminum root 237 ppm, calcium root 11,300 ppm, chromium root 8 ppm, cobalt root 129 ppm, iron root 810 ppm, magnesium root 1,700 ppm, manganese root 67 ppm, phosphorus root 2,111 ppm, potassium root 15,900 ppm, selenium root, silicic-acid leaf 40,000 ppm; root, silicon root 35 ppm, sodium root 3,510 ppm, tin root 6.7 ppm, zinc root 2.8 ppm.

Vitamins

Ascorbic-acid root 132 ppm, Beta-carotene root 660 ppm, carotenes plant 6,300 ppm, niacin root, riboflavin root 7.2 ppm, thiamin root 1.2 ppm. choline is a basic constituent of lecithin and is found in many plants e.g. hops, belladonna, strophanthus. It is almost classed as a vitamin.

Sugars and related compounds

D-mannose root, fructose root, glucose root, glucuronic-acid root, L-arabinose root, L-rhamnose root, sucrose root, mucilage root 290,000 ppm, mucopolysaccharides root 250,000 - 300,000 ppm, reducing-sugars root 51,500 ppm, xylose root,

The phytoactives

It is the presence of these remaining constituents that give comfrey its function and we can now begin to assign specific action to each of these chemical components
allantoin leaf 13,000 ppm; root 6,000 - 8,000 ppm, asparagine root 10,000 - 30,000 ppm, bornesitol root, caffeic acid root, chlorogenic acid root, consolicine root, consolidine root 17 ppm, echimidine root, echinatine root, GABA root, heliosupine-n-oxide root, hypoxanthine root, isobanerenol root, lasiocarpine root, lithospermic-acid root, lycopsamine root, octadecatetraenic-acid seed, pyrocatechins root 24,000 ppm, rosmarinic acid leaf 5,000 ppm, sitosterol root, stigmasterol root, symlandine plant, symphytine root, symphytocynoglossin root 21 ppm, tannin plant 80,000 - 90,000 ppm, viridiflorine plant.

Allantoin is described in Merck as a topical vulnerary and used in skin ulcer therapy. In the veterinary field it has been used topically to stimulate the healing of suppurating wounds and resistant ulcers. In the 5th edition 1941, it reported that it was used externally to stimulate cell proliferation in 0.3-0.5% aqueous solutions. It was recommended for ulcers, non-healing wounds, fistulas etc.

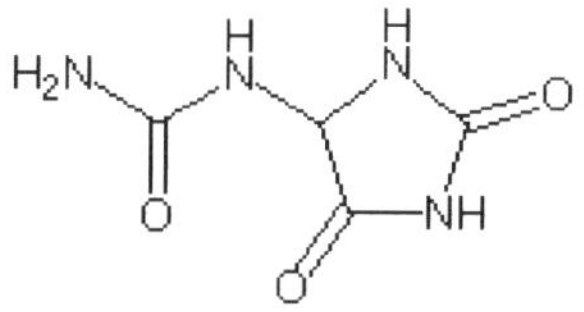

Asparagine is an amino acid that is frequently found in plant materials (e.g. mallow, chamomiles, liquorice, hops, sage and many *Pueraria* spp) most of which seem to be associated with long traditions of use for their soothing benefits.

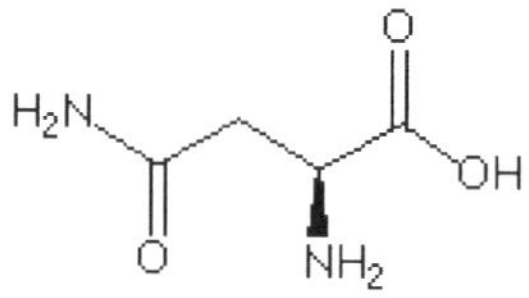

Bornesitol is very rarely found in plants, it has been found in plants like the Periwinkle or *Vinca minor* (a powerful anti-cancer drug) Borage or *Borago officinalis* and *Platycladus orientalis* none of which seem to have any relation to each other.

Caffeic acid is found in many fruits, vegetables and functional herbal materials like Horse chestnut (*Aesculus hippocastanum*), Burdock (*Arctium lappa*), Arnica (*Arnica montana*), Calendula (*Calendula officinalis*), Ivy (*Hedera helix*) and Self Heal (*Prunella vulgaris*) to name but a few. It is a selective inhibitor for leukotriene biosynthesis and can also inhibit arachidonate lipoxygenase activity. It is frequently associated with plants that are used for drainage and reduction of swelling and for the strengthening of cell membranes in the skin.

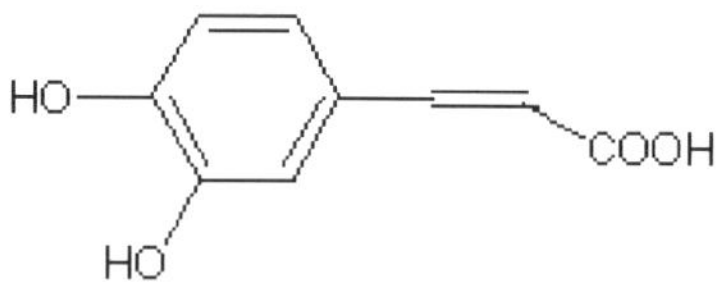

Chlorogenic acid – is a conjugated form of caffeic acid described above and has been cited for the stimulation of the immune system to stimulate T-lymphocytes. It often occurs in combination with caffeic acid. It is found in a multitude of plants that include Dandelion (*Taraxacum officinale*), Lime (*Tilea europaea*), St. John's Wort. (*Hypericum perforatum*), Golden Rod (*Solidago virgaurea*). Japanese Honeysuckle (*Lonicera japonica*) and Elder (*Sambucus nigra*) all respected herbal remedies for problem skin conditions.

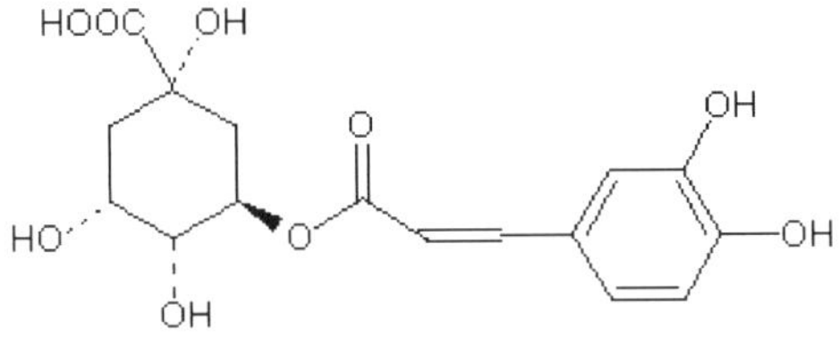

Consolicine and *consolidine* are extremely rare and we could only find them in Alkanet (*Alkanna tinctoria*). The effects appear to be a CNS-paralytic, curaroid and myoparalytic, but in truth little is known of the systemic or biological effects of this constituent.

Echimidine and *echinatine* are pyrrolizidine alkaloids and though this class of chemicals have been long recognized as potential skin irritants, they are definitely skin stimulants and at low levels could trigger a cellular response that was not detrimental. As so wisely stated by Paracelsus the father of toxicology "the dose makes the poison"

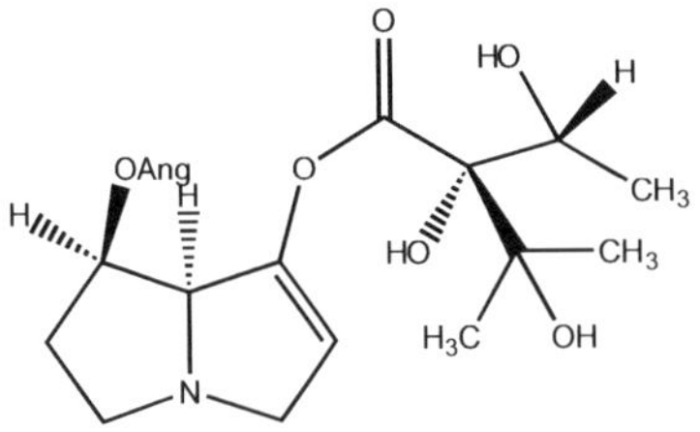

Echimidine

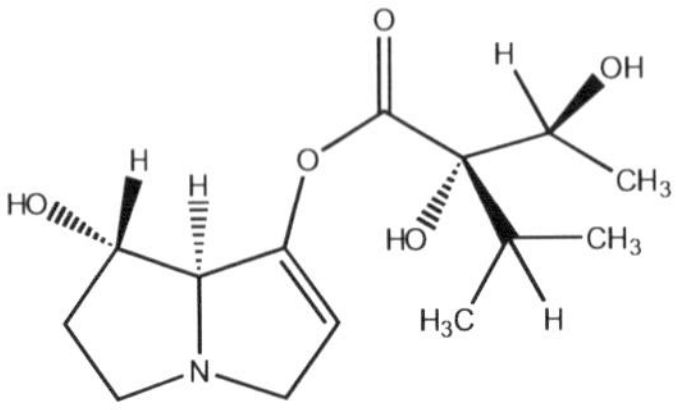

Echinatine

GABA is a multifunctional cellular stimulant and is anti-hepatotoxic, anti-varicose and anti-oedemic leading to cellular drainage and reduction in inflammation and swelling.

Heliosupine-n-oxide is a closely related pyrrolizidine alkaloids

Hypoxanthine found in Hound's Tongue, *Cynoglossum officinale*, *C. australe*, and *C. pictum*, in Viper's Bugloss, *Echium vulgare*, and in *Heliotropium supinum*

Isobanerenol appears to be unique to comfrey and has no reported properties.

Lasiocarpine is a related pyrrolizidine alkaloids and again seems unique to *Symphitum* species.

Lithospermic acid is a rare molecule (an analogue of caffeic acid) that is found in Blessed Thistle (*Cnicus benedictus*), Bugleweed (*Lycopus europaeus*) and *Lithospermum ruderale*. The material seems to suppress lactate dehydrogenase leakage particularly in renal cells.

Lycopsamine is a closely related pyrrolizidine alkaloids and found in Siam Weed (*Chromolaena odorata*) and *Echium* species.

Octadecatetraenic acid is found in Stoneseed (*Lithospermum officinale*). *Lappula squarrosa* and Chickweed (*Stellaria media*). This again is extremely rare in the plant kingdom and is reported to be extremely toxic.

Pyrocatechins or *catechol* is found in numerous plants like Tea (*Camellia sinensis*), *Ginkgo biloba*, Guarana (*Paullinia cupana*), and various *Pueraria* species. Catechol is a bioflavonoid and has an antioxidant and protective effect on skin cells. It is listed in Merck as an antiseptic.

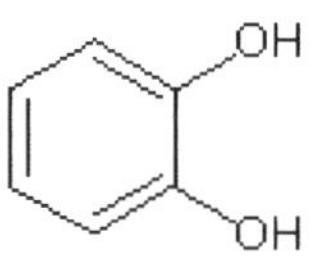

OH

OH

Rosmarinic acid is an anti-inflammatory, antioxidant and antimicrobial ingredient found (as you might expect) in Rosemary (*Rosmarinus officinalis*), Sage (*Salvia officinalis*), Lemon Blam (*Melissa officinalis*) and numerous other plants. It has also been attributed with antiviral, antithrombotic, antiplatelet and antihormonal activities and the suppression of cytokine-induced proliferation of murine cultured mesangial cells have also been reported for rosmarinic acid. It has been used topically in Europe as a nonsteroidal anti-inflammatory drug.

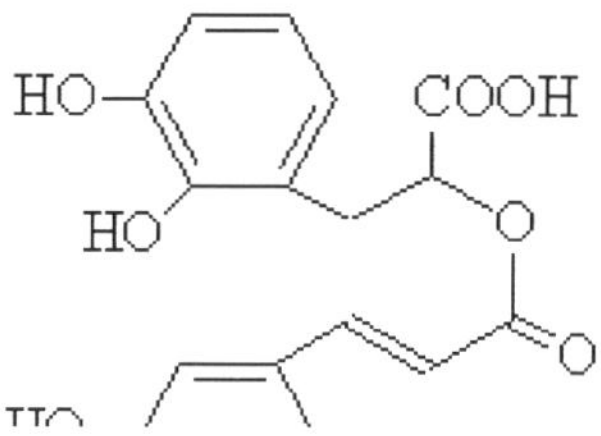

HO

COOH

HO

O

O

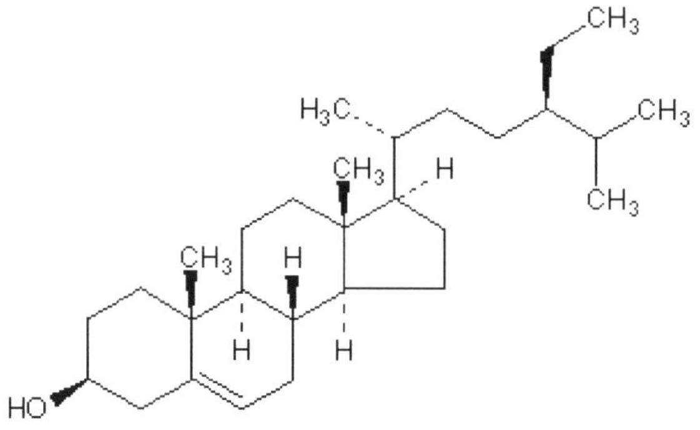

Sitosterol is a plant sterol and like its synthetic analogues hydrocortisone and corticosterone it has powerful skin properties including the reduction in skin erythema (skin redness), the reduction of pruritis (skin itching) and the reduction in inflammation.

CH_3 H_3C CH_3 CH_3 H CH_3 CH_3 H H H HO

Major plants that contain β-sitosterol: *Annona cherimola* MILL. [Annonaceae] Seed 10000-14000ppm, *Crataegus laevigata* (Poir.) DC [Rosaceae] Flower 6500-7800ppm, *Crataegus laevigata* (Poir.) DC [Rosaceae] Leaf 5100-6200ppm, *Nigella sativa* L. [Ranunculaceae] Seed 3218-3218ppm, *Oenothera biennis* L. [Onagraceae] Seed 1186-2528ppm, *Salvia officinalis* L. [Lamiaceae] Leaf 5-2450ppm, *Morus alba* L. [Moraceae] Leaf 2000ppm, *Senna obtusifolia* (L.) H.Irwin & Barneby [Fabaceae] Seed 1000-2000ppm, *Fagopyrum esculentum* MOENCH. [Polygonaceae] Seed 1880ppm, *Ocimum basilicum* L. [Lamiaceae] Leaf 896-1705ppm, *Zea mays* L. [Poaceae] Silk Stigma Style 1300ppm, *Salvia officinalis* L. [Lamiaceae] Stem 1214ppm, *Ocimum basilicum* L. [Lamiaceae] Flower 1051ppm, *Syzygium aromaticum* (L.) Merr. & L. M. Perry [Myrtaceae] Essential Oil 1000ppm, *Hippophae rhamnoides* L. [Elaeagnaceae] Seed 550-970ppm, *Glycine max* (L.) Merr. [Fabaceae] Seed 900ppm, *Nepeta cataria* L. [Lamiaceae] Shoot 900ppm, *Glycyrrhiza glabra* L. [Fabaceae] Root 500ppm, *Ocimum basilicum* L. [Lamiaceae] Root 408ppm, *Viola odorata* L. [Violaceae] Plant 330ppm, *Cnicus benedictus* L. [Asteraceae] Seed 243ppm, *Ocimum basilicum* L. [Lamiaceae] Sprout Seedling 230ppm, *Withania somnifera* (L.) Dunal [Solanaceae] Root 200ppm, *Serenoa repens* (W. Bartram) Small [Arecaceae] Fruit 189ppm, *Turnera diffusa* Willd. ex Schult. [Turneraceae] Shoot 33ppm, *Agrimonia eupatoria* L. [Rosaceae] Shoot 25ppm, *Medicago sativa* subsp. *sativa* [Fabaceae] Fruit 5ppm.

Stigmasterol is a closely related phytosterol and has similar properties to the sitosterol described above.

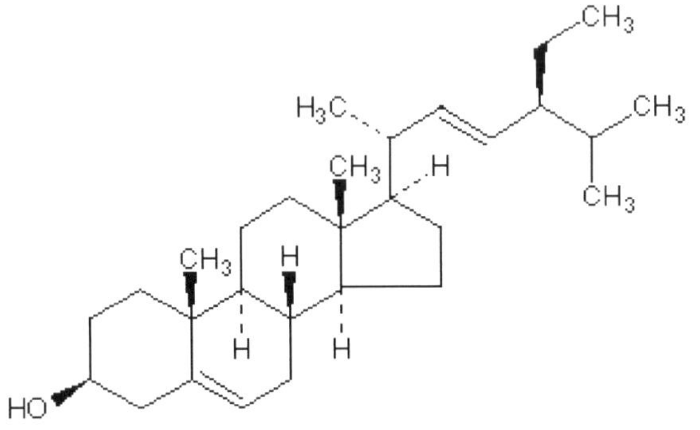

Other plants that contains stigmasterol (the main contenders): *Annona cherimola* MILL. [Annonaceae] Seed 3080-4000ppm, *Panax quinquefolius* L. [Araliaceae] Plant 500ppm, *Vigna radiata* (L.) WILCZEK [Fabaceae] Seed 230-230ppm, *Limonia acidissima* L. [Rutaceae] Fruit 150-150ppm, *Limonia acidissima* L. [Rutaceae] Leaf 120-120ppm, *Fagopyrum esculentum* MOENCH. [Polygonaceae] Seed 92ppm, *Centella asiatica* (L.) URBAN [Apiaceae] Plant 40ppm, *Medicago sativa* subsp. *sativa* [Fabaceae] Fruit 40ppm, *Salvia officinalis* L. [Lamiaceae] Leaf 5ppm.

Symlandine appears to be unique to comfrey and has no reported properties. It is a pyrrolizidine alkaloid.

Symphytine is another pyrrolizidine alkaloid.

Symphytocynoglossin is unique to comfrey and the properties are not known

Tannin is an astringent that firms microcapillary vessels

Viridiflorine is rarely found in the plant kingdom but is present in Hound's Tongue or *Cynoglossum officinale*. The function is unclear.

Conclusion

We have demonstrated that the choice of a plant for its anti-inflammatory properties is rarely dependant on one chemical entity present in the plant, but a symphony of individual components that work synergistically in order to bring benefit to the skin by acting as potentiators of a complex cellular metabolic pathway. In some cases the plant contains seemingly potent and toxic components, however, these are at such trace levels that they are more likely to act as triggers and initiators rather than toxins. A small spark that on its own would not be enough to cause a fire, but which in combination with other accelerants would cause a conflagration.

It has also been shown that specific chemical components occur in other plant genera and species. Comparison of the properties of one plant to another (where similar components are present) shows that these plants often exhibit similar properties and specific functional benefits are down to components that are unique to that plant. Examples are the bisabolol in chamomile, glycyrrhetenic acid in liquorice, allantoin in comfrey and asiaticoside in gotu kola.

Author's note: Thank you to my friends at Symrise for their help on this chapter.

Chapter 13

Natural Colors

Introduction

The chemistry of natural color is one that cannot fail to fascinate and intrigue. Nature provides a huge portfolio of possibilities and since one of our first papers on the topic [Natural ingredients for coloring and styling, *International Journal of Cosmetic Science* 24 5 287–302 (2002)], the availability of these materials has increased greatly.

Chemistry of Natural Dyes

Alkanin or Anchusin

Alkanna tinctoria, alkanet

In addition to the common name, it is called Orchanet and Spanish Bugloss and by the apothecaries it was known as *anchusa*. It has blue or reddish-purple flowers. Currently, alkanna root has no recognized medical uses except for its older use as an astringent. It is used as a cosmetic dye. The esteric pigments displayed excellent antibiotic and wound-healing properties in a clinical study on 72 patients with *ulcus cruris* (indolent leg ulcers).

Alkanet root contains a red coloring matter known as anchusin (alkannin or alkanet-red). Alkanet is an ancient dyestuff known throughout Europe.

Alkannin

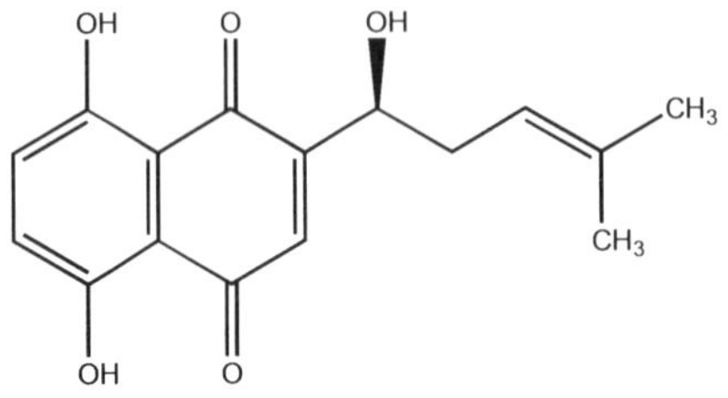

Annatto

Annatto, or norbixin, is extracted from the *Bixa orellana* or lipstick tree; it gives a yellow to deep orange color. The plant has entered into commercial cultivation for the production of this dye, which is used mainly in the food industry and for coloring dairy products such as butter and cheese, margarine and edible oils.

Bixa orellana, annatto

Annatto is a red to orange natural (golden yellow) pigment derived from the seed of the tropical bush *Bixa orellana.*

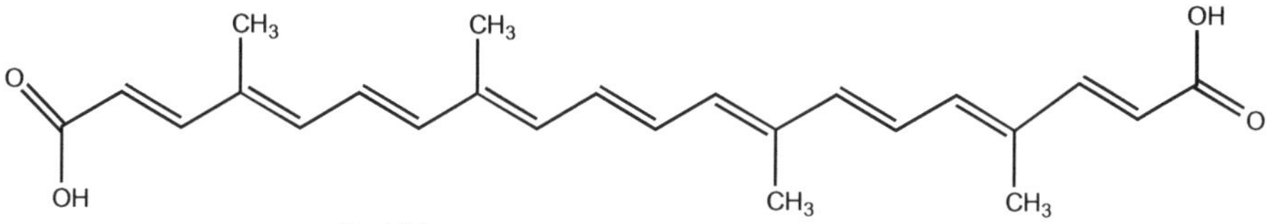

Norbixin

The major color present is *cis*-bixin, the monomethyl ester of the diapocarotenoic acid norbixin, which is found as a resinous coating surrounding the seed itself. Also
present, as minor constituents, are *trans*-bixin and *cis*-norbixin. The annatto bush is native to Central and South America where its seeds are used as a spice in traditional cooking.

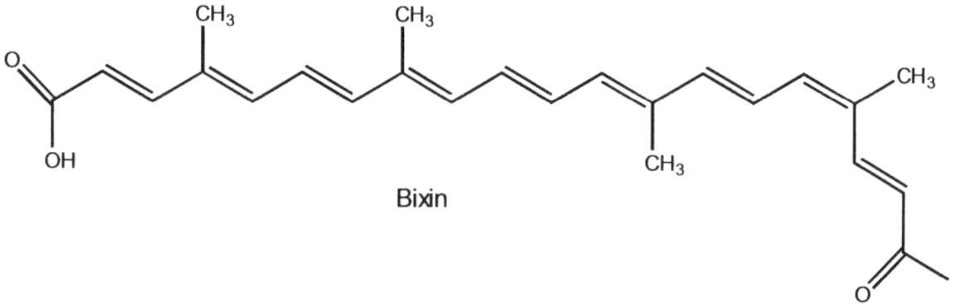

Bixin

It has been reported that the dye is also used in Brazil in pottery and as an insect-repellent, and in the Philippines in Boor, furniture and shoe polishes, nail varnish, brass lacquer, hair-oil, etc. It is further stated that Jamaica and South India have been major producers of the top quality product. From other countries, the dye has lacked the bright color required. In South America, the shrub is cultivated around villages, where it is native to and widespread throughout the neotropics. It is called *kiswe* or *kyswi* by the Waimiri Atroari. A red dye is obtained from the aril of the seed that is sometimes used for body painting. It is called *urucum* in the local vernacular.

Another chemical found in the plant that is responsible for some of the color is bixin. Bixin is one of the more stable natural yellow colors. However, it loses much of its tinctorial power gradually on storage, the process being accelerated by light and heat. Hence for manufacturing purposes fresh seeds are preferred. The tinctorial strength of bixin is comparable to that of ß-carotene, however, bixin is the more stable.

Anthocyanins

Anthocyanins are natural red, blue or violet plant pigments present in the cell sap of many flowers, fruits and vegetables. Anthocyanins are contained in cherries, plums, blackberries, black carrot, blueberries, cranberries, grapes, elderberry, mulberry, purple corn, rose hips, red cabbage and red currant.

They have the general structure shown below.

Anthocyanin

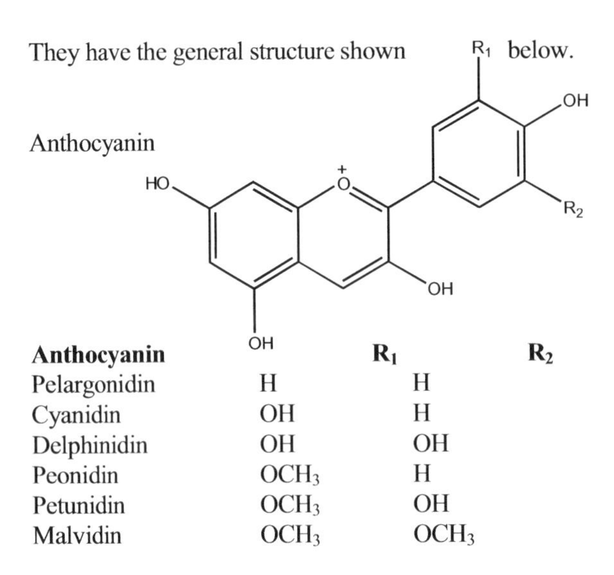

Anthocyanin	**R_1**	**R_2**
Pelargonidin	H	H
Cyanidin	OH	H
Delphinidin	OH	OH
Peonidin	OCH_3	H
Petunidin	OCH_3	OH
Malvidin	OCH_3	OCH_3

Cyanidin is found specifically in purple corn.

They are used as pH indicators and are heat-, oxygen- and enzyme-labile, but the stability is a major limitation for use. Red cabbage "juice" has been used in barbecue sauce and coloring pink lemonade.

Anthraquinones

Rubia tinctorum, roots of Madder

Madder is native to the Mediterranean and the Near East and was once widely grown as a dye plant and. was used in central Europe since around the ninth century. The common name madder comes from the Anglo-Saxon name *maedere* for "the plant." The generic name, *rubia* means "red" and the plant has been used as a source of a permanent red dye.

The two- to three-year old rootstock of the plants is used medicinally, which remains red when dried. The constituents include anthraquinone glycosides that are two red chemical entities derived from the roots and tubers, which are known as alizarin and purpurin. The isolation requires the prior hydrolysis of the glucoside precursor in the roots (Alizarin red from *Rubia tinctorum* occurs in nature as the glycoside and is called ruberythrinic or ruberythric acid and is its 6-O-β-D-xylopyranosyl-β-D-glucopyranoside). The Indian madder mainly yields purpurin (see Purpurin).

	R_1	R_2	R_3	R_4	R_5	R_6	R_7	R_8		
Alizarin			OH	PH	H	H	H	H	H	H
Purpurin			OH	OH	H	OH	H	H	H	H

In skin care it is considered an astringent, tonic, vulnerary and antiseptic. Madder is also used to cleanse open wounds and can remove skin blemishes by applying the bruised leaves externally. Externally, a decoction of madder can be used for skin problems, especially tubercular conditions of the skin and mucous tissue. The decoction can also be used as a bath additive.

Galium mollugo, White Bedstraw

The native Indians used bedstraw roots (*Galium mollugo*) to obtain a red color and the root of this plant is an important source of the dye alizarin.

Galium odoratum, Sweet Woodruff

Sweet Woodruff is similar to White Bedstraw and its roots can be used to produce a red/pink dye. It is the same family as madder.

Apigenin

Matricaria recutita, German Chamomile

This flavonoid, which occurs widely in plants, gives a dull, golden yellow and is usually obtained from German Chamomile or *Matricaria recutita.* Apigenin and lutolin were more active than the other flavonoids tested. The spasomolytic activity of chamomile has been attributed to apigenin, apingenin-7-0-glucoside and (-) - bisabolol, which have activity similar to papaverine and that Apigenin and lutolin were more active than the other flavonoids tested.

Apigenin or Natural Yellow 12

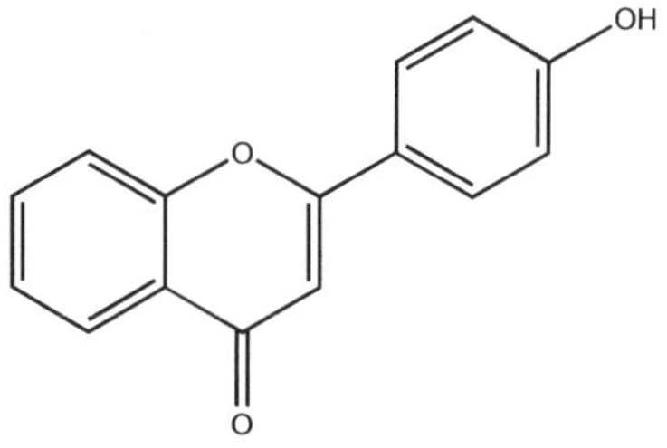

OH

O

O

Calendula officinalis, Marigold

It is also found in Marigold (*Calendula officinalis*), where it was shown using the mouse ear test that the flavonoids—and not the essential oil—were responsible for the activity and, of these, apigenin was more active than indomethacin in the test.

Artemisia inculta, Artemisia

Artemisia (*Artemisia inculta*) also contains apigenin and in a recent study was demonstrated to have anti-inflammatory activity.

Cuminum cyminum, cumin

Cuminum cyminum or cumin also contains apigenin and luteolin and their derivatives in addition to plants like Carrot (*Daucus carota*), Agrimony (*Agrimonia eupatoria*), Arnica (*Arnica montana*), Purple Coneflower (*Echinacea purpurea*) and Eyebright (*Euphrasia officinalis*)—all of which have demonstrated anti-inflammatory activity when used under the right conditions. Yellow apigenin is also found in parsley (*Petroselinum crispum*) and celery (*Apium graveolens*).

Arbutin

Shepherdia canadensis, Canada Buffaloberry

The Canada Buffaloberry (*Shepherdia canadensis*), also known as Canada Buffaloberry, Soopolallie, Soapberry or Foamberry, is one of a small number of shrubs of the genus Shepherdia bearing edible red berries. One recognized form however bears yellow fruits. The fruit produces a red dye. It contains ericolin glucoside, and chimaphilin which is a yellow naphthoquinone.

Arbutin (Ericolin)

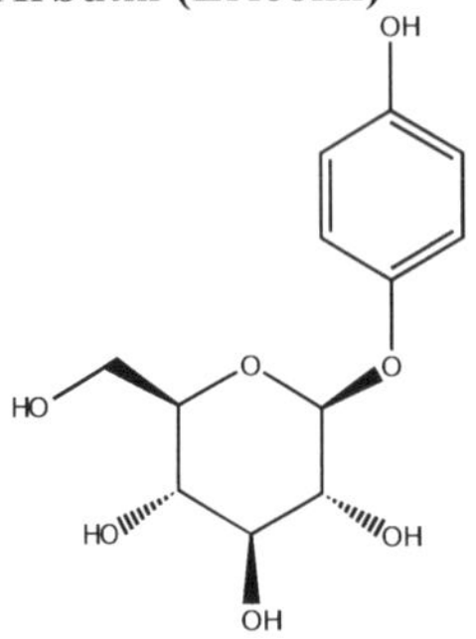

OH

O

O

HO

HO

OH

OH

Astacin

Astacin is found in the shells of crabs and lobsters *Homarus*, algae, sponges and fish. It is practically insoluble in water.

Astaxanthin

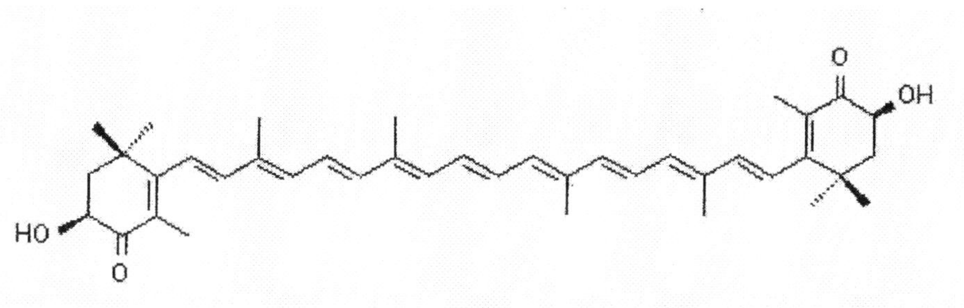

Azulene

Sadly this wonderful blue (blue-black) no longer seems to be available commercially. Azulene is a prime component of the essential oil of chamomile flowers, *Matricaria recutita* L. and related plant species. Products containing azulene generally also contain the other characteristic components of chamomile's essential oil. Azulene extracts are used in skin creams for reducing skin puffiness and wrinkles, and is also known for its anti-irritant and vulnerary properties.

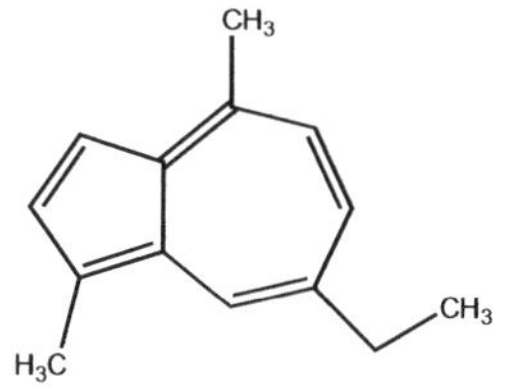

Chamazulene

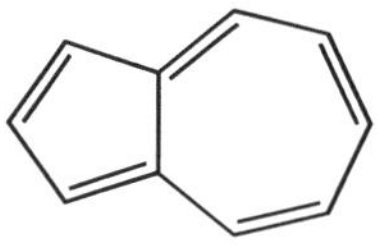

Azulene

It has never been clear whether the material was extracted from one of the chamomiles or from *Myoporum crassifolium* Forst or *Vanillosmopsis erythropappa* Schultz-Bip in which this material is also found. Maybe the source has become too scarce for there to be a reliable commercial supply.

Betalains and Betacyanins
Betanin

Dried beet juice contains 0.2–1.0% betanin and is most stable between pH 4.0–5.0. It is heat-, oxygen- and light-labile. Beets give rust/red or pink colors, but fade with time.

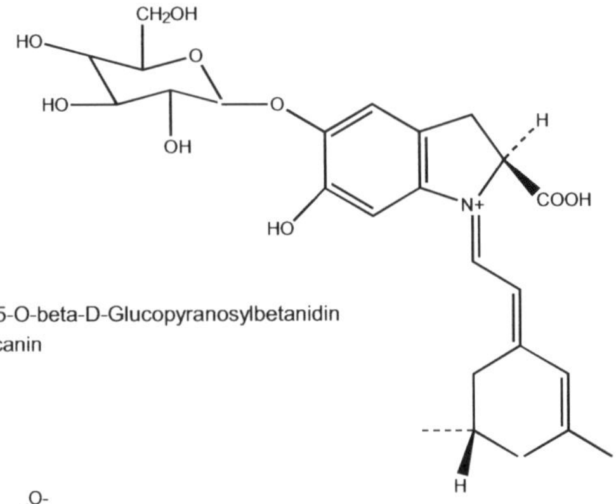

Betanin: 5-O-beta-D-Glucopyranosylbetanidin
Phytolaccanin

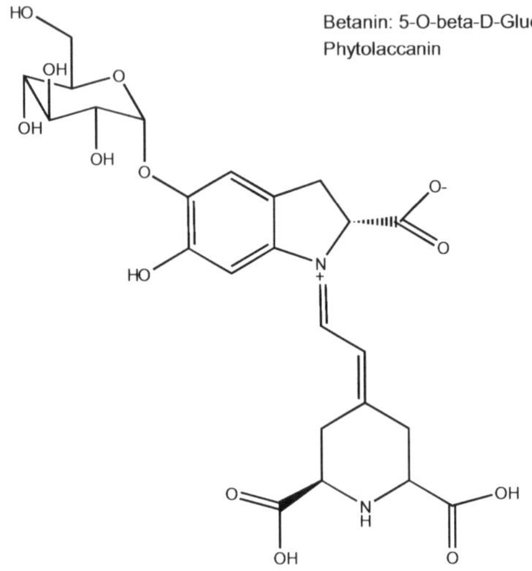

Betanin

This color occurs in the fruits of *Phytolacca americana* (Phytolaccaceae) and in *Portulacagrandiflora* (Portulacaceae). It is a purple pigment. Other places where it may be found are *Carpobrotus acinaciformis, Drosanthemum floribundum, Mesembryanthemum* spp. and *Opuntia bergeriana* and other *Opuntia* spp. (Cactaceae).

Brazilein, Brazilin

Caesalpinia echinata, brazilwood

The heartwood of American brazilwood Caesalpinia echinata contain a water-soluble compound called brazilin, which on oxidation converts to a red dyestuff called brazilein. As a mordanted dye, the colors can range from purple to bright red depending on the chemicals used.

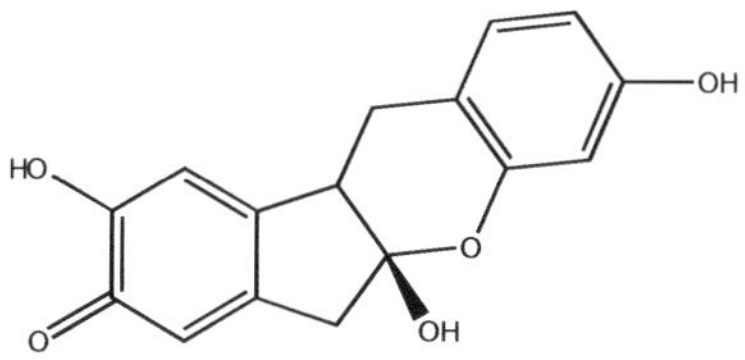

Brazilein

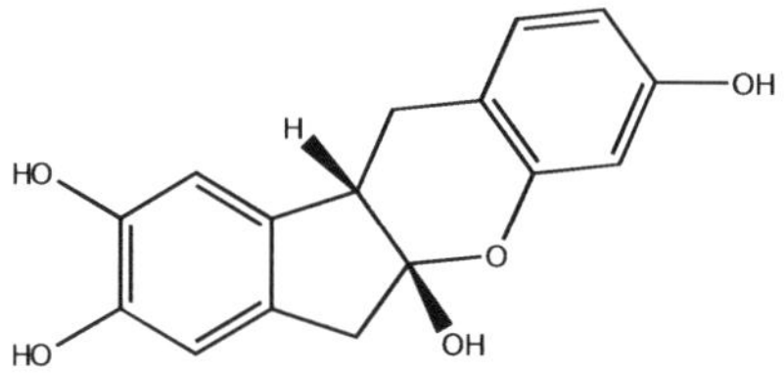

Brazilin, Natural Red 24

In an acidic solution, brazilin will appear yellow, but in an alkaline preparation it will appear red. Brazilin is closely related to the blue-black dye hematoxylin, having one less hydroxyl group.

Canthaxanthin

This is a carotenoid that naturally occurs in fungi but is more usually produced by "nature identical" synthesis. It is also a component of Spirulina-Dunaliella algae. The color can be yellow to an almost orange red. Canthaxanthin is a natural orange xanthophyll (see E161a) isolated from some mushrooms, crustacea, fish and also flamingo feathers. It is used to enhance the color of fish flesh, particularly trout and salmon. There is some evidence that prolonged ingestion of canthaxanthin in large amounts may cause spotting of the retina.

Capsanthin and Capsorubin

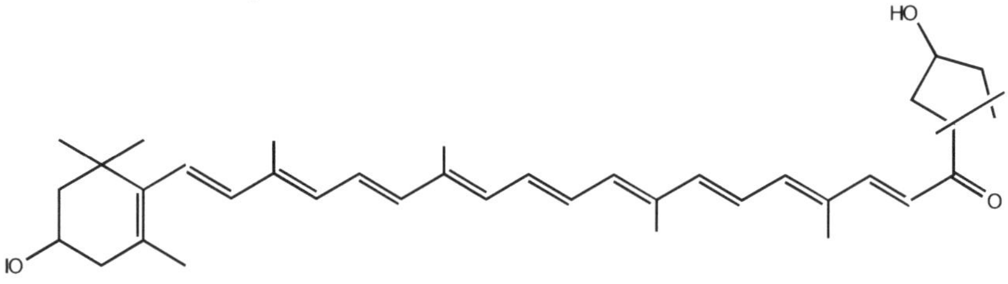

Capsanthin

Capsanthin and the related capsorubin are most commonly found in paprika or *Capsicum annuum.* It is rich in carotenoid pigments, including capsanthin, capsorubrin, carotene, luteine, zeaxanthin, and cucurbitaxanthin.

As well as being a dyestuff, it is also used in cosmetics in ointments, oils and emulsions for its stimulating effect and as a sports massage.

Capsicum annuum, Capsicum

It is also called Cayenne pepper, African Pepper, Chillies, Bird Pepper. It contains 0.1% capsaicin, capsacutin, capsico (a volatile alkaloid?) together with fixed oils. Capsicum in the form of an alcoholic tincture is sometimes incorporated into hair lotions as a stimulant, particularly in preparations designed for alopecia. The pigments present in paprika are a mixture of carotenoids, in which capsanthin and capsorubin dominate. These are oil-soluble, are stable to heat and pH variation, but may deteriorate in light.

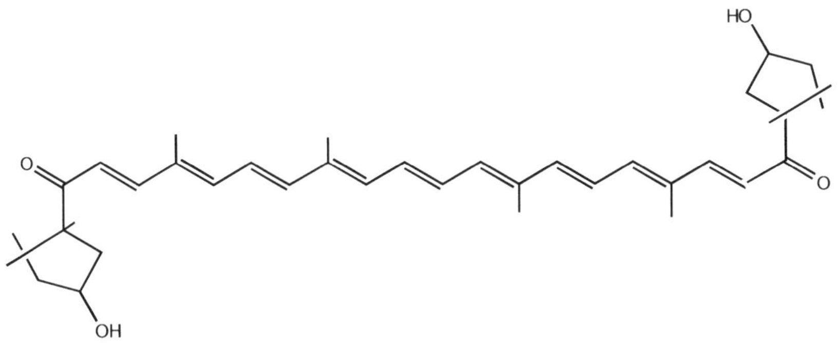

Capsorubin

It is employed in preparations to prevent chilblains and is the purest and most certain stimulant in *materia medica.* It produces natural warmth and equalizes the circulation. It is an important herbal remedy. It is used externally as a counterirritant, rubefacient and antiseptic. It is recommended for use in neuralgia, rheumatic pains, unbroken chilblains and is also used in cases of lumbago.

Capsicum is especially useful for providing counterirritation when applied to skin overlying an inflamed or irritated joint. Combined with myrrh, it makes a good antiseptic wash.

Caramel

Caramel coloring is produced by heat treatment of carbohydrates such as glucose syrup and sucrose, in the presence of ammonia, ammonium sulphate, sulphur dioxide or sodium hydroxide. The types of caramel color available include plain (spirit) caramel (prepared by controlled heat treatment of carbohydrates with or without an acid or base), caustic sulphite caramel (produced by heat treatment of carbohydrates with sulphur-containing compounds), ammonia caramel (heat treatment in the presence of ammonia) and sulphite ammonia caramel. Caramelized sugar or burnt sugar is formed by heating sugars without a catalyst.

Carminic acid

Coccus cacti, cochineal

This extract is associated with the protein material of the cochineal beetle and gives red, yellow and orange colors depending on the products and pH. *Coccus cacti* or scale insects are insects that live and feed on the prickly pear cactus (*Opuntia megacantha*) of Mexico and Central America. They increase rapidly in size and lose their original shape until they appear as protuberances of the plant. The dried pulverized bodies of these insects yield the red dyestuff cochineal, which the Aztec Indians used as body paint and for dyeing their fabrics a brilliant crimson. They also used cochineal for medicinal purposes.

It has been used as a pigment and as a coloring agent in cosmetics, paints and beverages, but it is expensive as it takes 70,000 insects to make one pound of dye. The homoeopathic tincture is prepared from the dried bodies of the female insects, which are larger than the male and have no wings. It is one of the main whooping cough remedies.

In both India and Australia attempts have been made to take advantage of this for the production of cochineal, for purposes such as dyeing soldiers' uniforms red. But chaos resulted. Either the insects fed so heartily that they wiped out the cacti; or the cacti multiplied so excessively that they became a real plague.

CI 45470
Carminic acid

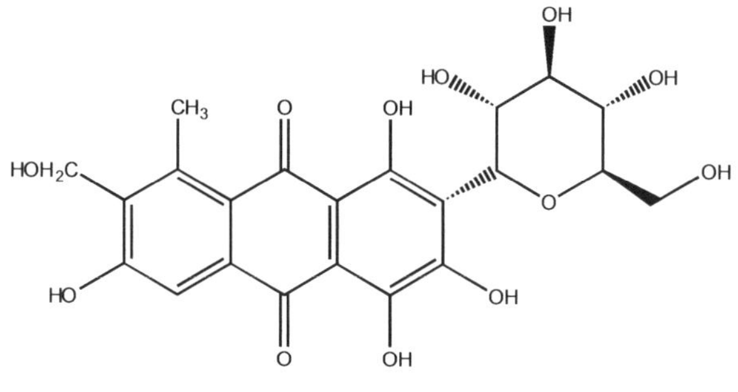

OH HO OH CH3 O OH HOH2C OH O HO OH O OH

This can be purified to yield carminic acid or reacted with alumina to produce the aluminium lake of carminic acid, which is referred to as carmine.

Carminic acid is a polyhydroxyanthraquinone acid or a red glucosidal hydroxyanthrapurin. It is water-soluble, acid-stable yielding orange to red shades depending on the pH. Its aluminium lake, carmine, is soluble in alkaline media and is very stable to heat, light and oxygen. In alkaline conditions, carmine provides a blue-red shade that becomes progressively less blue as the pH is decreased. Under acidic conditions below pH 3, carmine becomes insoluble.

Carotenes

Carotenes are fat-soluble and often synthetically produced natural pigments. The colors range from yellow to red. The pigment is sensitive to oxidation because of the conjugated double bonds and the molecule may be isomerized if involved in a heating processes. There are a number of related chemicals

ß-Carotene (0.6 ug = 1IU)
ß-Apo-8'-Carotenal (0.83 ug = 1IU)
Canthaxanthin (No vitamin A activity)
Bixin (Annatto Extract) (No vitamin A activity)
Lycopene (No vitamin A activity)

This is a group of yellow/orange colors extracted from such diverse sources as algae, carrots and palm oil. It is also available as a "nature identical product." Crystalline β-carotene is sensitive to air and light. Vegetable fat and oil solutions and suspensions are quite stable in products.

The carotenoids - apart from the chlorophylls - are the largest group of oil-soluble pigments found in nature. They consist of molecules with long chains made up of carbon, hydrogen and mostly oxygen (ß-carotene consists of only carbon and hydrogen). One of the carotenoid's characteristics is that the color varies (according to the type of carotenoid) from yellow to red-orange.

Carotenoids, like chlorophylls, exist in green plants. They are responsible for the yellow color of flowers and the pigments of many fruits and vegetables such as carrots, paprika and tomato. The first discovery of the color in carrots was the reason for the generic name of carotenoids. It is converted by mammals into vitamin A and as a result is called provitamin A.

ß-carotene is one of the major yellow colors used in the food industry and the largest use is by the dairy industry (butter, cheese and ice cream). The use of ß-carotene has almost entirely replaced E102 Tartrazine yellow, which for various reasons has received bad press in recent years.

The carotene family

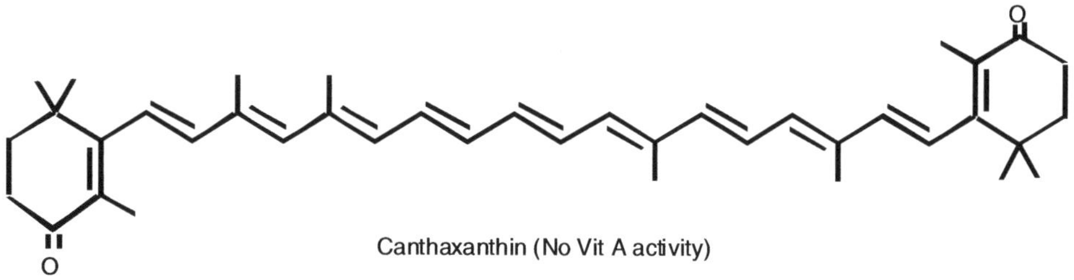

Canthaxanthin (No Vit A activity)

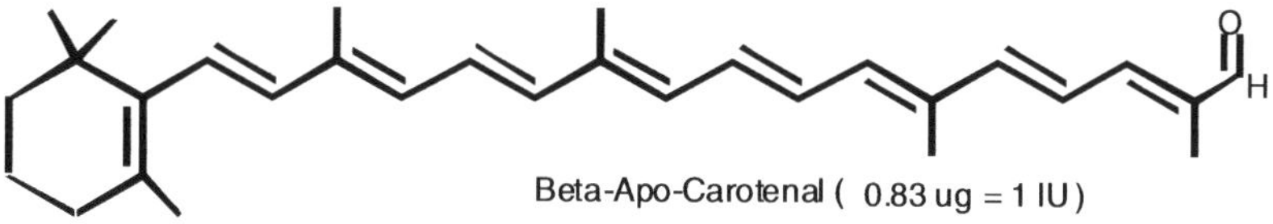

Beta-Apo-Carotenal (0.83 ug = 1 IU)

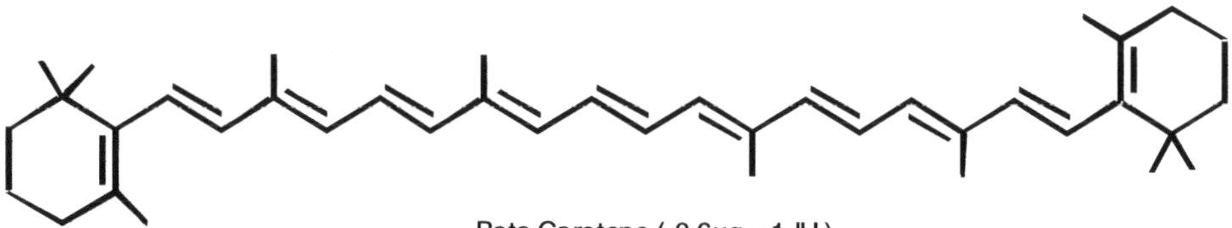

Beta Carotene (0.6ug = 1 IU)

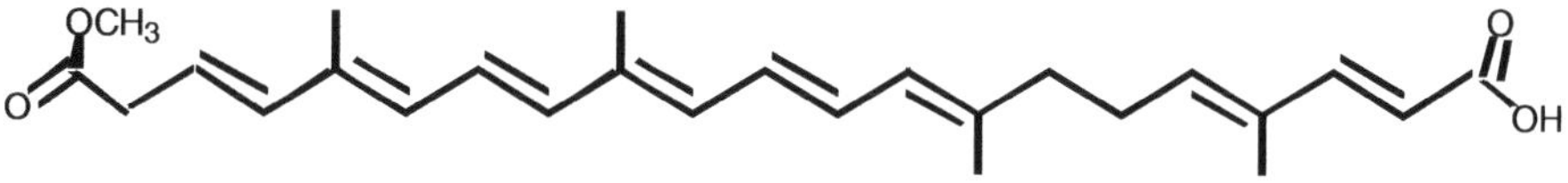

Bixin (Annatto)

ß-carotene is one of the popular free radical scavengers and antioxidants.

Carthamin

Carthamus tinctoria, Safflor

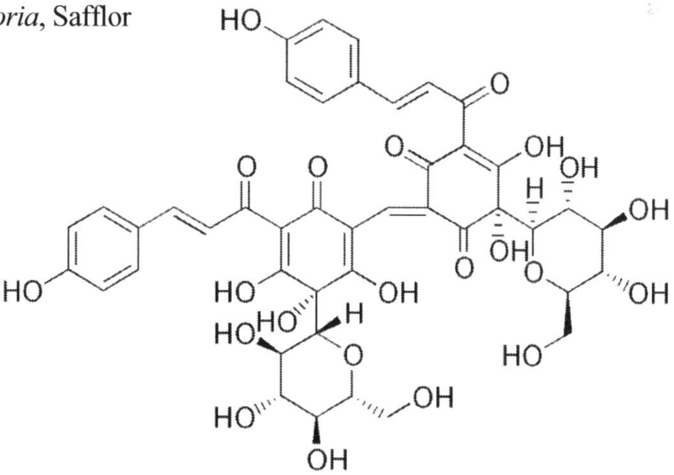

Carthamin is found in the flowers of *Carthamus tinctoria* or Safflor (Bastard Saffron), Dyer's Saffron, American Saffron, Fake Saffron, or *Flores Carthami*. It yields a pigment carthamin, which is a yellow-orange color. On closer examination it is shown to contain two coloring matters, one yellow the other red. Safflower (*Carthamus tinctorius* L.) was formerly used as a red dyestuff for textiles and is also currently used as a colorant by the food industry in small amounts.

The florets contain three major pigments, all of which are present as chalcone glucosides: the scarlet red carthamin is water-insoluble but the "safflor yellow" A and B or safflomin(e) A and B are both insoluble in water. The safflower florets are gathered just before they die off, and are used for dyeing wool, silk and leather. The flowers give a yellow dye, and the Chinese used an alkaline solution to produce the bright reds and purples for their silks. It also produces a pink dye that was used by the Indians to dye their official red tape used on legal documents. Mixed with talc, it was employed as rouge by actors. Apart from the seeds for oil and the flowers for dyeing, the Egyptians found the flower pleasing to the eye and included it in garlands laid on the mummies of their relatives. Remains of safflower were found in the tomb of Tutankhamun.

The seeds yield an oil that is used in India for burning and for culinary use. The flowers are used for their laxative and diaphoretic properties, also used in children's complaints of measles, fevers and eruptive skin complaints.

It was also the principal ingredient in Macassar Oil, the hair oil so very popular with the Victorian gentlemen. Certain oil paints are based on carthamin. Another little known fact is that it was used to impart a pink-red color to the ribbon used to bind legal documents (and still used to this day). Hence the expression "tied up in red tape."

Chimaphilin

Chimaphilin is a yellow naphthoquinone found in *Shepherdia canadensis* (see Arbutin) [Syn: *Hippophae canadensis*]. The fruit juice has been drunk in the treatment of digestive disorders and has also been applied externally in the treatment of acne and boils. The alternative name for the plant—*soopolallieis*—from the Chinook language for "soap" (*soop*) and "berry" (*olallie*) and the froth from the berries that contain a natural saponin or a jelly of the fruit, have been eaten as an insect repellent as it was said that mosquitoes were far less likely to bite a person who had eaten them.

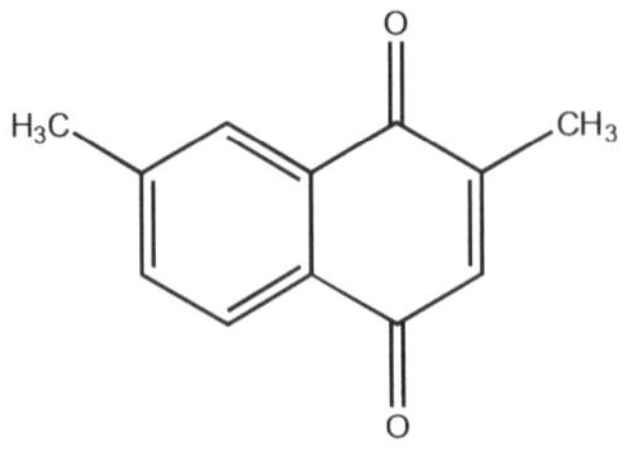

O

H_3C CH_3

O

Chimaphilin

Chlorophyll

Extracted from grass and alfalfa, this is present in all green plants and has always been a part of man's diet. It gives a moss green color. Naturally oil-soluble. It is also found in green vegetables such as spinach or *Spinacia oleracea* and the common stinging nettle or *Urtica dioica.*

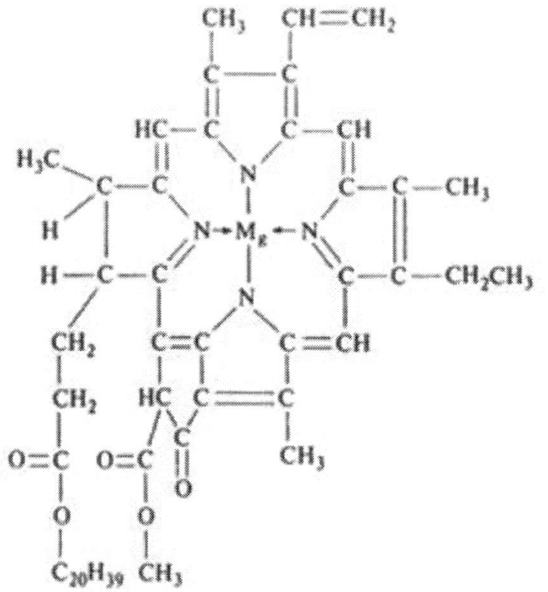

Chlorophyll

Copper chlorophyll

Derived from plants (as Chlorophyll) but gives a brighter more intense green color due to the replacement of the naturally occurring magnesium in the chlorophyll by copper. It is naturally oil-soluble.

This is produced as the copper chlorophyll but a saponification process renders this form water-soluble. The color is a bright green to green/blue.

Chrysophanic acid

Chrysophanic acid was discovered by Schrader in 1819, in *Parmelia parietina*, Linné, a common wall lichen. It was purified in 1843 by Rochleder and Heldt, who gave it the name chrysophanic acid, from its yellow color and later Schlossberger and Dopping decided the coloring matter obtained from rhubarb was identical. This coloring matter had been known under the names of rheine, rheumine, rhabarberic acid, rhubarb yellow, etc. Chrysophanic acid is chiefly obtained from a vegetable powder, called *Goa* or *Poh di Bahia*, which is the product of some unknown Brazilian plant probably *Andira araroba* (Leguminosae) and also known as Araroba. The powder contains from 70–80% of acid.

Chrysophanic acid has been chiefly employed as a local application in certain cutaneous affections such as eczema, various herpes conditions, psoriasis, acne and rosacea.

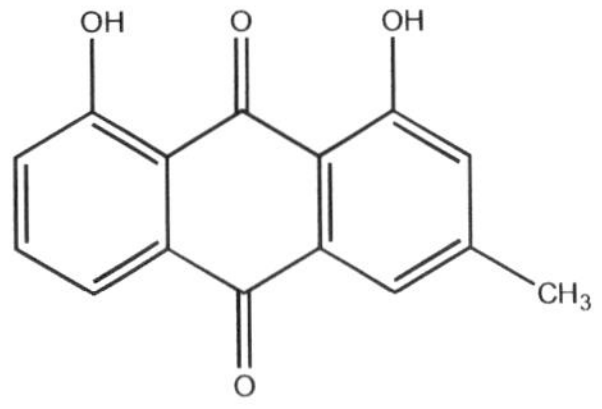

Chrysophanic acid

Rheum rhaponticum, rhubarb

Rhubarb roots make yellow, orange or red shades and was sometimes used for dyeing of hair. Rhubarb root has chrysophanic acid, a yellow dye, which will bind to keratin. Chrysophanic acid, as such, does not perhaps exist to any large extent in rhubarb, but is formed by the splitting up of chrysophane.

Citranaxanthin

This is a carotenoid pigment. There are natural sources of citranaxanthin, but it is generally prepared synthetically. It is used as an animal feed additive to impart a yellow color to chicken fat and egg yolks.

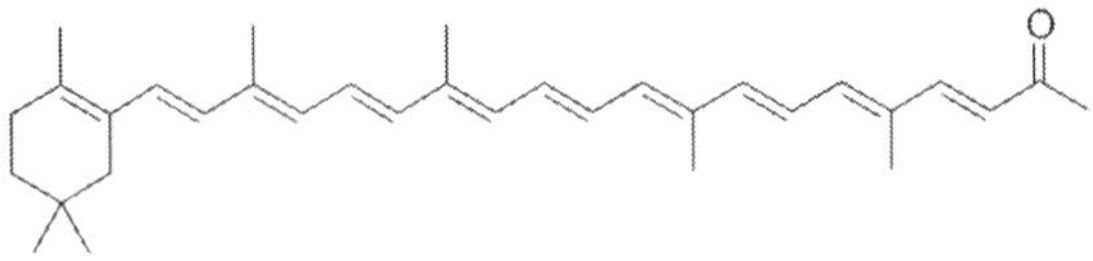

Citranaxanthin

β-Citraurin

β-citraurin from *Citrus sinensis* (Sweet Orange, Navel Orange) is a carotenoid pigment found in orange peel.

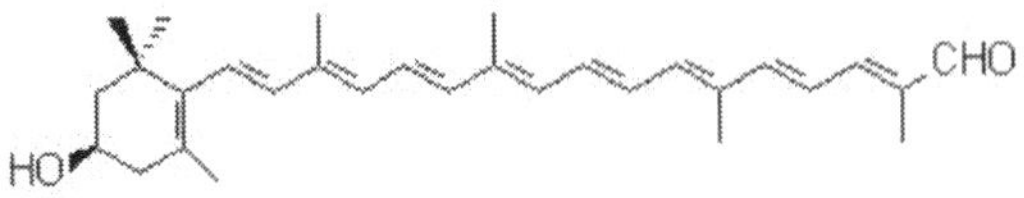

β-Citraurin

Crocetin

Crocus sativus, crocus

It is also known as CI Natural Yellow 6; CI 75100; Croci Stigma; Crocus; Safran. The dried stigmas and tops of the styles of the *Crocus sativus* contain crocines, crocetins and picrocrocine. They are delicate colors and should be protected from light.

Saffron is used as a food and cosmetic dye and flavoring agent. In some circles it is considered to be a food. It was once widely used for coloring medicines. There have been early reports of poisoning with saffron. This may be because of confusion with Autumn Crocus (*Colchicum autumnale*).

For years English saffron was the most highly esteemed (grown in Saffron Walden from about 1350), and most expensive, in Europe. Today, however, the best saffron comes from the barren plain of La Mancha in Spain. One kilogram of plant yields about 460,000 stigmas. The color is due to a pigment called crocin, so strong that one part crocin in a 100,000 parts of water is a deep golden color. The flavor comes from a related compound called picrocrocin. Saffron also seems to contain a substance that helps blood to clot. It is as a water-soluble dye that saffron probably found most use. The color is deep and rich, and most familiar today in the saffron robes of Buddhist monks.

Folkloric uses of saffron have included its use as a sedative, expectorant, aphrodisiac and diaphoretic. Anecdotal reports from the tropical regions of Asia describe the use of a paste composed of sandalwood and saffron as a soothing balm for dry skin.

The stigmas of *Crocus sativus* are rich in riboflavin, a yellow pigment and vitamins.

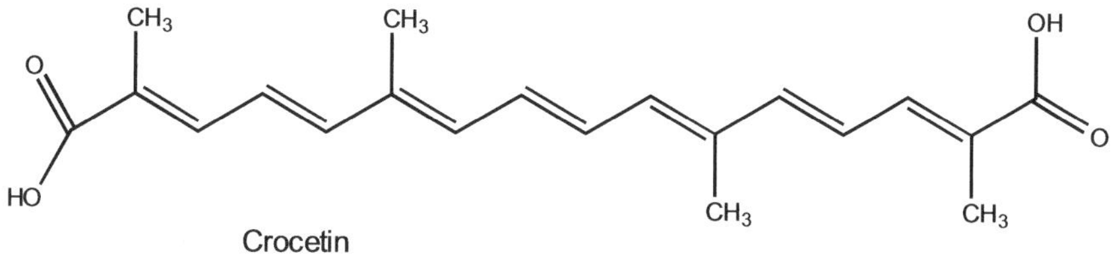

CH_3 CH_3 OH O HO O CH_3 CH_3

Crocetin

In addition, saffron contains crocin, the major source of yellow-red pigment. A hypothetical protocrocin of the fresh plant is decomposed on drying into one molecule of crocin (a colored glycoside) and two molecules of picrocrocin (a colorless bitter glycoside). Crocin is a mixture of glycosides: crocetin, a dicarboxylic terpene lipid, and alpha-crocin, a digentiobiose ester of crocetin. In addition, *cis*- and *trans*-crocetin dimethylesters have been identified.

Similar compounds have been isolated from other members of the family. A compound named gardenidin, obtained from gardenias, has been shown to be identical with crocetin.

Crocin

Crocin extract is the trade term for the yellow, water-soluble food colorant obtained from cape jasmine (*Gardenia jasminoides* L.) and from red stigmas of saffron (*Crocus sativus* L.). However, the extracts are not used interchangeably in all applications since saffron is valued as much for its aroma and flavor as for its coloring properties. It is an expensive spice that may prohibit its commercial use as a natural colorant.

Gardenia jasminoides, gardenia

A bright yellow color that has been in use for more than 1,000 years. Extracted from the fruit of *Gardenia jasminoides*. *Gardenia florida* [possibly a synonym for *G. jasminoides*] also contains crocin and is prescribed as antipyretic, hemostatic, antiphlogistic and in jaundice.

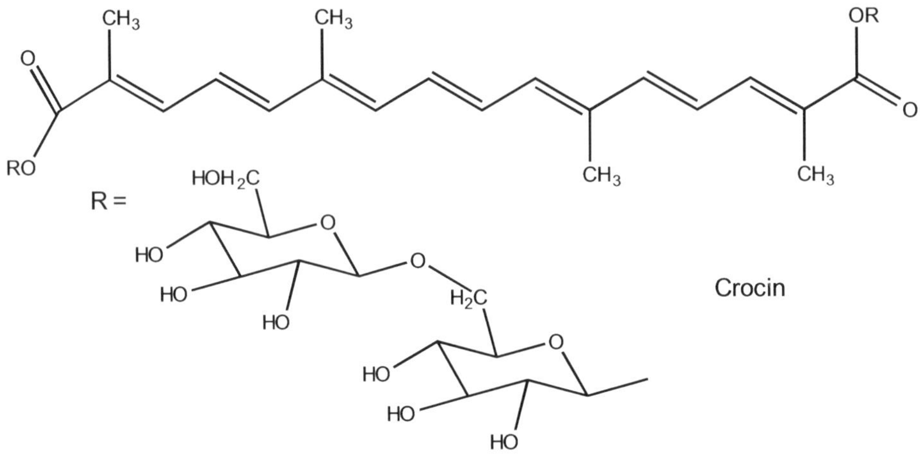

Crocin

A paste of the herb with flour and wine is used as a poultice on twists, sprains, strains, bruises and abscesses; very effective in injuries to tendons, ligaments, joints and muscles. Chinese medicine considers it to have anti-inflammatory, antipyretic, astringent, and hemostatic functions as well as use in the treatment of mastitis. The main component of its yellow pigment is crocin, which is now generally used as a natural yellow pigment.

It is also used for irritation, sore and swollen eyes and abscesses. *Fructus gardeeniae* is widely used in Chinese medicine. The fruits of *Gardenia jasminoides* also contain ursolic acid, which possesses hypothermic, sedative and anticonvulsant activities. The activity of this plant may, therefore, be in part due to the ursolic acid present

Crocus sativus, crocus

The characteristic yellow-orange color of saffron (*Crocus sativus* L.) comes from water-soluble pigment, the carotenoid crocin. Saffron carotenoids with ethanol-extractable mostly contain safranal as an antibacterial. It was used in Persian traditional medicine to treat some skin disorders. The extracted carotenoids from saffron, as an antioxidant, prevent many common diseases by taming harmful molecules known as free radicals.

Crocin is 8,8'-Diapo-ψ,ψ-carotenedioic acid bis (6-O-ß-D-glucopyranosyl-ß-D-glucopyranosyl) ester; α-crocin; di-gentiobiose ester of crocetin. It is also the coloring principle of saffron and also occurs in crocus.

Cryptoxanthin

Cryptoxanthin is a xanthophyll (see E161a) and is found naturally in members of the potato and tomato family [Solanceae], as well as in egg yolks, butter, the petals and flowers of plants in the genus Physalis, orange rind, papaya, corn, strawberries and apples,. It provides a natural yellow color but is not available for commercial coloring use may be because it is in too low a concentration to be commercially viable. Cryptoxanthin is closely related to beta-carotene, with only the addition of a hydroxyl group. In a pure form, cryptoxanthin is a red crystalline solid with a metallic luster.

Curcumin

Curcumin provides a water-soluble orange-yellow color. It is a natural extract obtained by solvent extraction from the dried rhizomes of turmeric (used in Indian cuisine as a flavoring agent). Curcumin may be used to compensate for fading of natural coloring in pre-packed foods. Recognized as an anticarcinogenic agent during laboratory tests.

This is the pigment of the spice turmeric and will give a range of color from yellow to a deep orange.

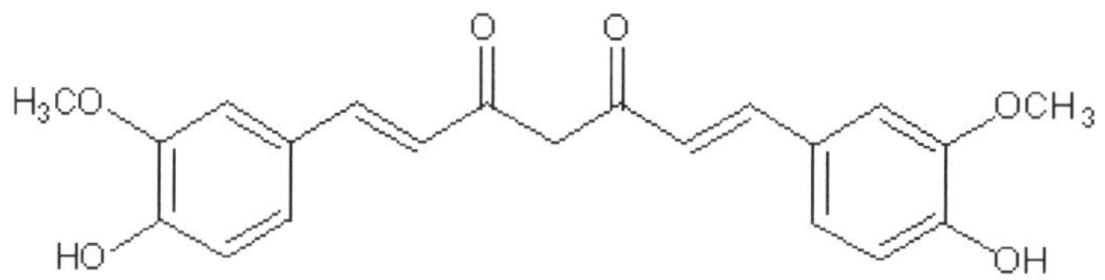

Curcumin

This has been in use as a food ingredient for more than 2,000 years. It also contains a closely related chemical called desmethoxycurcumin, where one of the methoxy groups is replaced with a hydrogen atom.

Curcuma longa, turmeric

The rhizome of *Curcuma longa* has been used as a medicine, spice and coloring agent for thousands of years. Turmeric was listed in an Assyrian herbal dating from about 600 B.C. and was also mentioned by Dioscorides. Among its many medicinal properties are its use as an antioxidant and anti-inflammatory. The anti-inflammatory activity of curcumin was first reported in 1971. In an extension of this work, it was reported that oral doses of curcumin possess significant anti-inflammatory action in both acute and chronic animal models. Curcumin was as potent as cortisone in the acute test (carrageenan edema), but only about half as potent as phenylbutazone in chronic tests.

The peoples of the Philippines utilize the fresh rhizome to treat recent wounds, bumps, bruises and leech bites. Mixed with gingelly oil (a locally produced oil), it is applied to the body to prevent further skin eruptions. Turmeric paste mixed with a little lime and saltpetre and applied hot is a popular application to sprains and bruises among other Filipino races. In smallpox and chicken pox, they make a thin paste of turmeric powder and apply it to the entire body in order to prevent pock marks from occurring by facilitating the scabbing process. It is also used for ringworm.

The peoples of Trinidad in the West Indies use turmeric rhizome poultices to reduce the swellings of sprains and pulled tendons, while the rhizome juice and oil is used for its antiseptic properties on various skin affections such as razor bumps, herpes lesions and venereal sores.

Among the races of India, turmeric has been used since time immemorial to treat skin problems. Both the Ayurvedic and the Unani practitioners have used a paste of powdered turmeric or its fresh juice made into a paste or a decoction of the whole plant as a local application in the treatment of leprosy and cobra bites. It is especially useful for indolent ulcers on the surface of the skin and gangrene in the flesh. A paste made from the powdered rhizomes along with caustic lime forms a soothing remedy for inflamed joints.

Turmeric is also used as an external application of "rouge" and is used by some women in India to suppress the unwelcome growth of facial hairs and upper lip moustaches. In Northern India, the rhizome is used by many natives for treating cuts, burns and scalds.

The natives of Samoa use powdered rhizome to sprinkle on newborn infants to help heal a recently cut umbilical cord, to prevent nappy rash from occurring, and to keep the skin continually soft and resilient. The powder is also used as a paste or poultice to treat skin ulcers and to help heal extensive skin eruptions. In parts of Africa, turmeric has been successfully tested for healing rashes due to allergies and psoriasis inflammation, and itching accompanying arthritis.

Datiscetin

This flavonoid has a yellow color and is found in Bastard Hemp (*Datisca cannabina*)

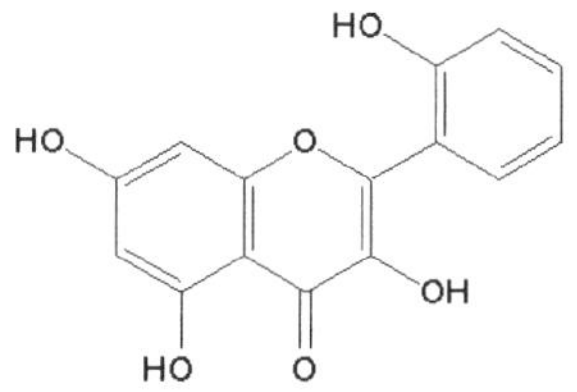

Datiscetin

Delphinidin

Delphinidin and is an anthocyanidin that gives colors from red wine to brown.

Deoxyisosantalin, Deoxysantalin

The heartwood (camwood) and roots of *Baphia nitida* yield a red dye called deoxyisosantalin that was used locally until recently to dye raffia and cotton textiles. The powdered heartwood is a used as red body paint and the paste is used as a cosmetic for the skin. As body paint, it is considered to have magic powers.

The leaves or leaf juice are applied against parasitic skin diseases. The leaves and bark are considered hemostatic and anti-inflammatory, and are used for healing sores and wounds. The powdered heartwood is made into an ointment with shea butter which is applied against stiff and swollen joints, sprains and rheumatic complaints.

Flavoxanthin

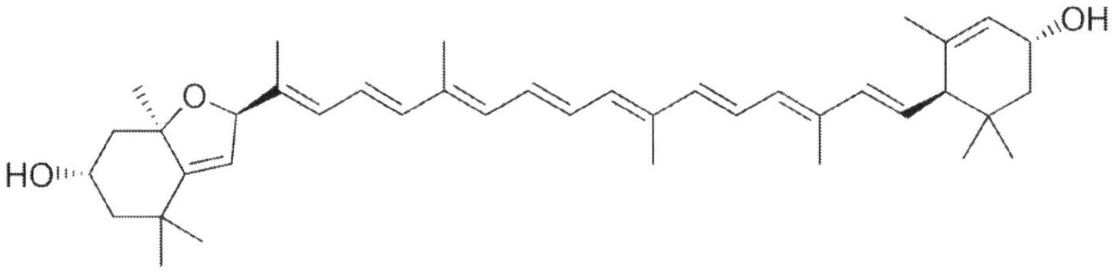

Flavoxanthin is a xanthophyll, providing a natural yellow color. Xanthophylls are mixtures of hydroxy derivatives of alpha-, beta- and gamma-carotenes, their natural epoxides and fatty acid esters.

Fumigatin

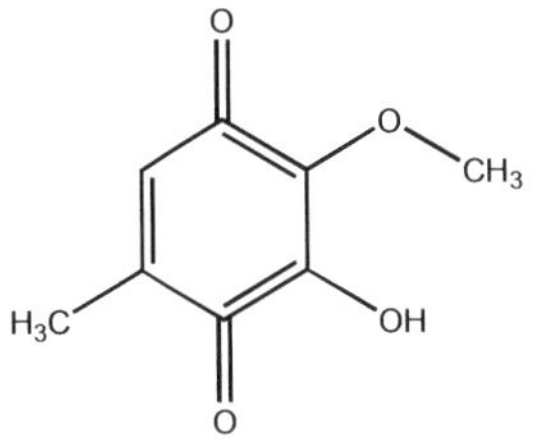

Fumigatin is isolated from *Aspergillus fumigatus* (a brown fungal toxin with antibiotic properties).

Fustin

Fustin is closely related to morin (see Morin) and these two flavones are found in the wood from *Chlorophora tinctoria* and *Morinda citrifolia* (Indian mulberry) whose pigments are mixtures of yellow to orange flavones similar in structure to fustin and morin. It is called maclurin, which has long been used as a yellowish-brown or khaki dye.

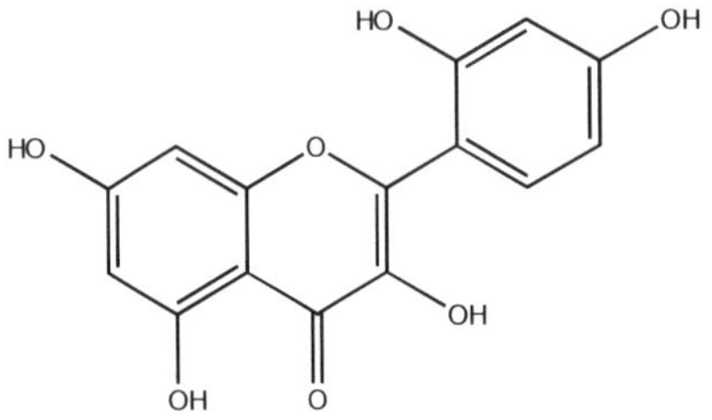

Genistein

Genistein is one of several known isoflavones. Isoflavones, such as genistein and daidzein, are found in a number of plants, with soybeans and soy products like tofu. Genistein functions as an antioxidant and many isoflavones have been shown to interact with estrogen receptors and cause an effect in the body similar to those caused by the hormone estrogen. It is a protein tyrosinase kinase inhibitor, both the estrogenic effect and the inhibition of kinases positively influences skin collagen content and is good for anti-aging products. In adipose tissue, it produces a lipolytic activity and is so useful in the treatment of cellulite.

Genista tinctoria, dyer's greenwood

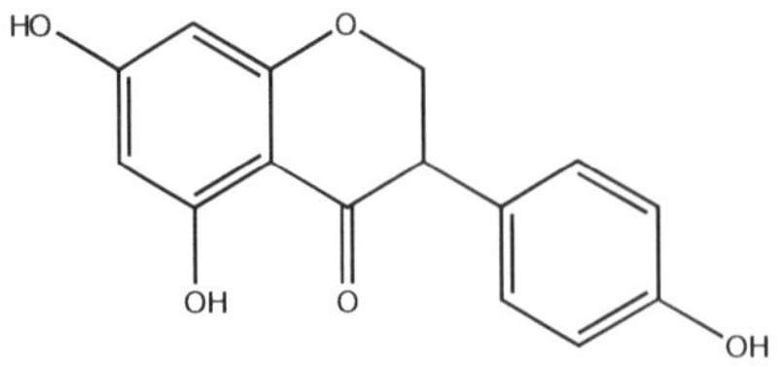

Genistein is a yellow dye contained as the glucoside genistin in dyer's greenwood, *Genista tinctoria.* An infusion is used against rheumatism, arthritis, dropsy and chronic skin disorders. In the past, the flowers were used for dyeing fabrics.

Gentisin

Gentisin is a yellow flavone found in the roots of *Genista tinctoria*, dyer's greenwood

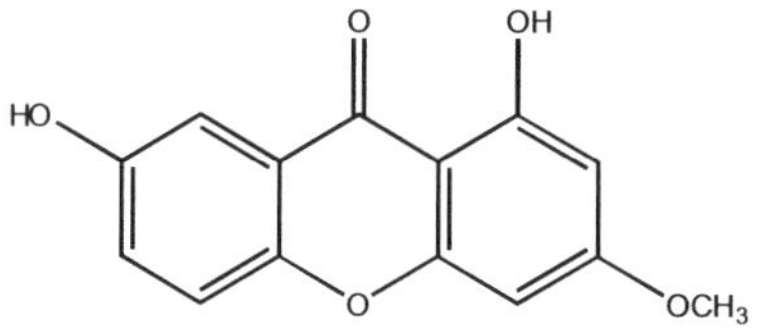

Haematoxylin

Haematoxylum campechianum, logwood

The heartwood of logwood (*Haematoxylum campechianum* L.) contains approximately 10% of a colorless compound called hematoxylin, which on oxidation transforms to a red to violet-blue substance hematoxein (also known as hematein).

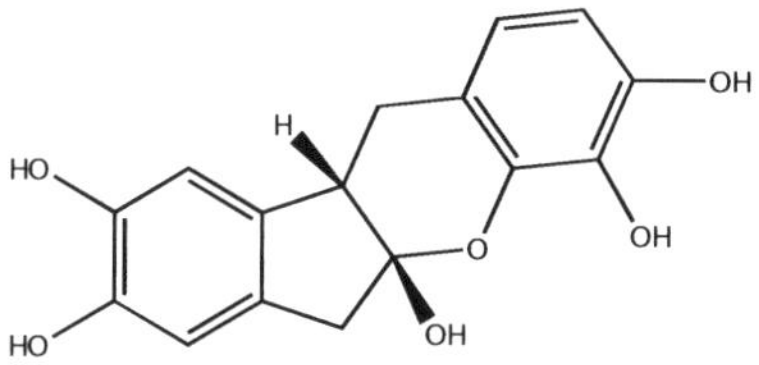

Haematoxylin, Hematoxylin, Natural Black 1 or C.I. 75290 (Figure)

An alcoholic solution of hematoxylin (0.2%) is used as an indicator. It has a range yellow to orange in acid solution and purple in alkaline solution. Solutions of hematoxylin are used in histology. Freshly prepared solutions have no staining powers; on keeping, however, the hematoxylin is oxidized to hematein, which is the actual coloring agent. Hematoxylin solutions for this purpose are prepared with ammonia alum, which appears to hasten the "ripening process." The tissues, after being stained red with hematoxylin, are washed in tap water, to change the color to blue. The mordants used to demonstrate nuclear and cytoplasmic structures are alum and iron, forming lakes, or colored complexes (dye-mordant-tissue complexes), the color of which will depend on the salt used. Aluminium salt lakes are usually colored blue-white while ferric salt lakes are blue-black.

The wood has medicinal properties. The infusion obtained when it is boiled in water has astringent properties. It is used in traditional medicine as a remedy for diarrhea and dysentery.

Hesperidin

Hesperidin is a flavanone glycoside (flavonoid) found in citrus fruits. Its aglycone form is called hesperetin. It acts as an antioxidant according to in vitro studies. Various preliminary studies reveal novel pharmaceutical properties. An animal study showed protective effects against sepsis. Hesperidin has anti-inflammatory effects. Hesperidin is also a sedative, possibly acting through adenosine receptors.

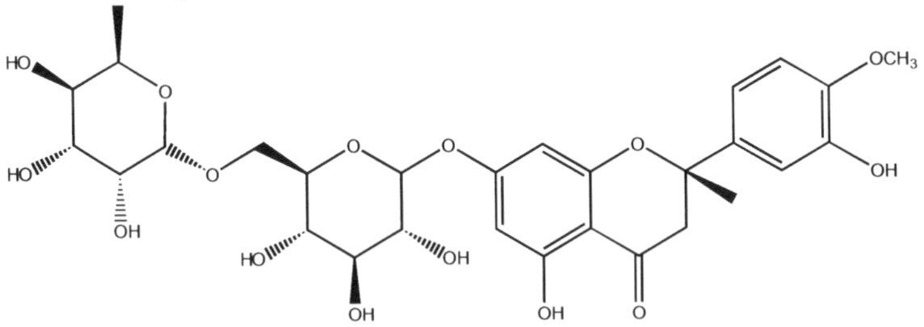

Hesperidin

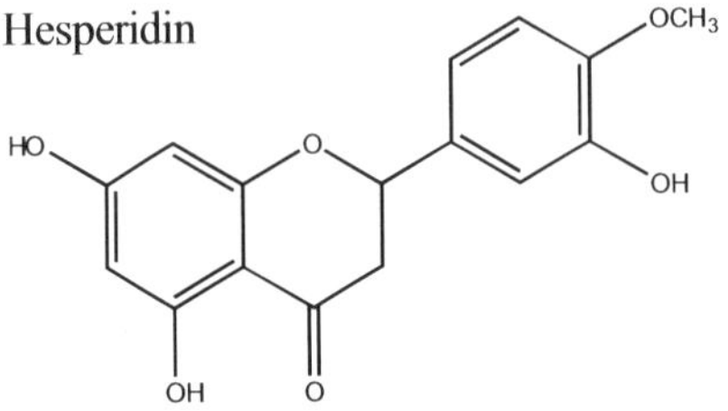

Hesperetin

Hypericin

Hypericum perforatum. St John's Wort

Hypericum is slightly astringent and is used for its healing application to wounds, sores, ulcers and swellings. Each year, the herb is supposed to show its red spots on the August 29, the day on which St. John the Baptist was beheaded. (Hence the name St John's Wort).

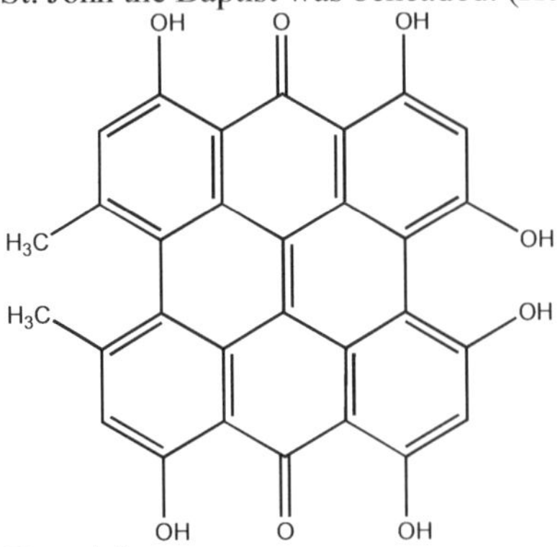

Hypericin

Herbalists combine it with hamamelis water as lotion for contusions and as an application to hemorrhoids. It is used as a mouthwash with calendula in parodontis and it is often combined with calendula as a lotion for contusions. The oil was much used in the past in the external treatment of wounds, burns etc

Indigotin

Indigofera tinctoria, indigo

Extracted from the fermented leaves of the plant *Indigofera* spp. This produces a blue to mauve color called indigotin (an indigoid structure). *Indigofera tinctoria* was also known as Pigmentum indicum or indigo. It was the deep blue dye used to color the denim of jeans and was first derived from plants such as woad and dyer's knotweed. Levi Strauss used the dye in the 19th century.

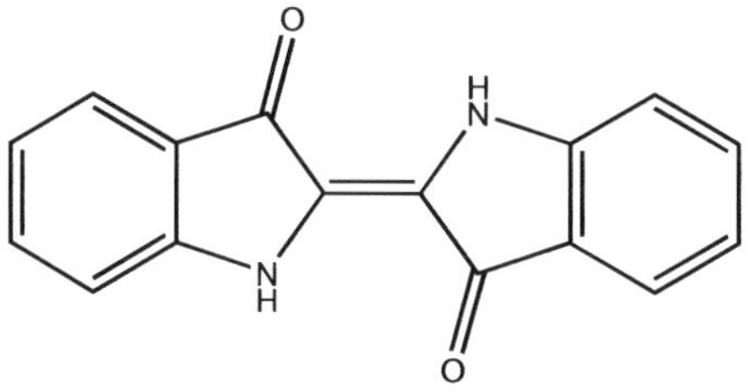

O
H
N
N
H
O

Indigotin

The blue dye is produced during the fermentation of the leaves, which is achieved using caustic soda or sodium hydrosulphite. A paste exudes from the fermenting plant material that is processed into cakes which are then finely ground. The blue color develops as this powder is exposed to air.

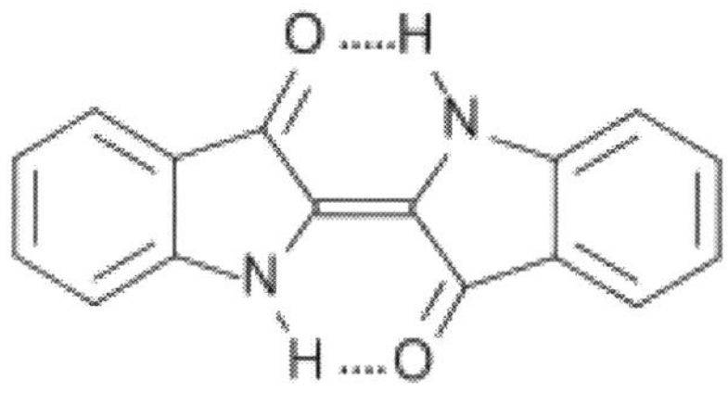

O ....H
N
N
H ....O

Indican

Indigo dye is a derivative of indican, a glucoside component of numerous *Indigofera* species and this is converted to blue indigotin using an enzyme process. This dye is quite colorfast and is combined with stabilizers and other compounds to produce a wide range of colorants. Today, almost all indigo used commercially is produced synthetically.

It is said to be good for mouth ulcers and externally the ointment will help infected ulcers and sore nipples.

Juglon

Naphthoquinones including juglone and juglandin, which are found in *Juglans cinerea*, *J. regia*, *J. nigra* (butternut, walnut, black walnut). The hulls of the nut contain dark colorant that is used as pink, brown to dark brown.

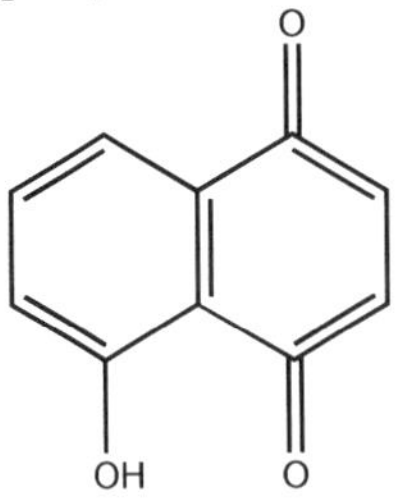

O

OH O

Juglon is a yellow-brown color and may be referred to as juglandic acid or nucin.

Kermesic acid

Kermes ilices

Kermes is one of the oldest dyes known, being mentioned in the book of Genesis (38:28) in the Bible and is a scarlet or crimson color. It is obtained from the bodies of an insect, *Kermes ilices* (formerly known as *Coccus ilicis*). It is chemically very similar to carmine and, as the older name indicates, the insects are related to those from which carmine is obtained.

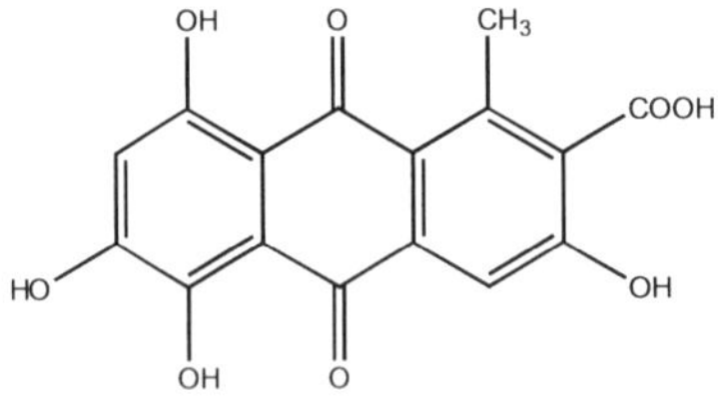

OH O CH_3

COOH

HO OH

OH O

Laccaic acid

Laccaic acids A, B, C and D, laccaic acid A being the most abundant crimson dye. Laccaic acids A, B and C are very similar, differing at a single point. Laccaic acid D is also called xanthokermesic acid, and closely resembles kermesic acid in structure.

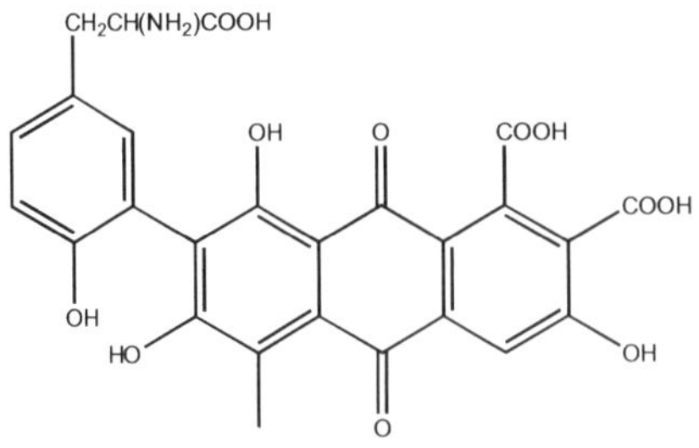

$CH_2CH(NH_2)COOH$

OH O COOH

COOH

OH

HO OH

O

Coccus laccae, lac insect

Lac is a yellow-red colorant obtained from the secretions of an insect found in India (*Coccus laccae*).

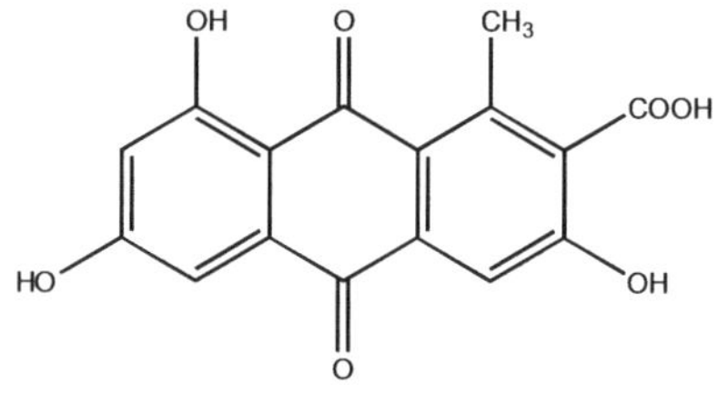

Laccaic acid D or xanthokermesic acid

Lapachol

It is reported to be useful for abscesses, ulcers and is anti-inflammatory, antiseptic, antiviral, bactericidal, fungicidal, insectifugal, pesticidal and viricidal.

Tabebuia avellanedae, Lapacho, Pau d'Arco, Taheebo

Lapachol is a yellow crystalline material from heartwood. Lapachol yellow is a naphthoquinone that was first isolated by E. Paterno from *Tabebuia avellanedae* (Bignoniaceae) in 1882.

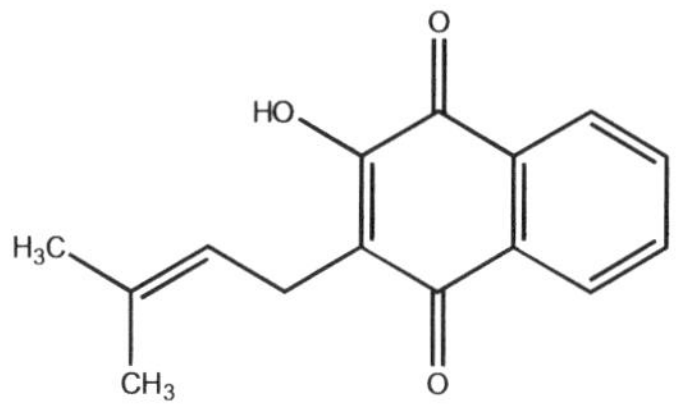

Lawsone

Lawsonia alba, henna

A color used frequently in hair care is from Henna or *Lawsonia alba*, *Lawsonia spinosa* and *Lawsonia inermis*, and is present in the leaves. It is the chemical lawsone that is responsible for the red color and is also found in *Juglans regia* or walnut.

This color has been used for nearly 5,000 years and was used by the ancient Egyptians for dying their hair and nails. Henna or Egyptian Privet was not only used by the Queens of Egypt to dye their hair, but was also used to decorate their skin as well. Mohammed is said to have dyed his beard with henna.

It is reported that, "*Lawsonia alba* is a shrub so laden with heavily scented white and yellow flowers that its perfume can be smelt a long way off. Every part of the tree is used by Eastern women to increase or restore their beauty. The powdered leaves make the best hair dye that is known and give the auburn color that is so admired. Sudanese women make a paste with catechu, cover their heads with it and leave it overnight."

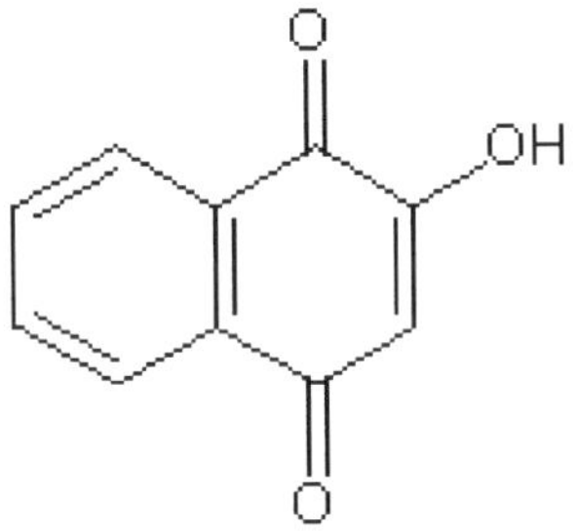

Lawsone [chinone molecule]

It is reported that, "*Lawsonia alba* is a shrub so laden with heavily scented white and yellow flowers that its perfume can be smelt a long way off. Every part of the tree is used by Eastern women to increase or restore their beauty. The powdered leaves make the best hair dye that is known and give the auburn color that is so admired. Sudanese women make a paste with catechu, cover their heads with it and leave it overnight."

The Malays are said to have a special Henna dance at weddings and use the leaves medicinally, externally to relieve tired feet. The plant has alterative, astringent and sedative properties.

It is reported that henna is also used as a hair brightening rinse for certain shades of hair (particularly in those cases in which it is desired to bring up a chestnut or auburn tint). Its content of lawsone makes it a substantive dye for keratin in acid solutions, where the dye is formed within the hair and not as a coating as in the case of metallic dyestuffs. Henna is quite innocuous and very few cases of allergy have been reported in its use. To all intents and purposes, danger from pure unadulterated henna is non-existent.

It has been reported that UV absorption properties can depend on the fact that chemical components are capable of reacting with the skin and so have UV absorption properties. In this respect, these "interactive extracts" are similar in function to DHA or dihydroxyacetone. The best known in the category is henna, although walnut extract has also been used successfully. There are water- and oil-soluble versions available.

Juglans regia, walnut

This extract is made from the fresh green shells of English walnut *Juglans regia*. The aqueous extract has been shown to be particularly effective as a self-tanning sunscreen agent. Its most important component is juglone (5-hydroxy-1,4-naphthaquinone) a naphthone closely related to lawsone (2-hydroxy-1,4-naphthaquinone). The self-tanning properties of henna are a result of the interaction between lawsone and the skin. Likewise, juglone is known to react with the keratin proteins present in the skin to form sclerojuglonic compounds.

These are colored and have UV protection properties. It is likely that these reactions, as with DHA, are the result of Maillard and Browning reaction sequences. Unlike with DHA, the sclerojuglonic compounds have more of a red-brown color, rather than the yellow color associated with a keratin protein/DHA reaction.

Henna has also been used as a medication for scurf and superficial wounds and has been used to treat skin infections such as tinea and lawsone is also known to be antibacterial.

Lutein

Lutein is a xanthophyll (see E161a) that provides a yellow-red color. It is related to carotene (E160a) and is normally found in green leaves so is available as a natural plant extract.

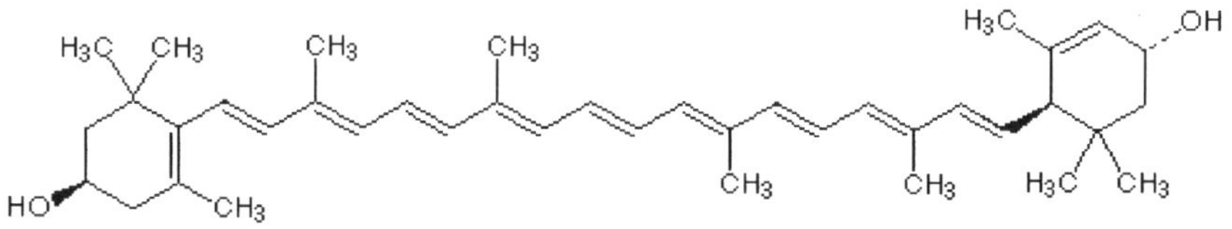

Tagetes erecta, Aztec marigold

Lutein is found in plants like dandelion (*Taraxacum officinalis*) along with violaxanthin, St. John's Wort (*Hypericum perforatum*). Marigold (*Calendula officinalis*) and Orange peel (*Citrus aurantium amara*). Interestingly, lutein is often found in combination with violaxanthin. It has also been found in Mexican tarragon (*Tagetes lucida* Cav.)

Luteolin

Reseda luteola, dyer's rocket or weld

The color luteolin is found in Dyer's Rocket (also known as Weld) or *Reseda luteola*. It is one of the oldest yellow dye plants and is found in many parts of Central Europe. The leaves and seeds are used, which contain more dye than the stems. An infusion of the plant has been used for treating wounds.

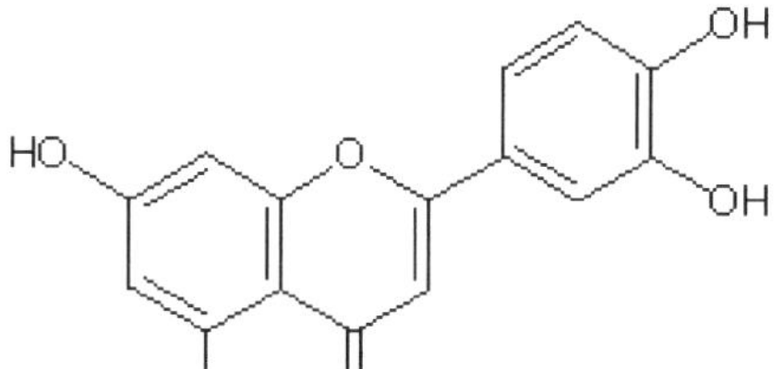

Luteolin

Genista tinctoria, Dyer's Broom, Dyer's Greenweed

This dye is also present in Dyer's Broom, Dyer's Greenweed or *Genista tinctoria*, where the color is a more green-yellow. An infusion of the plant has been used for chronic skin disorders. It has anti-inflammatory and antibacterial properties.

Antirrhinum majus

The 7-glucoside and 7-glucuronide is found in the petals of *Antirrhinum majus* (Scrophulariaceae). The 7-galactoside and 7-rutinoside occur in *Capsella bursa-pastoris* (Cruciferae) and the 3'-glucoside in *Dracocephalum thymiflorum* (Labiatae).

Lycopene

An extract from tomatoes that gives a red to orange color and is also reported in other plant sources such as *Rosa rubiginosa* (rose hip), *Taxus baccata* (yew), *Calendula officinalis* (marigold) and *Citrullus lanatus* (watermelon), though clearly at much lower levels. It is insoluble in water.

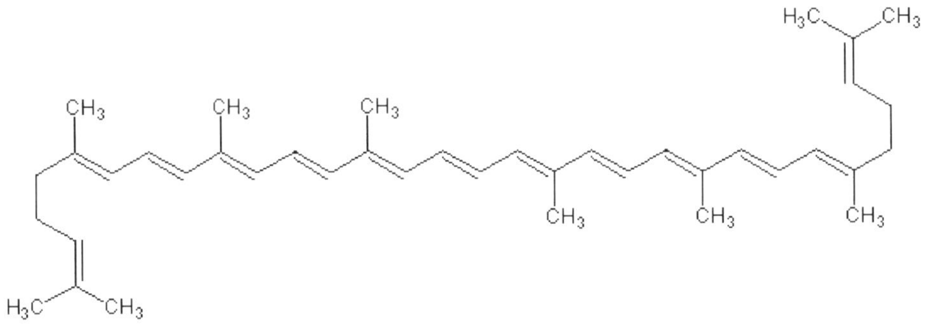

It has a similar structure to the other carotenes. It is a powerful antioxidant and free radical scavenger.

Monascin

The use of *Monascus* microorganisms is also a rich source of natural color and produces chemical species that give a red color. These include monascin, ankaflavin, rubropunctatin and monascorubrin, which have the following molecular skeleton. This is traditionally grown on rice in the Orient and is said to have an antibacterial effect. The color is currently available in purple and in red, but a new yellow-orange is close to being finalized commercially.

Since 800 A.D., red yeast rice has been employed by the Chinese as both a food and a medicinal agent. The therapeutic benefits as both a promoter of blood circulation and a digestive stimulant were first noted in the traditional *Chinese Pharmacopoeia, Ben Cao Gang Mu-Dan Shi Bu Yi,* during the Ming Dynasty (1368–1644).

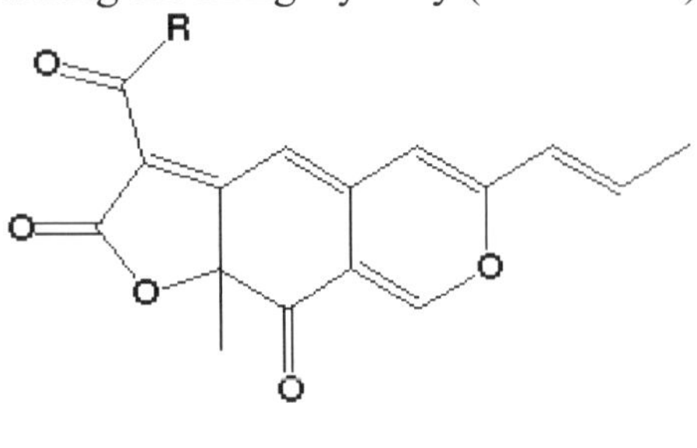

R=C_7H_{15} **Monascorubrin**

Chinese traditional medicine practitioners utilize red yeast rice to treat abdominal pain due to stagnant blood and dysentery, as well as external and internal trauma. It is known as Hung-chu or Hong-Qu in Chinese.

Monascus purpureus, monascus

Results suggest that pigments from *Monascus purpureus* could be an acceptable substitute for traditional colorants (tests performed in strawberry yoghurt). While the color does change slightly during various production stages, the change is comparable to that of FD&C Red#40 and is less than the changes found using carmine or beet juice powder.

Morin

The heartwood of *Artocarpus heterophyllus* (Jackfruit) and the heartwood of *Maclura pomifera* tree contain morin. The dye from *Alnus jorullensis* (Mexican Alder) contained important amounts of luteolin, quercetin, and morin, making it rather unusual in the family of the yellow flavonoid dyes.

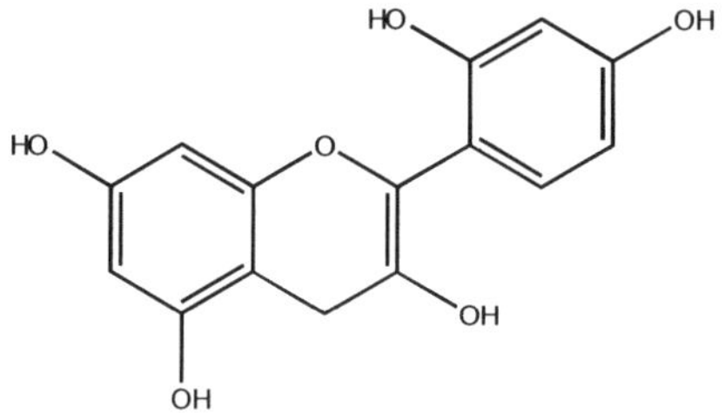

Morus alba, mulberry

Morus alba L. has been used for inflammation, as an astringent, emollient in common folk remedies and medicine. The main ingredients of the Mulberry (*Morus alba*) root extract are flavonoids (mulberrin, morin, mulberrochromen), triterpenoids (α-,β-amyrin, cytosterol) and acids (palmitic acid, stearic acid). Mulberry Root Extract shows high tyrosinase inhibition effect and high absorption of UV at region A, B, which consequently exhibits an excellent whitening effect for the skin.

Myrtillin

Hibiscus sabdariffa,

The swollen red calyces of *Hibiscus sabdariffa* (often referred to as the hibiscus flower itself), which form the outer covering of the flower buds, are dried and used to make a rosy citrus flavored tea. *Hibiscus sabdariffa* is a member of the Malvaceae family. The water-soluble anthocyanins malvin (malvidin-3,5-glucoside) and myrtillin (different delphinidin 3-glucosides) belong to the ingredients of Malvaceae. It is also the deep purple found in *Viola tricolor* (Heartsease).

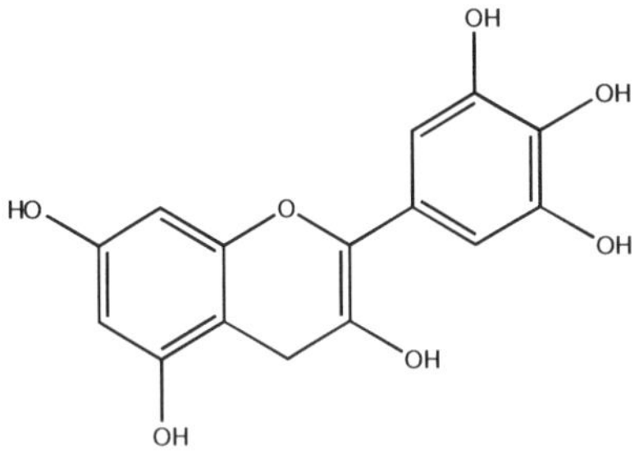

Orcein

Rocella tinctoria, archil

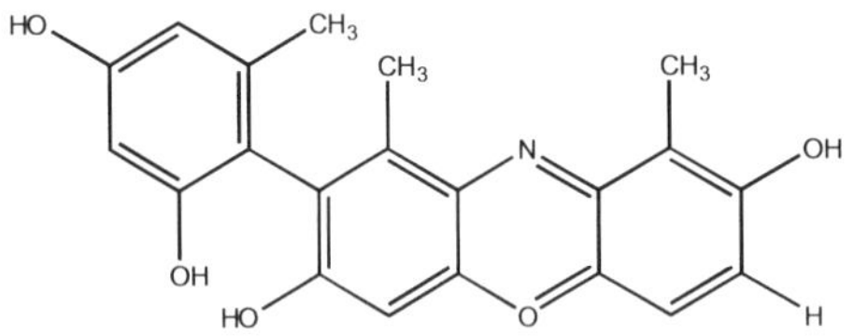

Orcein and orchil are colorings derived from archil, the lichen *Rocella tinctoria*. Orcinol is derived from the lichen and then converted to orcein (a reddish-brown dye, also used as a microscopical stain) by the action of aqueous ammonia and air. Orchil is a purple-blue dye. Orcein is a mixture of compounds with a phenoxazone structure, composed of hydroxy-orceins, amino-orceins and amino-orceinimines (see molecular diagrams).

Pelargonidin

Pelargonidin is found in *Pelargonia* spp., especially red pelargoniums.

Raphanus sativus, radish

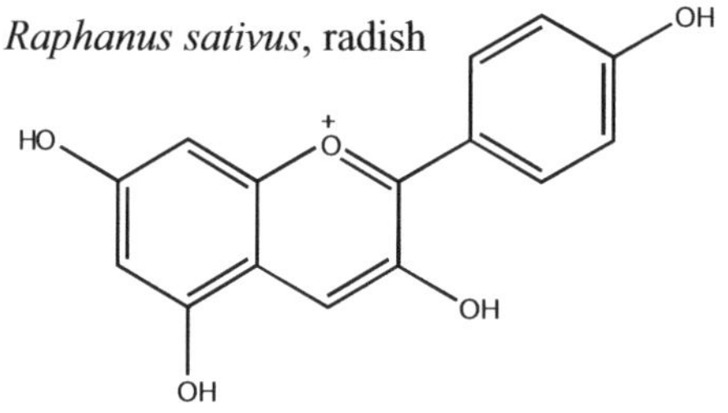

HO OH OH OH O^+

Pelargonidin-3-sophoroside-5-glucoside, mono- or di-acylated with cinnamic and malonic acids

Phycocyanobilin

Phycocyanobilin is a blue phycobilin, i.e. a tetrapyrrole chromophore found in cyanobacteria and in the chloroplasts of red algae, glaucophytes and some cryptomonads. It is also extracted from blue algae, often from spirulina. It is a color shade similar to the blue gardenia. The chemical species responsible for this color is one of the phycocyanobilin molecules, one of the structures of which is shown below.

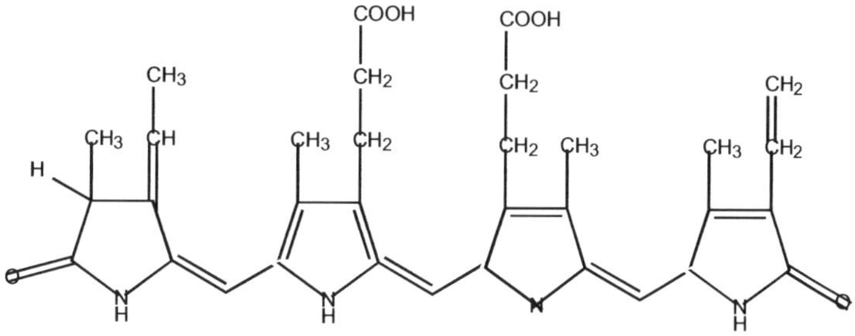

COOH COOH CH3 CH2 CH2 CH2 CH3 CH CH3 CH2 CH2 CH3 CH3 CH2 H N H N H N N H

It is a rich blue color that is obtained from the fermented leaves of the plant *Isatis* spp (probably better known to most as woad). It was used for many years for dying fabrics and has been used in conjunction with herbs for coloring use. This has identical structure to the indigotin found in indigo. There are similar structures called indirubinoids, specifically 6,6'-dibromoindigotin, which can be obtained from whelks (*Murex trunculus*, *Murex brandaris* and *Thais haemastoma*), normally common to the Mediterranean region. The color is a regal purple.

Phycobiliproteins

Phycobiliproteins are red or blue pigments that are found in Rhodophyta, Cyanophyta and Cryptophyta algae. They are built up of bilins, which are open-chain tetrapyrroles classified as blue phycocyanins, red phycoerythrins and pale blue allophycocyanins.

Polyporic acid

Polyporic acid is a dark violet color obtained from *Polyporus vidulans* which is a parasitic mushroom found on oak trees. It has also been isolated from *Lopharia papyracea* (a fungus) and from the lichen *Sticta coronata*. It is a toxic constituent of the mushroom *Hapalopilus rutilans*.

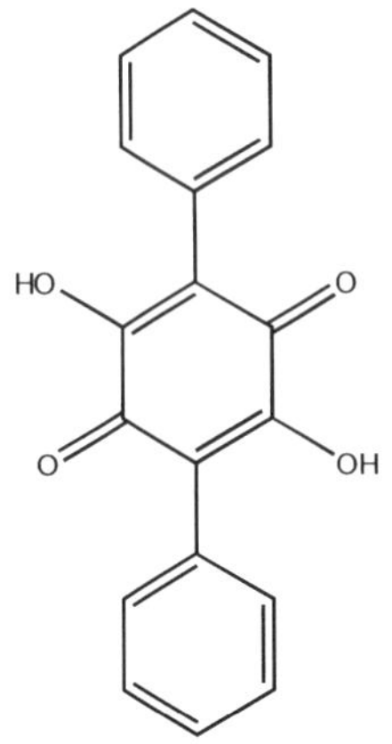

HO O O OH

Pratol

Trifolium pratense, clover

From clover or *Trifolium pratense* one can obtain a natural colorant called pratol (7-hydroxy-4'-methoxyflavone) which is a dull, golden yellow. There are a number of flavonoids that can be used from plant sources. Clover has been traditionally used for eczematous skin conditions, especially where the skin is pruritic. It is also useful for boils and pimples. A lotion of red clover used externally can give relief from itching in skin disorders, acne, boils and similar eruptions, as well as eczema and skin problems with irritation.

Melilotus officinalis, Melilot

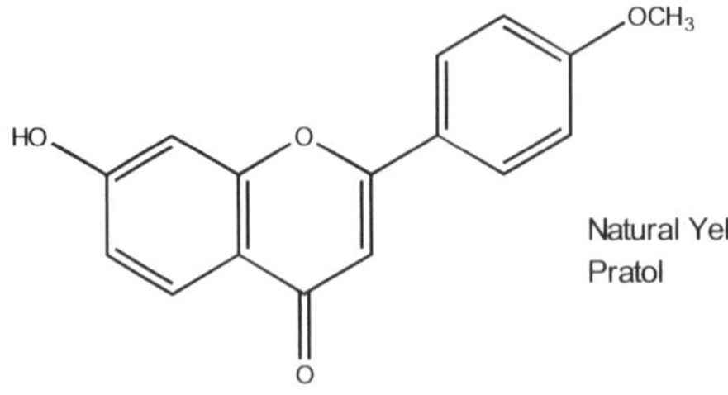

OCH_3 HO O O

Natural Yellow 10
Pratol

According to another source it is also found in Yellow Sweet Clover or Melilot (*Melilotus officinalis*), where flower heads are used externally in the treatment of ulcers, burns, sores and skin complaints. Used internally to treat chronic skin conditions such as psoriasis and eczema.

Purpurin

Rubia tinctorium, Madder

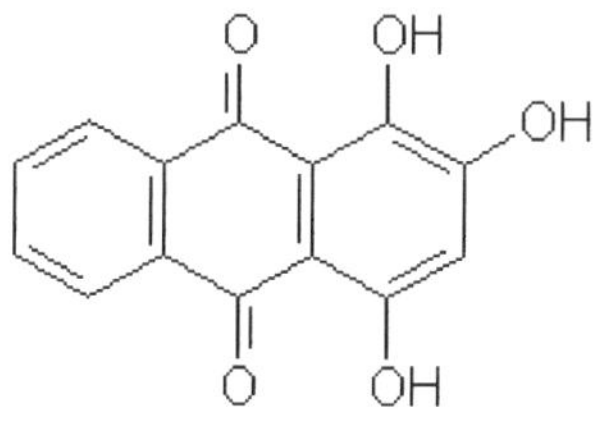

Purpurin

Quercetin

Quercetin is widely distributed in the plant kingdom and is the most abundant of the flavonoid molecules. It is found in many foods, including apple, onion, various berries, hops, oak bark, tea, horse chestnuts and brassica-type vegetables. It is also found in many seeds, nuts, flowers, barks, and leaves. The medicinal botanicals include *Ginkgo biloba*, *Hypericum perforatum* (St. John's Wort), *Sambucus canadensis* (Elder), and many others. The inner bark of *Quercus velutina* (oak tree) contains quercetin. It is a tetrahydroxyflavonol and is an orange-brown color.

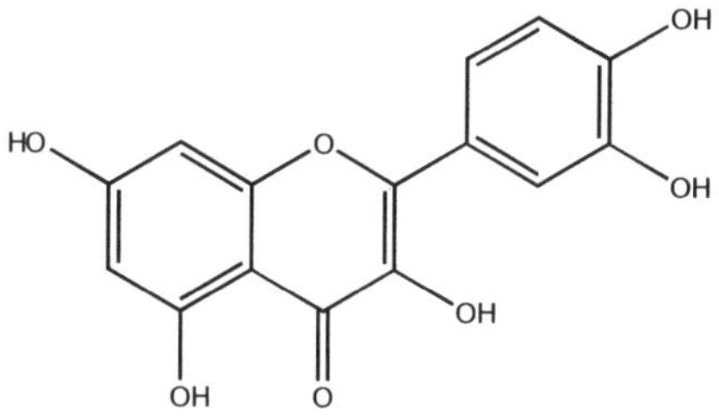

Quercetin has anti-inflammatory activity that appears to be because of its antioxidant and inhibitory effects on inflammation-producing enzymes such as cyclooxygenase and lipoxygenase with the subsequent inhibition of inflammatory mediators, including leukotrienes and prostaglandins. The inhibition of histamine release by mast cells and basophils also contributes to quercetin's anti-inflammatory activity. Thus quercetin is indicated in any inflammatory condition.

Rhodoxanthin

Rhodoxanthin is a xanthophyll (see E161a) found naturally in yew tree seeds. It is yellow in color. Not commercially available.

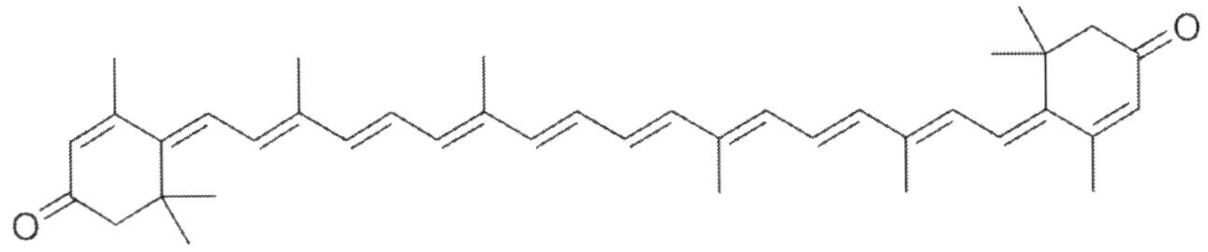

Rhodoxanthin is an orange-yellow and found in autumn leaves and also in the seeds of the yew tree (*Taxus baccata*) as well as in the feathers of the downy-legged pigeon

Riboflavin

Riboflavin is an essential dietary requirement, as it aids in the metabolism of fats, carbohydrates and proteins. It is also needed for other functions including red blood cell formation, respiration, antibody production and general well-being. Activation of vitamin B6 and folic acid require riboflavin.

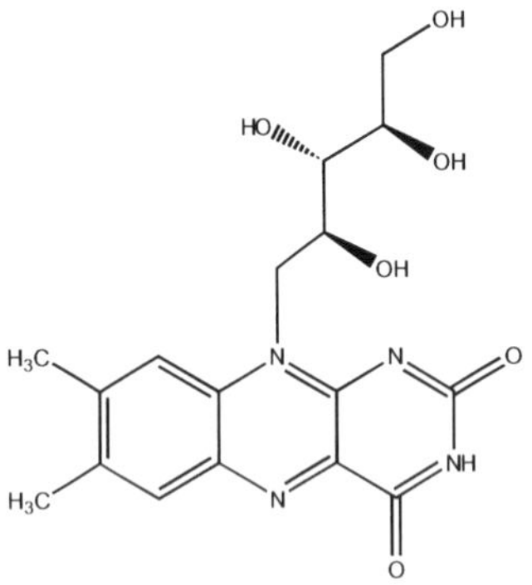

Riboflavin has been found to aid in the treatment of eye disorders, e.g. cataracts. It is found naturally in liver, kidneys, eggs, milk but is destroyed upon exposure to light. It is manufactured industrially using yeast or other fermenting organisms, used as a yellow coloring and as vitamin fortification, but is difficult to incorporate into most foods due to poor solubility.

Rottlerin

Light reddish-brown plates or needles with golden luster from ethyl acetate, also reported as brownish-yellow plates from toluene. It is practically insoluble in water.

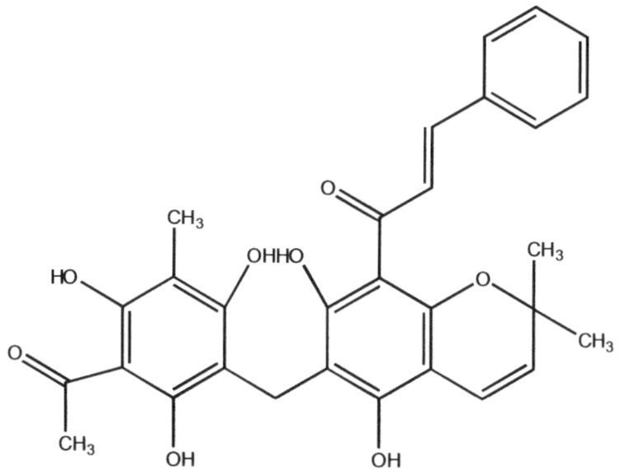

Mallotus philippinensis, kamala

Kamala is largely used in India externally for cutaneous troubles, and is effective for scabies.

The principal phenolic component of kamala is an anthelmintic dye obtained from *Mallotus philippinensis* (Lam.) Muell. Arg. (also known as *Rottlera tinctoria* Roxb.), Euphorbiaceae. The root of the tree is used in dyeing, and for cutaneous eruptions, also used in solution to remove freckles and pustules. It has activity as protein kinase inhibitor.

Rubixanthin

Rubixanthin is a xanthophyll (often described as E161a), that provides a natural yellow color in foods consumed as part of a normal diet, however it is not commercially available. This widely used description is likely wrong and E161a should be described as Flavoxanthin and Rubixanthin should be referred to by the E number E161d.

Rosa canina, dog rose

Rubixanthin, 3-hydroxy-γ-carotene or natural yellow 27, is also described as being the red-orange color found in fairly large amounts in the anthers and styles of dog rose *Rosa canina* and is insoluble in water. It is likely that γ-carotene was the precursor of this pigment. As a food additive it used under the E number E161d as a food coloring. It seems to be a very rare material in the wild and rubixanthin seems to be restricted to *Rosa* spp. Although it has been identified in several other plants, namely, *Cuscuta salina*, *C. subinclusa* and *Gazania rigens*.

Sanguinarine

Sanguinarine is a quaternary ammonium salt from the group of benzylisoquinoline alkaloids. It is extracted from some plants, including *Sanguinaria canadensis* (bloodroot), *Argemone mexicana* (Mexican prickly poppy), *Chelidonium majus* and *Macleaya cordata*. It is also found in the root, stem and leaves of the opium poppy.

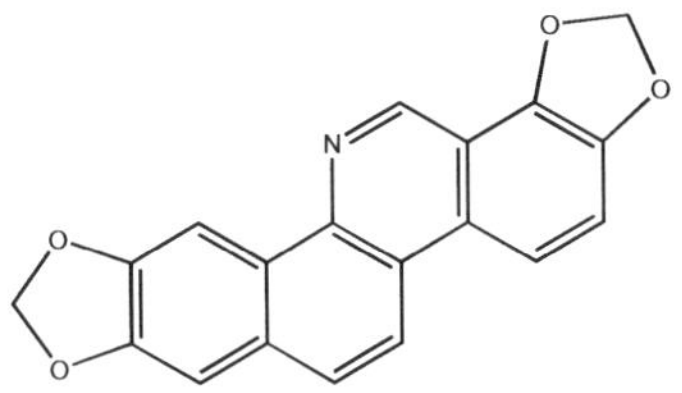

If applied to the skin, sanguinarine kills cells and may destroy tissue. In turn, the bleeding wound may produce a massive scab, called an eschar. For this reason, sanguinarine is termed an escharotic.

Sanguinaria Canadensis, bloodroot

It is also known as Red Puccoon, Indian Plant, Tetterwort, Sanguinaria. The dried rhizome is used. Sanguinarine, an alkaloid extracted from *Sanguinaria canadensis*, is used as an anti-plaque agent in toothpaste and mouthwash preparations. Sanguinarine is effective in vitro against a number of common oral bacteria, some of which are considered to play an important role in plaque formation. Bloodroot extracts have been shown to reduce glycolisis activity in saliva. Sanguinaria has been classified by the U.S. Food and Drug Administration as a herb that is unsafe for use in foods, beverages or drugs.

Santal

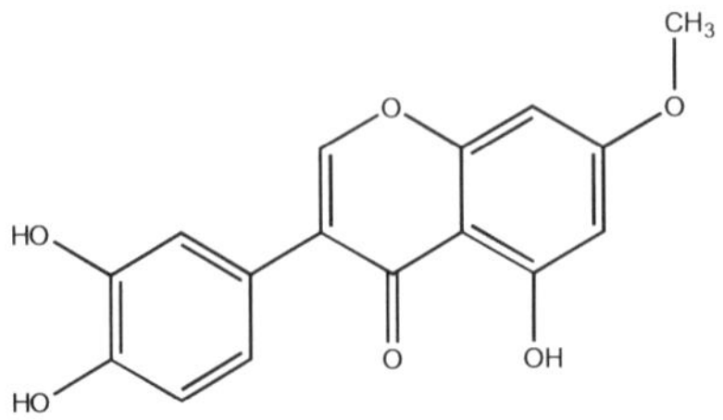

Santalin

Pterocarpus santalinum, red sandalwood

The red obtained from *Pterocarpus santalinum* or red sandalwood is a complex molecule known as santalin. There are a number of forms of this basic structure, which all give quite intense red colors. The stability of this red is quite good compared to the others. It has been traditionally used for many centuries.

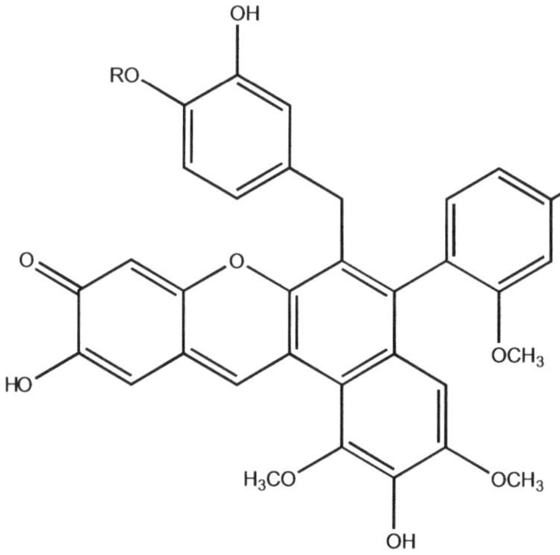

Santalin A R = H
Santalin B R = CH_3

The red pigments are called santalin A (9,10,12-tri-o-methylsantalin) and B (9,10,12,4'-tetra-o-methylsantalin), while the yellow pigments are santalin Y and AC. Santalin A and B are soluble in organic solvents and alkalis but not in water. A yellow isoflavone pigment, santal (see santal) is also present. The properties are said to be astringent, cooling and tonic, with possible benefits as an anti-inflammatory, anti-allergic, and for skin diseases (traditional use).

Sappanin

Caesalpinia sappan, sappan wood

Sappan wood or East Indian red wood (*Caesalpinia sappan*) is a multipurpose tree used primarily as a source of medicine and dye. It was first called brezel wood in Europe.

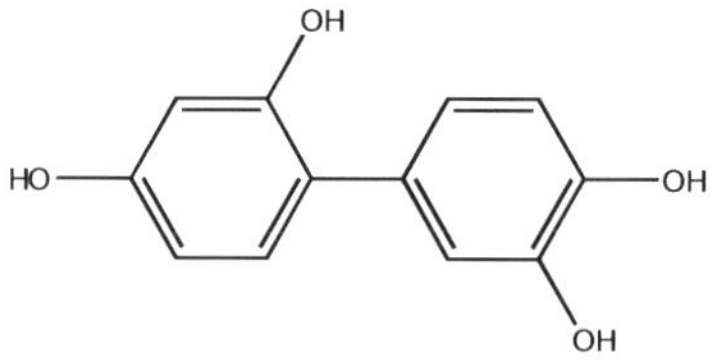

The plant possesses medicinal properties as an antibacterial and for its anticoagulant properties. It also produces a valued type of reddish dye called brazilin, used for dyeing fabric as well as making red paints and inks. The wood is somewhat lighter in color to Brazil wood (see brazilein, brazilin) and its related species, but contains the same the active dyestuffs.

Sappanin (Figure)

Chloroform, n-butanol, methanol, and aqueous extracts of the *Caesalpinia sappan* showed antimicrobial activity against standard methicillin-sensitive *Staphylococcus aureus* (MSSA) as well as MRSA.

Spinulosin

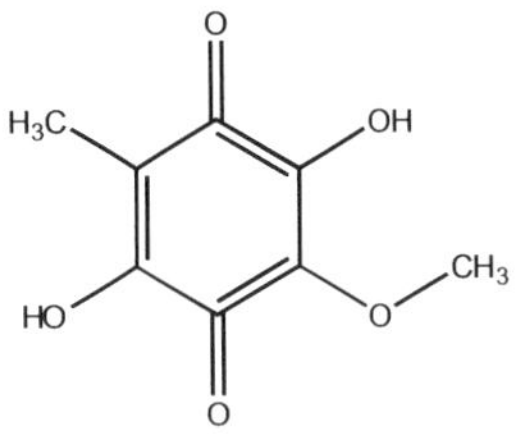

O
H_3C OH
CH_3
HO O
O

Spinulosin is a produced by *Penicillium spinulosum* and has a bronze purple color.

Violaxanthin

Violaxanthin is a xanthophyll (E161e) that provides a natural yellow to orange color, however it is not commercially available. Found in abundance in yellow pansies where it is responsible for the color of the petals.

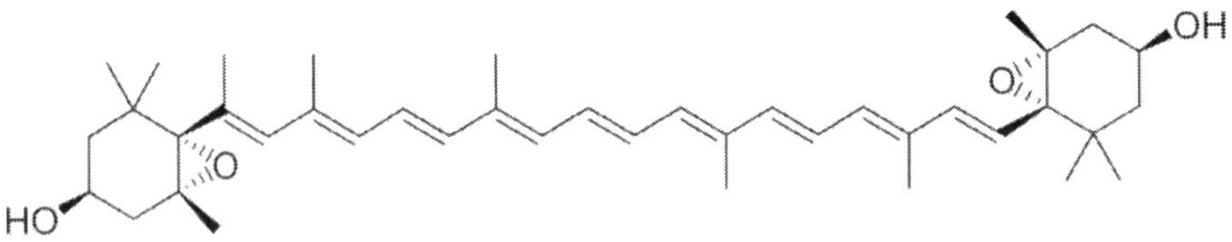

OH
O
O
HO

Zeaxanthin

Zea mays, corn (maize)

Zeaxanthin from corn (maize); *Zea mays* is insoluble in water.

Conclusion

Natural colours are not easy to work with and the environment must be free of metallic ions, traces of bleaching agents and peroxides. Many of them are colour indicators and will react strongly to pH shifts, so buffered systems and a stable pH are vital. The rewards for the hard effort of research are the benefits that these materials bring to the skin. Some are anti-inflammatory, others antioxidant and most will confer some beneficial effect on cellular function. This is the reason for using these materials as most of them are not in Annex IV of the European Directive, nor are they approved by the FDA for colour use. It is just unfortunate that the favoured antioxidant is bright orange or the skin soothing additive is blue!

Chapter 14
Marine Extracts and Marine Margin Plants

The growth in thalassotherapy and related spa treatements makes this topic interesting and offers huge scope to the marketer and formulator alike.

A

Acantophora orientalis. It is one of the red algae in the group described as Rhodophyta.

Alaria esculenta. Dabberlocks or Edible Kelp. Commercially as Wakame. It is found on exposed shores, attached to the rocks at the lowest tidal level and below into shallow water. It is common in the North Sea and the northern Atlantic. Note the yellowish-olive stalk continues as the midrib of the long, thin, fragile, yellow-green frond, with wavy edges. This midrib is what distinguishes it from the *Laminaria* species, which it resembles. The holdfast consists of spreading, rootlike segments. The main stalk has many, short-stalked, swollen-ended outgrowths in which the spores are produced. It is known commercially as Wakame, the fresh fronds and midribs are eaten in soups or salads or used as a thickener in many parts of the world. Kalpariane (ex Biotech Marine) is described as an energy and wrinkle filler that acts to protect against the weather (wind, sun, cold) and skin stresses (air conditioning, cigarettes smoke, pollution, etc). It is a balanced and rich composition for skin nutrition, protection and suppleness, skin moisturizing that stimulates hyaluronic acid and prevents stretch marks.

Alsidum helminthocorton. [Syn. *Fucus helminthocorton*] Known as Corsican Moss. It is found in the North Atlantic and growing on the Mediterranean coast, especially on the island of Corsica. It grows in tangled tufts of slender, brownish white, cylindrical threads. It has a cartilaginous, tufted, entangled frond, with branches marked indistinctly with transverse streaks. The lower part is dirty-yellow, the branches more or less purple. The taste is saline, the odor, that of seaweed. The whole plant is used medicinally as an anthelmintic and vermifuge that acts very powerfully on lumbricoid intestinal worms. The dose is 10–60g taken with honey, treacle, syrup or made up in infusion. Supplier: Alban Muller as Corsican Moss HS.

Anatheca dentata. Dentate Anathec. This is described as being an emollient, hair conditioner and moiosturiser with tonic properties. Greensea.

Armeria maritima. Thrift or Sea Pink. Traditional use: Used for its antibiotic action, antiobesic. It cannot be employed as an antiseptic poultice as it may cause dermatitis or local irritation. Rarely used even in folk medicine. Folklore: Sea Pink. Nervine, tonic, as a tea to alleviate mental depression. A standard brew of the flowers is used. It is described as antimicrobial, purifying, moisturising and regenerative. Supplier: A&E Connock Cosflor Sea pink, Cosflor Thrift. Greensea—Marsh Daisy.

Artemia salina. One of the Oleazooplancton and contains polyunsaturated fatty acids. It has regenerating and hydrating action. There are commercial sources of the extract of the various zooplankton. Suppliers: Vincience as GP4G and also Inductylor.

Artemisia maritime. Sea Wormwood. This is described as having antimicrobial and tonic properties. Supplier: Greensea.

Ascophylum nodosum. Also known as Tangle or Knotted Wrack. It grows in abundance on the west-facing coasts of Europe that are exposed to the Gulf Stream, as well as in North Temperate zones on both the Atlantic and Pacific coasts of North America. It is found attached to rocks and boulders at the upper and middle tidal levels and can be very abundant on sheltered, rocky shores and estuaries. It has disc-shaped holdfast from which short, tough, dark-green stalks grow, developing into long, flat, leathery branches that fork repeatedly and produce egg-shaped air bladders. These are tough and not easily "popped." Minerals from sea water are concentrated in the body of the plant. It is usually taken for its iodine content. It is also harvested in the Norwegian fjords. This seaweed had a long history of use in cosmetics, dating back to Roman times when it was used as a coloring agent. Knotted wrack contains algin and is widely used as an emulsifying or thickening agent in soups, puddings and jellies. A brown alga is part of wrack found on European rocky coasts. Formerly, in Brittany, women from the seashore used to rinse their hair with its decoction. As a result, their hair was supple and shiny thanks to the sheathing and revitalizing properties of the alga. Suppliers: Bottger: ACP 941. Active Organics: Actiphyte of Ascophyllum. Atrium: Aldavine, Homeo-Age, Homeo-Soothe. Algea: Algea Esthe. Codif: Algowhite. Alban Muller: Ascophyllum HS. Gelyma: Gelyol AN 55. Codif: Pheofiltrat Ascophyllum HG. Laboratoires Sérobiologiques: Phytofirm LS, Vitaplex LS. Provital: Polyplant Marine, Polyplant Oily Seaweed, Polyplant Seaweed. E.U.K.: Sea Extract Ascophyllum, Grau: Sea-Weed Extract HS 2449. Greensea : Ascophyllum.

B

Blendingia minima. [Syn. *Enteromorpha minima*] Found on rocks, timber and other algae at the upper tidal level. It is frequently found in the North Sea, English Channel and the Atlantic. It has soft, delicate, green fronds that may be simple or slightly branched and sometimes inflated.

C

Cakile maritima. Sea Rocket. Traditional use: Can replace cod liver oil. Use for scrofula, lymphatics, after malarial attack. The youngest and most tender shoots of Sea Rocket are used as a flavoring. In Canada, the root is pounded, mixed with flour and eaten when there is a scarcity of bread. The leaves are fleshy and crisp and have dietetic properties similar to watercress. In skin care, it might be used more for the marketing name than the benefits. Suppliers: Active Concepts, Solabia.

Calliblepharis ciliata. This seaweed is found attached to rocks in pools at the middle tidal level and below into shallow water, and washed ashore in the spring. It is frequently found in the English Channel and the Atlantic, south from Ireland. It has a holdfast comprising stout, branched rootlets and a short stalk that widens into a flat, bladelike, dark red, pointed frond that sometimes divides in larger specimens. Suppliers: Codif: Extrait de Calliblepharis ciliate.

Callophyllis lacinata. Found attached to rocks, stones or *Laminaria* seaweeds in shallow water and is also washed ashore. It is found in the North Sea, English Channel and Atlantic. It has fairly thick, flat, forked, crimson fronds that broaden out from a very short stalk. In summer, reproductive bodies form in tiny growths at the frond's margin or over the whole surface.

Caulerpa peltata. This species has not been exploited (as far as we know) for use in cosmetics at the present time, but would be expected to be sun protecting.

Caulerpa racemosa. It is a green algae –Chlorphyta -that is sometimes called Sea Grape. It is a skin conditioning agent, humectant, sun protecting on the hair and an antioxidant. Supplier: Greensea - Sea Grape, Noevir - Umibudou Ekisu.

Chlorella vulgaris. Chlorella, This isdescribed as being an antioxidant, moisturising and antioxidant. Greensea. Chlorella.

Chlorella sorokiniana. This is another species of chlorella that is described as beingantioxidant, regenerative and with tonic properties. Greensea - *Chlorella sorokiniana*.

Chondrus crispus. Irish Moss, Carrageen Moss, Carragheen Moss, Pearl Moss that is found in European and North American coasts. It is not a moss but a small procumbent seaweed with fan-shaped fronds. It is found attached to rocks and stones at the middle tidal level and below into shallow water. It is rare in the Baltic, common in the North Sea, English Channel and Atlantic. Note that this seaweed can vary in form and color, but usually has either a distinct flat stalk or else a very short stalk with wide, flat, wedged-shaped fronds. These are generally purple-red, with lighter, iridescent tips, but turn green in sunlight. The margin of the frond is never rolled inward and this is the way to distinguish it. *Gigartina stellata* (see below). Irish Moss is often described as the dried thallus of two red algae, *C. crispus* and *G. mamillosa* growing on the shores of the North Atlantic. The whole plant is used. The alga is green to purple when fresh but is bleached by watering and exposing to the sun. Chondrus contains 55-80% of a mixture of polysaccharides, consisting of carbohydrate ethereal sulfate occurring as sodium, potassium and calcium salts. On hydrolysis 31-33% of galactose and smaller amounts of glucose and fructuronic acid are produced. In addition, the alga contains 10% of protein, carrageenan giving the seaweed the properties of being relaxing, antiviral, reduces tightening, protects the protein macromolecules (e.g. collagen, elastin and keratin). It is also high in nutrients and trace elements, carrageen is one of the most widely used edible seaweeds. There are three types of carrageenan in *C. crispus*, identified as kappa-, lambda- and iota-, forms are extracted. It is a nourishing food and fresh or fried, it is used as a vegetable in soups, stews or as a thickener in mousses, jellies and desserts. It is demulcent, antitussive and emollient. Topically, it is used as a lotion in chapped and roughened hands and in dermatitis. It is also demulcent, pectoral and nutritious. Used in chronic coughs, bronchitis etc. It is used extensively as an emulsifier and stabilizer in toothpaste, as a binder in tablets and other powdered products. It may also be used as a thickener in hand lotions and creams. In the past it was used as a fixative in hair setting lotions. Supplier: A&E Connock: AEC Carrageenan Gel. CP Kelco: Genugel, Genu-1. Carrageenan. There are numerous mixtures that contain *Chondrus crispus*. Laboratoires Sérobiologique: Seanamine BD, Seanamine SU, Phytofirm LS, Vitaplex LS. Other suppliers would include Active Organics, BiotechMarine (Gelaig, Phycol CC), Cosmetochem, Crodarom, Grau. Alban Muller (Irish Moss), Bertin, Greentech, E.U.K., Vege-Tech etc.

Chorda filum. Sea Lace. Is found attached to rocks and stones in shallow, gravel-bottomed water down to 20m. It is found in the Baltic, North Sea, English Channel and Atlantic. It has long, whiplike stems, unbranched and slimy, that grow from a small disc-shaped holdfast. Young plants are covered in fine hairs, while the adult stems are hollow, tough and filled with air.

Chroococcidiopsis thermalis. This is described as being able to coat the hair and give a restructuring effect while at the same time it provides a sun protection effect. Greensea.

Chylocladia verticillata. Ii is found during late spring and summer attached to stones or rocks in pools at the middle tidal level and below. It is frequent in the North Sea, English Channel and the Atlantic. It is pink to mauve in color, but it turns yellow when exposed to the sun.

Cistus monspeliensis. Montpellier Rockrose. This marginal plant encourages the release of beta endorphins, is anti-inflammatory and has an anti-stinging property to protect the skin. Supplier: Codif - ArEAUmat Cistacea.

Cladophora fascicularis. Biosorption is an effective method to remove heavy metals from waste water. In many treatment plants reeds are used (*Phragmites* spp) to help clarify and purify water. They do this by absorbing water into the roots and absorbing the heavy metals to produce water that is more potable. In salt water and on the fringes of where fresh water enters the sea there exist seaweeds that perform a similar task. *Cladophora fascicularis* has been investigated as a function of time, initial pH, initial Pb(II) concentrations, temperature and co-existing ions. At a time when raw material suppliers are looking for new ideas, this seaweed might have benefit as a solution to skin pollution and for use in products that detox the skin.

Cladophora rupestris. Found at all times of the year, growing on rocks and beneath large, brown seaweeds at the middle tidal level and below into deep water. Several dense layers of cellulose and a chitin-like material protect the alga from drying out. It is frequently found living under *Fucus* where it remains moist at low tide. The alga is dark because of a wide range of pigments that can absorb enough light even in the shade of the large brown algae. Most green algae live in the upper shore, but this species can survive at almost any level, in the middle down to the lower shore where it is among the kelp. It is common in the Baltic, North Sea, English Channel and Atlantic. It has densely tufted, dark green fronds with numerous, erect branches that look and feel rather harsh and wiry. It might be of benefit as a moisturizer for dry and parched skin.

Cladophora glomerata. Meekong Weed. In Laos, *Cladophora* species are commonly eaten as a delicacy and they are usually known in English as "Mekong weed." The algae grow on underwater rocks and thrive in clear spots of water in the Mekong River basin. They are harvested 1 to 5 months a year and most often eaten in dry sheets much like Japanese *nori*, though much cruder in their appearance. Mekong weed can also be eaten raw, in soups, or cooked—as in a Lao preparation called *mók kháy*.

Cladosiphon okamuranus. Mozuku is a type of edible seaweed naturally found in Okinawa, Japan. Most of the mozuku is farmed by locals, and sold to processing factories. The main use of mozuku is as food, and as source of one type of sulfated polysaccharide called Fucoidan, which is used in cancer treatment aid as health supplements. It is also classified as a skin conditioning agent. Supplier: Yaizu Suisankagaku—Fucoidan YSK.

Codium fragile. A dark green siphonous alga that lives in the intertidal zone. It is a fuzzy-looking patch of tubular "fingers" that hang down from the rocks when the tide is out, giving it the common name Dead Man's Fingers. It is also called Chonggak or Miru, and is widely eaten either fresh or dried in Korea and Japan.

Codium tomentosum. Found on mud, sand, rocks or in deep pools at the middle tidal level and below to 20m. It is common in the North Sea, English Channel, Atlantic and Mediterranean. It has cylindrical, dark, yellowish-green, felt-textured fronds, with dichotomous branching, and the spongy holdfast that comprises a mass of closely woven filaments. Supplier: BiotechMarine—Codiavelane, Hydra Concept

Colpomenia sinuosa. It has been found to have antioxidant activity. Six fatty acids were found in the plant—methyl myristate, methyl palmitate, methyl behenate, methyl lignocerate, methyl oleate and methyl tetradecatrienoate, a sterol, stigma-5,23-dien-3β-ol and a metabolite, *tris* β-butyrolactone have been isolated. These last two materials have not been identified in seaweed before. They may have potential in restructuring skin care.

Corallina officinalis. This seaweed is found attached to roots just below the surface of rock pools in shady places at the middle tidal level. It is rare in the Baltic, but common in the North Sea, English Channel, Atlantic and Mediterranean Seas. Supplier: BiotechMarine—Phycocorail, Phyco Algae 100. Codife—Concentre Coralline. Yves Rocher.

Crithmum maritimum. Crest Marine, Sea Fennel, Samphire. Traditional use: Antiobesity agent, antiscorbutic, diuretic. It was an effective protection against scurvy. It may be taken in infusion and an item of food, often pickled. Modern examination has shown skin benefits. Folklore: Gerard described its cultivation in 1598 in England and Quintyne described it in France in 1690. It was dedicated to St Peter (whose name in Greek signifies a rock), the patron saint of all fishermen. Since the plant chose to grow only on the rocks of the seashore, it was called St Peter's herb, the herb of Saint-Pierre or Samphire. It was cried in London streets as crest marina. *C. maritimum* contains dillapiol, eugenol, crithmeme. It has a revitalizing action, energizing action on the cells. It has a bacteriostatic, anti-inflammatory action and has been used to cleanse and purify the epidermis. Supplier: Active Organics, Codif—ArEAUmat Samphira, A&E Connock, Solabia, BiotechMarine, Alban Muller, Greentech, Greensea, Provital, E.U.K., Vege-Tech.

Cystoclonium purpureum. This seaweed is found attached to rocks and other seaweeds in pools at the lower tidal level and below into shallow water. It is common in the Baltic, North Sea, English Channel and the Atlantic, south from Ireland. It has a rootlike holdfast from which grows a thick, round stem about twice the diameter of the numerous branches that divide from it then subdivide. All the branches narrow at their bases and taper to fine points at their tips. The whole plant is a dull, purplish-red that looks pinker in water.

D

Delesseria sanguinea. It is found attached to rocks or *Laminaria* seaweeds in deep, shady pools at the lower tidal level and below into shallow water. It is quite rare in the Baltic, but common in the North Sea, English Channel and Atlantic. It possesses crimson, wavy edged, leaflike fronds with clearly defined midribs and conspicuous pairs of veins, it grows in up to 20m depths and is harvested only by diving. The "leaves" are attached to branched, rounded stalks growing from a small holdfast. In winter the fronds disintegrate and small spore outgrowths and leaflets develop along the midrib. Supplier: Codif - Rhodofiltrat Delessaria.

Desmarestia aculeate. Attached to rocks or stones in pools at the lower tidal level and below into shallow water; sometimes washed ashore. It is uncommon in the English Channel and north Atlantic. It has a strong main stem bearing numerous alternately arranged side branches that are bare in winter but covered with delicate little tufts in summer. It is bright green in color, turning brown as the plant ages.

Dictyopteris membranacea. Sea Fern. is a brown alga of the European Atlantic coasts belonging to the Dictyoptales family. In order to provide a sustainable source the algae collected in the sea are placed in hatcheries where they produce the spores which ensure their perpetual reproduction. The small algae are later transferred to the open sea, grown on a longline system and collected for extraction after a year. The aromatic note of *Dictyopteris membranacea* has a very characteristic odour due to the presence of dictyopterene pheromones which have been utilized in perfumerywhere it is known as Parfum d'Anthée. Dictyopteris Oil encourages an adipocyte nourishing action that increases their volume and so results in lips that are replumped, have fewer lines and roughness. Dictyopteris Oil also has a volume-enhancing effect on the breasts. Supplier: Codif.

Dunaliella salina. This algae has the properties to protect the hair and is also a moisturiser, with nourishing and regenerative properties. Greensea,

Durvillea antarctica, Bull Kelp. It is described as moisturising, purifying and soothing, Supplier: Greensea

E

Ecklonia maxima. Sea Bamboo is a species of kelp native to the southern oceans. It is most typically found along the southern Atlantic coast of Africa, from the very south of South Africa north to Namibia. In these areas the species dominates the shallow (up to 8 m) temperate water kelp forests offshore. From the holdfast attached to a rock or the large holdfast of another kelp, a single long stipe rises to the surface waters, where a single large pneumatocyst holds a tangle of blades at the surface.

Ecklonia radiate. A species of kelp found in the Canary Islands, the Cape Verde Islands, Madagascar, Mauritania, Senegal, South Africa, Oman, southern Australia, Lord Howe Island and New Zealand. It is described as a hair and skin conditioning agent. Supplier: Southern Cross Botanicals - Australian Kelp.

Eisenia arborea. Southern Sea Palm. This is described as being regenerative, a seboregulator and veinotonic. Supplier: Greensea.

Enteromorpha intestinalis. Nori or Tiger Moss. It grows on gravel and rocky bottoms. At low tide, a part of them tend to bend and float on the surface. It is quite common in rock pools, especially those that have a high position, 5–10 m above sea level, and get splashed during storms. It is pale green in color. The shoots that grow from a common root are tubular, often gas-filled, intestinelike and devoid of branching. It is described as beingoxygenating, nourishing and moisturizing. Supplier: Greensea - Gut Weed.

Enteromorpha compessa. Aonori. From very ancient times *Enteromorpha* (green edible algae) have been used in powder form as a seasoning in China. This green alga grows in temperate and cold seas. It contains amylopectine-like proteins and has a regenerating action, is sebostatic and anti-inflammatory. Supplier: Provital—Homeostatine, Pronalen Sport. Atrium - Homeoxy. BiotechMarine - Phycol EC. E.U.K. - Sea extract AO-Nori.

Enteromorpha flexuosa. It is described as a hair conditioning agent. Supplier: Innovalg - Proenteine.

Enteromorpha intestinalis. *Nori* or Tiger Moss is found throughout spring and early summer in rocky pools at the upper tidal level, in estuaries, salt marshes and brackish waters. The dead, bleached fronds are often found on the beach in late summer and autumn. It is found in the Baltic, North Sea, English Channel, Atlantic and Mediterranean. It is known in Japan as green *Nori* and in China as Tiger Moss, and is valued as an excellent edible seaweed and used in numerous dishes.

Enteromorpha linza [Syn. *Ulva linza*]. It is found attached to stones, rocks and other seaweeds in pools from the upper to lower tidal levels. It is frequent in the Baltic, North Sea, English Channel, Atlantic and Mediterranean. It has bright green, spirally twisted, unbranched fronds, with crinkled edges, which appear flattened, but are in fact hollow.

Eryngium maritimum. Sea Holly. Traditional use: Most commonly used in uterine irritation, bladder diseases, painful micturirion with painful and ineffective attempts to empty the bladder. Folklore: The roots are fleshy and have the flavor of chestnuts. They were candied in the 16th and 17th centuries and called Kissing Comforts in allusion to their aphrodisiac properties. Supplier: A&E Connock, Cosmetochem.

Eucheuma cottonii. It has been patented as a stabilizer for milk products that was comprised of a carrageenan extract taken from *E. cottonii* seaweed. The largest producer is the Philippines, where cultivated seaweed produces about 80% of the world supply. The most commonly used are *E. cottonii* and *E. spinosum*, which together provide about three quarters of the world production. These grow at sea level down to about 2 metres. The seaweed is normally grown on nylon lines strung between bamboo floats and harvested after three months or so when each plant weighs around 1 kg. It is described as being oxygenating, film forming, moisturising and antioxidant by the supplier: Greensea.

Eucheuma spinosum. Described as a skin conditioning agent.

F

Fucus ceranoides. Found attached to rocks and stones between the upper and lower tidal levels, in brackish water, and in estuaries and land-locked bays. It is common in the North Sea, English Channel and Atlantic. It has thin, olive-green, dichotomously branching fronds with a distinctive midrib and clusters of forked, swollen tips that contain the reproductive bodies.

Fucus distichus. Common Rockweed. It is up to 10 cm long with a short stout cylindrical stipe, branching dichotomous, flat and with a mid-rib. It is quite rare in Britain but has been recorded from: Orkney Islands, Fair Isle, St Kilda and the Outer Hebrides in Scotland; in Ireland it had been recorded from Counties Clare, Donegal and Kerry. Two subspecies of *F. distichus* (subsp. anceps and subsp. edentatus) have been described from the British Isles.

Fucus laminaria. Found around the shores of the North Sea and Atlantic (particularly Brittany) and Pacific Oceans. It contains iodine and bromine compounds, mucilaginous substances and carbohydrates. It is used in slimming and massage preparations.

Fucus serratus. Toothed Wrack, Kelpware, Cutweed is found attached to rocks at the lower end of the middle tidal level. It is very common in the Baltic, North Sea, English Channel and Atlantic. It has tough, short-stalked olive-green, branching fronds with thick midribs and serrated edges. In winter the tips have flattened, fruiting bodies growing from them. Small, white, spiral tubes of the worm *Spirobis* are often attached to older specimens. *F. serratus* has no air vesicles, while *F. nodosus* has the single air vessel (not in pairs). It is a deobstruent and used for "anti-fat" products. It influences the kidney and acts as an alterative. It was used in the past for the extraction of iodine, but the process is no longer economical. It also contains a substance called algin that is a used as a dressing for calico. Supplier: BiotechMarine - Dentactive. Atrium - Homeo-Shield. Gelyma - Seavie.

Fucus spiralis. Spiral Wrack or Twisted Wrack is found attached to rocks at the upper tidal level, often forming a distinct zone of about 1 metre. It is frequently seen in the North Sea, English Channel and Atlantic. It has broad, tough leathery, olive-green fronds that partially twist and have a distinctive midrib. The paler tips of the branches are swollen and contain the reproductive bodies but these do not extend to the rim, which is sterile. The "measles" are the reproductive organs, which contain either sperm or eggs.

Fucus vesiculosis. Bladderwrack, Sea-Wrack, Black Tang, Sea Wave, Kelp, Fucus, Quercus marina, Cutweed, Bladder Fucus, Blasentang, Seetang, Meeriche. Traditional use: The skin characteristics of some of the seaweed workers' hands were much improved in respect of smoothness, luster and hydration. Calluses were reduced and skin tone was clearly better. Many cosmetics that have a basis of bladderwrack make use of its local fat-dissolving properties. Folklore: Numerous products have used this property to moisturize the skin, soften the hands and body, and produce smoothing facial masks and shampoos. Historically it was used for obesity (excellent source of iodine) and cellulites where it has decongestive and stimulating properties. It has found use as a wash for psoriasis, is good for sprains and bruises and has been employed in products to stimulate the hair and scalp. The virtues of this alga were well-known by Brittany girls who traditionally mixed it with other marine plants to make a softening beauty milk that protected their skin from the effects of suntan and sunburn. Supplier: A&E Connock, Active Concepts, Active organics, Alban Muller, Arch, Bio-Botanica, Grau, Ennagram, Gattefosse, Greensea, Greentech, Ichimaru Pharcos, Indena, Laboratoires Sérobiologiques, Phytochim, Phytocos, Prod'Hyg, Provital, Solabia, Symrise.

G

Gelidium amansii. Japanese Isinglass. It is an economically important species of red algae commonly found in the shallow coast of many East and Southeast Asian countries. Japan, is the best variety and other sources are Ceylon and Macassar. This algae is used to make agar, and is sometimes served as part of a salad in the regions that produce it. The seaweed is boiled to extract mucilage that is then dried.

Gelidium cartilaginenium. It is also known as agar-agar produced from red algae. Supplier: BiotechMarine—Rhodysterol.

Gelidium sesquipedale. Described as a skin protectant. Supplier: Gelyma—Gelyol GS45.

Gigartina acicularis. This is a species that has not been used in cosmetics but might be worthy of investigation.

Gigartina canaliculata. Small quantities are harvested in Mexico (Baja California), and is available from south of Ensenada to Punta San Antonio. It is harvested from May to September by fishermen who pull it by hand during low tide.

Gigartina chamissoi. This is a species that has not been used in cosmetics but might be worthy of investigation.

Gigartina pistillata. This is a species that has not been used in cosmetics but might be worthy of investigation.

Gigartina skottsbergii. Southern Moss. This red alga is becoming increasingly valuable as a resource to providing the raw material for the carrageenan industry established in Chile and elsewhere. It is described as being nourishing, film forming, cooling, regenerative and oxygenating. Supplier: Greensea—Southern Moss.

Gigartina stellata. False Irish Moss. Found attached to rocks and stones at the lower tidal level. It is common in the North Sea, English Channel and Atlantic. It has flat, tufted, dark red-brown fronds that fork six or seven times and are attached to a disc-shaped holdfast. These fronds have inrolled edges that form a channel. This, and the small, fruiting "pimples" that are dotted along the older specimen in summer and winter, distinguishes it from Carragheen, *Chondrus crispus*. It is edible and can be used like Carragheen to thicken soups, stews and desserts. Supplier: BiotechMarine—Hydrane BG. Vege-Tech—Vege Plex VP-1297.050WB, Vege-Tech VT-357. Greensea.

Gracilaria sp. This is a red seaweed from which polysaccharides are extracted that are known as agarose. Supplier: Biowhittaker—BMA Agarose CLE. Cambrex Bio Science—Litex Agarose HSB-LV 2500.

Gracilaria confervoides. This is a species that has not been used in cosmetics but might be worthy of investigation.

Gracilaria vermiculophylla. This is described as a skin conditioning agent. Supplier: Koei Kogyo.

Gracilaria verrucosa. This is described as a skin conditioning agent, humectant and skin protectant. Supplier: Radiant Inc.

H

Halichondria panicea. Breadcrumb Sponge.

Halidrys siliquosa. Sea-oak is found attached to rocks at the middle tidal level and below into shallow water. It is uncommon in the North Sea, English Channel and the Atlantic.
It has tough, leathery, rather flattened, olive-brown fronds with regularly alternating side branches. The tips of the branches have long, pointed, oval-chambered air bladders that are rather like seed pods.

Himanthalia elongata. Button Weed, Sea-thong or Thong-weed. Button Weed is an alga that belongs to the order of Fucales. It can be found in renewed waters of exposed coast areas. It is the second alga most harvested in France for food purposes coming just after *Laminaria digitata.* Supplier: Codif—Pheofiltrat Himanthalia, Alban Muller—Phytami Himanthalia. E.U.K.—Sea Extract Himanthalia. Greensea.—Himanthalia or Thingweed. Vege-Tech—Vege-Tech VT-339.

Himanthalia lorea. Sea-thong or Thong-weed is found attached to rocks or growing in colonies in large pools at the lower tidal level and below into shallow water. It is common in the English Channel and the Atlantic. They have olive-green, mushroom-shaped buttons that develop above the holdfast out of the center of which grow long, brown, leathery, straplike branches that are forked and hang downward. The spotty, reproductive parts of the plant develop on these in scattered pits. Sometimes the buttons will exist without the rest of the frond. As they become older, they look like tops until they hollow out on the upper surface like a funnel.

Hizika Fusi-formis. It is described as beingnourishing and regenerative. Supplier: Greensea.

Hypnea musciformis. Supplier: BiotechMarine—Biorestorer is used for hair restructuring, Hypneane described as a moisturizer. Assessa-Industria—Ormagel and Quiditat ranges. Chemyunion—Hydrahair Sphere.

I

Inula viscose, [Syn: *Dittrichia viscose*]. It is a Mediterranean perennial marginal plant rich in flavonoids that secretes a resin to protect its leaves from UV radiations, In skin care it gives daily protection and provides strong antioxidant properties, protective and repairing effect towards UVAs and UVBs and anti-elastase activity. A dye can be extracted from the roots. Supplier: Codif—Hydrofiltrat Paniopsis G.

L

Laminaria angustata. Supplier: Orient Stars—Oristar FCD. Technoble—M-034, Marinoble-AN.

Laminaria cloustini. Supplier: BiotechMarine—Phyconnexine (was shown to protect against cutaneous aging through a firming and plumping effect. The activity was through cell communication on an epidermal, dermal and hypodermal level that was restored and protected with the material. Bottger—ACP 941.

Laminaria digitata. Sea Tangle, Oarweed, Sea Girdles or Tangle. Traditional use: When applied to the skin, it is considered to have antiseptic and moisturizing properties. Folklore: Fried or dried Oarweed is high in nutrients and trace elements, and is often labelled Kombu when sold commercially (however, this is incorrect as this is the fungus that grows in a tea of seaweed. It contains alginic acid, mannitol, vitamins, minerals and iodine. When applied to the skin, it is considered to have antiseptic and moisturizing properties. It makes an excellent addition to soups and stews. Supplier: Active Organics. BiotechMarine—Aquaphycol LD, Bio-Energizer, Phycol LD. Codif—Complex Algomarin. Pheofiltrat Laminaria, Phycojuvenine. Alban Muller—Devil's Apron. Greentech. Greensea. Chemyunion: Energilium, Hidrahair AV. Provital: Pronalen Slimming, Polyplant Marine, Polyplant Slimming. E.U.K.: Sea Extract Laminaria. Laboratoires Sérobiologique—Vitaplex LS.

Laminaria groenlandica, Arctic Sea-Girdle. This is described as being nourishing, hair conditioning, moisturising and regenerative. Greensea.—Arctic Sea-Girdle.

Laminaria hyperborean. Rough-stalked Kelp. It is described as nourishing, antioxidant, purifying, cooling and veinotonic, Supplier: Crodarom—Laminaria Gel BG. BiotechMarine—Phycoboreane. Greensea—Oar-Weed.

Laminaria japonica. Supplier: Active Concepts—AC Hydrating Seaweed Complex. E.U.K.—China Extract Lung Xu Cai.

Laminaria ochroleuca. Supplier: BiotechMarine—Antileukine 6, Laminaine Marine, Phycolanine. Doosan—DS-Cerix5 Plus.

Laminaria saccharina. Sugar sea-belt, Sea Belt, Sugar Kelp, Sugarwrack, Poor Man's Weatherglass, Neptune Kelp is found in rocky pools or attached to small stones, shells or rocks on muddy or sandy flats at the lower tidal level and below to 20m. It is a brown alga that has the characteristic of emerging only at spring tides. Its name *saccharina* stems from the sugars that crystallize on the surface of the alga as it dries out. It is very common in the North Sea, English Channel and Atlantic. It has smooth, round, sandy-yellow stalks that expand into a single, long, flat undivided frond with wavy edges and crinkled undulations all over it. The holdfast comprises several layers of branching fibers at the stalk's base. The dried fronds have a whitish deposit on the surface that has a distinctly sugary taste and they are highly valued for their edibility, particularly in China and Japan. This is the seaweed used by amateur weather forecasters (hence the alternative name Poor Man's Weatherglass) as a guide to the humidity of the atmosphere. The plant contains laminarine, marine iodine, fucosterol, alginic acid. It has a slimming effect, normalizes, rebalances and re-equilibrates the epidermis, has sebum-controlling action (sebostatic action). Supplier: BiotechMarine—Laminaria Saccharina DJ, Phlorogine. Alban Muller—Phytami Devil's Apron. Bioland—Recelderm-503. Ennagram—Laminaria Saccharina Extract. Greensea.—Neptune Kelp.

Laurencia pinnatifida. Pepper Dulce is found attached to rocks and in crevices at the middle tidal level and below into shallow water. It is common in the North Sea, English Channel, Atlantic and Mediterranean. The plant has a strong distinctive smell and taste, as the name implies.

Leathesia difformis. Found growing on rocks and other seaweeds, such as *Corallina*, at the middle tidal level and below from March to the end of September. It is very common in the North Sea, English Channel and the Atlantic.

Limonium vulgare. Sea Lavender. Traditional Use: The powdered root is applied to old ulcers or made with a soothing ointment for piles. A wound herb, it encourages the rapid healing of damaged tissue. External use: Crush the flowers and apply directly to the wound. Supplier: A&E Connock—Cosflor Sea Lavender. Greensea.

Limonium narbonense. Swamp sea-lavender. Perennial marginal seaside plant from Camargue in France that grows in salt marshes and has developed a remarkable protection system to fight against free radicals caused by UV irradiation. In skin care it provides photo-protection by providing strong antioxidant activity, with a protective and repairing effect that is proven to delay skin ageing in skin exposed to UV. Supplier: Codif—Hydrofiltrat Statice G.

Lithothamnium calcarum. Maerl. This is a sedimentary red alga that lives affixed on the sea bed and can be found abundantly on Brittany coasts. Maerl is a calcareous alga consisting of mineral substance (up to 95%) and therefore looks like a rock and is often mistaken as coral. Its skeleton is mainly composed of calcium carbonate and magnesium carbonate. These two elements represent 35% of the plant's dry weight and Maerl concentrates them from the sea. Supplier: Yves Rocher—Maerl MP. Provital—Polyplant 5043, Polyplant Marine Plus. E.U.K.—Sea Extract Lithothamnium.

Lomentaria articulata. It is found attached to rocks and other seaweeds in pools at the upper tidal level and below. It is common in the North Sea, English Channel and Atlantic. The shiny, transparent plant is dull-purple to crimson red in color.

Lythophyllum incrustans. This is found covering rocks on exposed shores at the middle tidal level and below. It is common in the North Sea, English Channel, Atlantic and Mediterranean. It has a thick, chalky, rough-textured, purple to pink or mauve crust that adheres closely to the surface on which it grows. Eight other similar but hard to differentiate species are found around Britain.

M

Macrocystis integrifolia. Perennial Kelp. This seaweed is described as purifying, sun protecting, moisturizing and cooling. Supplier: Greensea.

Macrocystis pyrifera. Kelp, Giant Kelp. Traditional use: It is reported that kelp, like many seaweeds is rich in materials that provide mucilaginous and soothing effects when applied topically. Folklore: Kelps refer to species of *Laminaria* and *Macrocystis* although kelp is often used in reference to species of *Fucus*. The traditional uses of kelp in obesity and goiter are presumably attributable to the iodine content. It is mainly used for its thickening, gel-forming and stabilizing properties. Macrocystis has the distinction of being the largest seaweed in the world; the largest attached plant recorded was 65 m long and the plants are capable of growing at up to 50 cm per day. It is described as being cooling, moisturizing, firming and regenerative. Supplier: Active Concepts—AC Hydrating Seaweed Complex. Active Organics—Actiphyte of Sea Kelp. BiotechMarine—Macrocystis Pyrifera. Arch—Marine Plasma Extract III. Vege-Tech—VT-027 Extract of Kelp. Greensea—Giant Kelp.

Maris sal. Dead Sea Minerals. Traditional use: The Dead Sea area is well-known throughout history for its health benefits. The Dead Sea is very rich in minerals that provide a good source of skin conditioning. It is also known for its restorative power to cure a number of skin diseases and surgical wounds. Folklore: Dead Sea water contains 345g of mineral salts per liter and has a salinity about 10 times that of ocean water, from which it differs markedly in composition. The major salts are magnesium, sodium, potassium and calcium. The proportion of sodium to total salts, however, is much less than that of ordinary sea water. The high specific gravity gives the water a remarkable buoyancy.

Monostroma fuscum. Found in rock pools at the lower tidal level and below into shallow water. It is frequently found in the North Sea and Atlantic, north from the Irish Sea. Often mistaken for Sea Lettuce (*Ulva lactuca*).

Monostroma nitidum. Supplier: Carrubba—Aonori Extract.

O

Odonthalia dentate. This seaweed is found attached to rocks, or sometimes other seaweeds, in pools at the lower tidal level and below into shallow water or cast up onto the beach. It is common in the North Sea and the Atlantic, north from Ireland and northern England. The main stem is with solid, flattened, dark red, with irregularly branched fronds tipped by shorter, toothed sub-branches and branchlets. The fronds have a distinct peppery smell.

P

Palmaria palmata [Syn. *Rhodymenia palmata*]. Sea Parsley or Dulse. It is found on rocks and on the stalks of *Laminaria* and other seaweeds at the middle tidal level and below into shallow water. It is very common in the North Sea, English Channel and the Atlantic. It has tough bladelike, dark purplish-red, lobed and divided fronds growing directly out of the wide, disc-shaped holdfast.

Traditional use: It is cited for its anti-irritant properties. Folklore: Sea parsley is a variety of red seaweed (rhodophyta). High in vitamins, and thought to have medicinal properties, dulse has a delicious slightly salty, nutlike taste. It is described as a skin conditioning agent. Supplier: Lessonia—Palmaria Micronized. Active Organics—Actiphyte of Dulce. Bio-Botanica—Dulse Extract. Atrium—Homeoxy. Silab—Whitonyl. Ennagram, E.U.K., Alban Muller, Vege-Tech, Greensea and Codife also have extracts.

Pancratium maritimum. Sea Daffodil. It is described as firming, soothing and antioxidant. Supplier: Greensea.

Pelvetia canaliculata. Channelled Wrack is found attached to rocks at the upper tidal level, and above, where it forms a distinct belt, often only moistened by spray. It is common in the North Sea, English Channel and Atlantic. It has tough leathery, branching fronds that roll in on themselves along the margins and so enables moisture to be retained. The divided, swollen tips contain the reproductive bodies. It is olive to blackish-green when dried. The plant contains fucane, which activates the microcirculation, and is said to be a tonic to the blood supply (veinotonic). BiotechMarine—Bio-Energizer, Pelvetiane, Phycol PC. Gelyma—Efficiensea. Alban Muller—Phytami Pelvetia.

Phyllacantha fibrosa. It is a brown algae proven to significantly reduce the waistline (clinical study) at 1%. It inhibits adipocyte differentiation, stimulates lipolysis and causes a decrease in perilippins. Supplier: Codif—Pheoslim.

Phymatolithon calcareun. This seaweed is found either by itself or sometimes encrusting small stones at the bottom end of the lower tidal level and below into shallow water. It is found in the North Sea, English Channel, Atlantic and Mediterranean. It has a patch of thick, bumpy, chalky, violet-red growth that becomes erect with nodular branches as the specimen ages and so resembles red coral.

Plocamium cartilagincum, ***Plocamium coccineum***, ***Plocamium vulgare***. These are found attached to rocks or other algae in pools at the lower tidal level and below into shallow water; it is also washed ashore. It is common in the North Sea, English Channel, Atlantic and Mediterranean.

Polysiphona lanosa or ***Polysiphona fastigata***. These seaweeds are found growing on *Ascophyllum* seaweeds, occasionally on *Fucus* seaweed and very rarely on rock at the upper and middle tidal levels and in estuaries. Tiny red outgrowths on the plant are the algal parasite, *Choreocolax polysiphoniae*. It is described as a skin conditioning agent. Supplier: Gelyma—Extract of Polysiphona lanosa Sun'Ytol.

Porphyra tenera. Described as a skin conditioning agent. Supplier: Biobank—OM-X Extract.

Porphyra umbilicalis. Purple Laver, Laver or Nori is found attached to rocks or stones in sandy places at all tidal levels. Natural *Porphyra* has been a foodstuff and enjoyed by the Japanese for more than a 1,000 years. It is very common in the North Sea, English Channel, Atlantic and Mediterranean. *Porphyra* sp. occurs not only on Japanese coasts, but also in many other East Asian areas such as China coasts. It has delicate membranous fronds that grow in irregular, lobed leaflike clusters, attached at one point to a minute, disc-shaped holdfast. The rosy-mauve or purple color turns olive green as it withers, then black and brittle when dry. The plant is cleaned, boiled until tender, then cooked in oatmeal and fried in bacon fat. Laver bread is a Welsh delicacy eaten all over the British Isles. It is cultivated for consumption in Japan by placing bundles of bamboo in shallow off-shore water. Once the seaweed is established the laver-covered bamboo is transferred to brackish water estuaries where it continues to grow and then produces more lush softer fronds. They are cultivated in numerous forms and serve for making sushi or other delicacies. Described as a skin conditioning agent, moisturiser and antioxidant.. Supplier: Mibelle—Helioguard 365. Gelyma—Helionori. Alban Muller—Phytami Porphyra, Porphyra HS. Provital - Polyplant Marine. E.U.K. Sea Extract Nori. Vege-Tech—Vege-Tech VT-351. Greensea.

Porphyra yezoensis. Supplier: Lessonia—Nori Micronized.

Prasiola stipiata. It is found on rocks at the top end of the middle tidal level. It is common in the North Sea, English Channel and Atlantic.

R

Rhodymenia palmata. Edible Dulse. The thallus, fresh or dried, is used, which has a delicious flavor and is easily digestible. Hot rhodymenia lemonade is a satisfying evening drink in winter and used as a remedy for chills and colds. Supplier: New Age Botanicals.

S

Saccorhiza polyschides [Syn. *Sacchoriza bulbosa*]. Furbelows. Found attached to rocks at the lowest tidal level and below 20m. It is common in the North Sea, English Channel and the Atlantic. It has a large, flat, wavy-edged stalk, twisted at the base but rather stiff toward the upper part, which grows out of a huge, knobbly, hollow holdfast that has small attachment roots. The stalk expands into a massive fan-shaped frond, divided into numerous ribbons. This is the largest seaweed found around Britain and is most common along the south and west coasts, although it only lives for one year.

Salicornia bigelovii. Supplier: Seaphire—Seaphire Extract.

Salicornia europaea. Glasswort or Marsh Samphire. This plant is found on open sandy mud in salt marshes and estuaries, sometimes on sandy soils inland and occasionally on gravel foreshores. The stems become rather woody by summer and the plant turns from bright green to reddish. Lightly boiled in salt water, drained and served with butter and seasoning, the marsh samphire is a delicious vegetable to serve with fish, eggs or meat. It is described as being tonic and antioxidant. Supplier: Greensea

Salicornia herbacea. Salicorne is a hardy coastal plant that has adapted to grow in soils with a low to very high salt content. This achievement is linked to the presence of water and ammonium ion transporters, which play an essential role in protecting the plant from dehydration and the high salt content of their environment. In the 14th century, the ashes of Salicorne were used to produce soap and glass. The ashes of Salicorne are still used today in the manufacture of Alep soap that has been made exclusively with natural raw materials since antiquity. Supplier: Technoble—Coral Grass. Codif—Cire de Salicorne. Natural Solution-Incheon—Hamcho Extract. Yves Rocher—Extrait de Salicorne. There is also an oily extraction that stimulates AQP8 and AQP3, stimulates urea synthesis in the epidermis. These actions lead to a reinforcement of the lipidic barrier, decreases in TEWL and an increase in skin hydration. Supplier: Codif—Hydrasalinol.

Sargassum filipendula. A brown alga. Supplier: Assessa-Industria—Ormagel range and Seaweederm, Seaweedex.

Sargassum fulvellum. Described as a skin conditioning agent. Supplier: Biospectrum—Jeju Fucoxan. Bioland—Sargassum Fulvellum Extract.

Sargassum fusiforme. I.D. Bio—Sinominceur.

Sargassum horneri. It is described as a skin conditioning agent. Supplier: Sansho Seiyaku—Akamoku Extract. Maruzen Pharmaceuticals—Sargassum Horneri Extract.

Sargassum muticum. Supplier: Gelyma—Phyactyl.

Sargassum pallidum. It is described as an antifungal and antioxidant. Supplier: Fragrance Oils Intl.—Gulfweed Extract 101873.

Sargassum vulgare. It is described as a skin conditioning agent. Supplier: Assessa-Industria—Quidgel BRM.

Scytosiphon lomentarius. It is found attached to rocks, stones, shells and other seaweeds in rock pools at the middle tidal level and below into shallow water. It is found in the Baltic, North Sea, English Channel, Atlantic and Mediterranean. It has shiny, green-yellow or brown, tubular unbranched fronds, is rather slimy to touch, with periodic constrictions that make them resemble a string of sausages. Short stalks join them to a disc-shaped holdfast. Several generations are produced each year, growing lower and lower down the beach as the weather gets warmer.

Spirulina maxima. It is a blue algae and is a source of pigment that grows wild in the alkaline lakes in Mexico and Chad and known as Tecuitlatl. This microscopic seaweed belongs to the Cyanophyceae. Spirulina is rich in proteins (65-70% dried matter), nucleic acids (4% dried matter), unsaturated fatty acids, chiefly linoleic acid and gamma linolenic acid. It also contains natural pigments phycobilins (blue) specifically phycocyanins, carotenoids (red) and chlorophyl (green). Phycocyanin has been reported as being between FD&C Blue No.1 (Brilliant Blue) and Blue No.2 (Indigo Carmine). However, the exact spectral properties of all phycobiliproteins are dependant on their algal source, state of purification and physiochemical environment. The plant contains a wealth of amino acids, lipids and vitamins as well as the minerals and trace elements. The history of Spirulina dates as far back as the Aztec civilization: Emperor Noctezuma was fond of Tecuitlath (Spirulina) and used to send his people to pick them from the lakes of the area. The ancient Aztecs considered it to be a source of sacred power and priests and warriors sustained themselves on dried spirulina wafers. Spirulina was also harvested by natives of the Sahara Desert, where it was known by the name *dihe*. For centuries it has been known for its remarkable energizing and rejuvenating properties. A Japanese philosopher (Toru Matsui) who lived in a retreat on Mount Hakone, near Tokyo, was said to have had a diet consisting entirely of spirulina for more than 15 years. Supplier: Greentech—Selenium. Active Organics. Phytochim. Provital. Exsymol—Protulines. Solabia. Grau. Greensea.—Blue Alga or Spirulina. Alban Muller. Vege-Tech. Lessonia—Microzest Spirulina.

Spirulina platensis. Supplier: Bottger—Algemara. BiotechMarine—Blue Algae. Greentech. Mibelle—Iso-Slim Complex. Collaborative laboratories—Spirox. Cyanotech—Spirulina Pacifica. Greensea.

Spirulina subsalsa. This is a species that has not been used in cosmetics but might be worthy of investigation.

T

Tetraselmis suecica. Gold Microalgae. Contains polyunsaturated fatty acids and has topical action. It is described as a skin conditioning agent. It is described as being soothing. Supplier: Eclosarium—Plancton Marin Sa Eclate. BiotechMarine—Oleaphytoplancton, Phycol Phytoplancton. Greensea— Gold Microalgae.

Thymus carnosus. Portuguese Thyme. This plant just made it into the sea margin category, It is described as astringent, antimicrobial and antioxidant. Greensea.

U

Ulva lactuca. Sea Lettuce, Green Laver, Aosa. It is found at all times of the year attached to rocks or stones at the upper tidal level and below to shallow water, or floating free in pools or washed up. It has wavy green, leaflike fronds that vary in shape, but often grow in bunches, giving rise to its common name. It has been used for a long time as a sea vegetable for its dietary supply. High in nutrients and trace elements, sea lettuce is sometimes eaten in Britain as a substitute for laver and highly valued as a culinary ingredient in many parts off the Caribbean, South America and Far East. In Australia, sea lettuce is used in salads and pickles, or even added as an ingredient of stir-fry recipes. It is also eaten by the Canadian tribes Kwakwa-ka'wakw, Heiltzuk and other Northwest Coast groups. *U. lactuca* has a regenerating action, and has been reported as protecting and hydrating to the skin. In Cuba it used to be taken in dry decoction as a vermifuge. In Brittany, France it was applied as a dressing on wounds to prevent infections. *U. lactuca* remains on skin surface and captures molecules responsible for the creation of free radicals. The presence of zinc increases the self defense system of the skin against internal aggressions. Supplier: Codif—Chlorofiltrat Ulva. Yves Rocher—Extrait D'Ulva Lactuca, MCEH 100, Ulva Lactuca ME PG 40. Secma (BiotechMarine)—Phycol UL. IdB Holding—Phytelenes of Ulva EG 746 Liquid. Laboratoires Sérobiologiques—Sveltonyl. Libiol—Biomeduline. E.U.K.—Sea Extract Ulva. Greensea. BioBotanica—Sea Lettuce. Lessonia—Aosa Micronized, Microzest Ulva.

Ulva pertusa. Described as a skin bleaching agent. Supplier: Technoble—Marinoble-AN.

Undaria pinnatifida. Sea Mustard, Wakame, Japanese Kelp. The algae has resistance against harsh environmental stress (UV light, water movements and abrasions) that is related to its content of a special sulphated polysaccharide called fucoidan that protects the algae's body wall from losing integrity and stability. It contains proteins, lipids, micronutrients, vitamins and sugar that lead to a revitalizing and energizing action on skin cells. It has been reported that fucoidan has radical scavenging and hyaluronidase inhibitory properties. It is these properties that make Wakame a useful antiaging active ingredient in cosmetics. Supplier: Crodarom produces a glycerin/aqueous extract of *U. pinnatifida* (Phytessence Wakame) and has shown that the protection of hyaluronic acid degradation by inhibition of hyaluronidase enzyme activity prevents deterioration of skin tissues, decrease of dermal thickness and improved firmness. Somaig—Wakamine. Gelyma—OM-X Extract. Codif—Pheofiltrat Undaria. BiotechMarine—Phycol UP. Greensea.—Wakame or Sea Mustard. Provital—Polypalnt Marine Plus. E.U.K.—Sea Extract Wakame. Marinova Pty—GFS10, GFS75. Lessonia—Wakame Micronized. Sako—Mekabumato.

Valonia fastigiata. This is a species that has not been used in cosmetics but might be worthy of investigation.

Valoniopsis pachynema. This is a species that has not been used in cosmetics but might be worthy of investigation.

Chapter 15

Gemmotherapy - The Life Force and Vitality of Buds

Introduction

Gemmotherapy is derived from the Latin *gemma*, meaning "bud", and the Latin *therapīa* or Greek *therapeia*, meaning "medical treatment." These medicines, which sit astride the boundary between homeopathy and herbal medicine, have strong energetic qualities, and thus are considered low potency homeopathics and may be taken along with other low potency liquid homeopathic medicines. They are used for their tonic, anti-inflammatory and particularly for their strong drainage properties.

It is an offshoot of traditional homeopathy, whereby a homeopathic mother tincture (potency 1x) is prepared from the fresh sap, buds, seeds, etc. that the plant produces in the spring. They are made principally from the embryonic tissue of various trees and shrubs (the buds and emerging shoots), and also from the reproductive parts (the seeds and catkins) and from newly grown tissue (the rootlets and the cortex of rootlets) when the tree or shrub's annual germination is at its peak in order to capture the various nutrients, vitamins, plant hormones and enzymes that are released during this process, and which in some cases are only present in the plant at this time. In a small number of cases they are made from the early spring rising sap. The only exception is with the seeds that are taken in autumn.

History

In 1970, Pol Henry (1918–1988), a Belgium medical doctor, laid the foundations for *phytoembryotherapy* (treatment using embryonic vegetal tissues). His laboratory work revealed the therapeutic properties of buds and young shoots. These germinal tissues contain all the components and the genetic blueprint of the future plant (*totum*) because this package of ingredients and information will unfold into the final plant.

Several years later, the French homeopath Max Tétau, MD, invented the term *gemmotherapy* for treatment using buds and young shoots.

Gemmotherapy became an accepted form of herbal medicine in France (entering the Pharmacopée Française in 1965). Today, it is the health care practitioners and general public in Italy who best know gemmotherapy, where the remedies are used as a complementary or a first-choice treatment in a variety of common conditions.

Until recently it has been virtually unknown outside these two countries, with the partial exception of the United Kingdom, the United States, Germany and the Czech Republic. The first companies to introduce a product based on this technique into the cosmetic industry were Gattefossé and Greentech, but now EUK has delivered a full range that will enable a complete range of products to be developed.

Philosophy

Gemmotherapy is one of the most important subsections of phytotherapy. The essence of this method is the medicinal application of germinal plant tissues present in buds, and young shoots. This vital energy of the trees and shrubs is at its highest point when the new leaves, branches and flowers begin to emerge and it is believed that it is concentrated in these parts and remains as an active force that shows in the potency of the preparations.

The extracts therefore contain exceedingly valuable substances not present in other parts of the plant and include natural hormones, minerals, trace elements, nucleic acids, amino acids, vitamins and enzymes.

Production

Gemmotherapy is a cellular, energetic and global phytotherapy. The manufacturing process of gemmotherapy is based on maceration of plant buds (or young shoots) in a blend of glycerine and ethanol during three weeks to dissolve the most important part of the active principles.

Now, researchers are continuing to study the therapeutic properties of buds and young shoots. Naturopaths and homeopaths are using them increasingly.

The following example is taken from the European Pharmacopoeia, 6th edition, European Directorate for the Quality of Medicines and Healthcare, Strasbourg 2007 [on Wikipedia]:

A young shoot of ***Rubus fruticosus****, the Bramble or Blackberry.*

Producers follow the protocol laid out in the Pharmacopée Française and subsequently in the European Pharmacopoeia.

As soon as possible after being picked (and with a delay of not more than a few hours), the buds, young shoots or other parts to be used are cleaned and weighed. A sample is set aside so that the dry weight of the plant material can be determined. This is done by heating it at 105°C in an oven until its weight does not reduce any further.

The rest of the fresh plant material is macerated in an equal quantity of alcohol and vegetable glycerin, to a combined total weight of 20 times the equivalent amount of the dried sample. The proportion of fresh buds to excipient is about 1:9, and of young shoots, about 1:4. [Another method described the procedure as follows: when the buds and embryonic plant tissues are harvested, they are soaked in a mixture of glycerin and alcohol, in a ratio where the basic medium corresponds to 1/20 of the dry weight of fresh plants used].

The mixture is left to stand for one month in a cool, shaded environmen, and is agitated at intervals. It is then decanted and filtered under constant pressure. After standing for a further 48 hours, it is filtered once more. The resulting liquid is known as the souche, *or stock. It consists of equal parts of glycerin and alcohol, and 10–25% plant material, depending on the water content of the plant used.* [Another method describes the technique as follows: After three weeks, the macerate is filtered and diluted to 1/10 with a mixture of water, alcohol and glycerin].

To prepare the remedies in their final bottled form, one part of the souche *is diluted with nine parts of a mixture of 50% glycerin, 30% alcohol and 20% water, and this is lightly succussed (shaken rather than struck) 30 times to bring it to the 1DH (1X or 1:10) homeopathic potency.* [It is this 1/10 solution, Hahnemann's first decimal, that is the medicinal form of gemmotherapy remedies.]

If properly stored in lightproof bottles and a cool, preferably dark, environment, the therapeutic qualities of gemmotherapy remedies do not diminish significantly in time. However, European regulations stipulate that they should be used within five years of the date on which the plant material from which they were made was picked.

One or two producers use modes of production at variance to that described in the European Pharmacopoeia; for example by diluting the remedies in the final stage to the 2DH homeopathic potency, or by not diluting them at all and offering them as 'concentrated' gemmotherapy remedies. Concentrations stronger than 1X or 1DH, risk causing some intolerance phenomena according to another source.

Gemmotherapy Remedies—Traditional Uses

Aesculus hippocastanum (Horse Chestnut bud) as Vegebud*
It is indicated for most cases in circulation disturbances (chronic redness of the face, varicose ulcers, hemorrhoids). It is particularly valuable for varicose veins and related disorders.

Alnus glutinosa
Coronary artery blockage, cerebral infarction, anticoagulant, chronic inflammatory conditions, angina, angeoneogenesis, phlebitis, adrenal gland, migraine headaches

Betula alba (Birch bud) as Vegebud
It is very efficient for those suffering from retention of water, rheumatic diseases, articular pain, cellulitis, and kidney stones.

Castanea sativa (Sweet Chestnut bud) as Vegebud
It is indicated in varicose veins and poor circulatory conditions such as venous insufficiency. It has a positive effect on the lymphatic system and acts in depth on edema and cellulitis.

Cornus sanquinea
Thyroid drainage, general thyroid disorders, endocrine system drainage, anti-necrotic action, treats hemorrhages due to anticoagulants.

Corylus avellana (Hazel bud) as Vegebud
It is often used for certain circulation problems such as varicose ulcers. It is also helpful in liver conditions such as cirrhosis.

Crataegus sp (Hawthorn bud) as Vegebud

It is recommended in cases of tachycardia, in low blood pressure or in hypertension. It is a sedative for the central nervous system (anxiety, stress, and depression).

Ficus carica
Stomach and duodenum, antibacterial action, damaged mucous membrane, gastritis
peptic ulcers, neurovegetative dystonia, dysbiosis, digestive system (Crohn's disease, irritable bowel syndrome, colitis), obesity, warts.

Fraxinus excelsior (Ash tree bud) as Vegebud
It may be used to treat inflammation and injury of the feet, ankles, legs and knees.

Juglans regia (Walnut bud) as Vegebud
It is indicated for an imbalance of the intestinal flora. It also has beneficial properties in the treatment of certain skin infections (psoriasis, acne, and eczema).

Juniperus communis (Juniper young shoot) as Vegebud
It is a general stimulant of the internal organs and helps to eliminate uric acid, urea and cholesterol from the system.

Lonicera nigra (Honeysuckle bud) as Vegebud
It is indicated in coughs and mild asthma.

Olea europaea (Olive young shoot) as Vegebud
It is an excellent remedy for the circulation system particularly at the cerebral level (memory loss). It is indicated in atherosclerosis and cerebral arteriosclerosis.

Pinus montana
Deformative arthritis (osteoarthritis, rheumatoid arthritis), joint drainage, cartilage regeneration, liver/lymphatic drainage.

Pinus sylvestris (Pine bud) as Vegebud
It is useful in the treatment of various forms of osteoarthritis. It remineralizes and fights wear and destruction of the articular cartilage (also called the hyaline cartilage).

Plantanus orientalis
Acne, vitiligo.

Prunus amygdalus dulcis (Almond bud) as Vegebud
It lowers the level and the rate of formation uric acid and to an extent urea. It is recommended in hypertension.

Quercus pedunculata (Oak bud) as Vegebud
It is useful in masculine senescence (loss of libido), sexual debility and tiredness.

Quercus pedunculata* L. syn. *Quercus robur L. (Oak, Fagaceae family) buds have adaptogenic and immunostimulant properties. According to the analogical biological model proposed by Pol

Henry, the Belgian physician who is regarded as the father of gemmotherapy, as Oak is a dominant species in the mixed forests of the temperate regions of the globe, thanks to its capacity to adapt to different soils, in the same way the gemmotherapy extract, obtained from fresh buds, may stimulate in a general way the body's reactions to restore homeostasis. This kind of response suggests that *Quercus* acts on the hypothalamus-pituitary axis (see Asian Ginseng); therefore *Quercus* modulates the stress response, the basal metabolism, the immunity system and also sexual hormones. *Quercus* gemmotherapy extract, having a nonspecific action, is indicated as general tonic and adaptogen, to restore any kind of imbalance, stressful situations, and asthenia, to treat immune deficiency conditions and in particular geriatric patients. [We have included the Italian interpretation of this plant from EPO (Estratti Piante Officinali) S.r.l.–Istituto Farmochimico Fitoterapico, where the indications are extremely similar and give good comparison.]

Ribes nigrum (Blackcurrant bud) as Vegebud
It is a powerful anti-inflammatory (especially for rheumatism and tendonitis). It is recommended for allergies and certain skin infections (eczema, psoriasis). It is used also in asthma, chronic bronchitis, emphysema and allergic rhinitis.

Rosa canina (Dog Rose young shoot) as Vegebud
It is an important remedy for the shere of oto-, rhino-, laryngology (ORL) or more simply ENT (ears, nose and throat) and the associated conditions of rhinitis, tonsillitis, otitis and nasal-pharyngitis. It is also indicated in certain cutaneous problems (eczema, furunculosis, and herpes).

Rosmarinus officinalis (Rosemary young shoot) as Vegebud
It is an effective hepatic-protector. It is also a *désintoxicant* (detoxifying agent is the nearest translation) of the body. It normalizes the levels of cholesterol, triglycerides and urea.

Rubus fructicosus (Blackberry young shoot) as Vegebud
It is recommended in chronic respiratory insufficiency: emphysema, chronic bronchitis.

Rubus idaeus (Raspberry young shoot) as Vegebud
It is often recommended in female senescence, dysmenorrhoea, and amenorrhoea. It increases ovarian secretion and at the same time controls the levels of estrogen and progesterone.

Syringa vulgaris
Stimulation of coronary artery-collateral circulation, angina pectoris/myocardial problems, antispasmodic. Must be off of aspirin before use

Tilia tomentosa (Silver Linden bud) as Vegebud
It is indicated in cases of insomnia. Its antispasmodic properties suggest its use for cardiac palpitations, and spasmophilia (an abnormal tendency to convulsions, tetany, or spasms from even slight mechanical or electrical stimulation). It is detoxifying and lowers the levels of cholesterol, and uric acid.

Vaccinium vitis idaea (Cowberry young shoot) as Vegebud
It is useful for menopausal women to counteract the uncomfortable symptoms of these hormonal changes (e.g. hot flushes).

Viscum album
Breaks down fibrous tissue and enhances elimination, breaks down cysts and scar tissue anywhere in the body, epileptic syndrome. It is used after operations to get rid of scar tissue.

Vitis vinifera (Grape Vine bud) as Vegebud
It is recommended in intestinal inflammations. Its anti-inflammatory ownership is interesting at circulation level (hemorrhoids, phlebitis). But it is in most cases indicated in rheumatic pain and osteoarthritis.

Conclusion

In a world that seeks new ideas and new directions it is not often easy to find a totally new concept, but it is believed that the concept of vitality and germinating life forces could be irresistible to some marketers.

* *Vegebud is available from EUK*

Author's note: my huge thanks to Jean-Pierre Cavin for his help on this chapter.

Chapter 16

Gums, Gellants, Bulking Agents and Thickeners

Thickeners are used to control phase separation, prevent syneresis, extend shelf life, add volume, slow down or eliminate crystal growth, help suspend particulate materials, form gels and have a positive effect on product application, e.g. spreadability, delivery.

Acacia senegal. Gum Arabic, Cape Gum, Egyptian Thorn

It is found in North Africa from Abyssinia to Senegal. It is found in the Sudan, particularly the province of Kordofan, in Central Africa and West Africa. The gum exudes from the tree naturally, but larger yields are obtained by making incisions. The main component is arabin, the calcium salt of arabic acid. The structure of the gum is complex and has not yet been fully elucidated. The gum is built upon a backbone of D-galactose units with side chains of D-glucuronic acid with L-rhamnose or L-arabinose terminal units. The molecular weight is 200,000 to 600,000 Daltons. Acacia gum is demulcent, and soothes irritated mucous membranes. Consequently, it is widely used in topical preparations to promote wound healing and as a component of cough and some gastrointestinal preparations. Domestically it is also used as an adhesive. It does have slight thickening properties but is too sticky to be helpful. The blending with tragacanth gum helps reduce the tackiness.

Acacia Seyal Gum

Itis the dried, gummy exudate obtained from Acacia seyal, Fabaceae. Acacia seyal, the Red acacia, known also as the shittah tree (the source of shittim wood). Acacia seyal is, along with other acacias, an important source for gum arabic, a natural polysaccharide, that drips out of the cracks of the bark and solidifies. The gum is used as an aphrodisiac, to treat diarrhoea, as an emollient, to treat hemorrhaging, inflammation of the eye, intestinal ailments and rhinitis. The gum is used to ward off arthritis and bronchitis. In view of the medicinal use, there is little risk of adverse reactions when this material is applied topically to the skin.

Algae Extract

The term "algae" is normally used to denote a complex mixture of many seaweeds, which include the red, brown, green and blue classes. Seaweeds contain much higher levels of minerals than terrestrial plants and also contain a rich selection of trace elements, amino acids, short chain proteins and polysaccharides. The polysaccharides are of particular importance, since these natural sugars can have a profound benefit on the skin, especially in the areas of moisturization, skin smoothing and skin softening.
There is a gellant obtained from selected classes of red and brown seaweed (algae) collected in unpolluted tropical waters of the Brazilian Northeastern coast. In water this gives a white opaque gel that is quick to break and non-tacky. It is said to be highly substantive to hair and skin, provides good moisturizing properties, and helps to protect skin, hair and scalp.

Algenic Acid or Algin

The brown algae are multicellular and have differentiated structures that, in some species, bear a superficial resemblance to the roots, stalks, and leaves of more advanced plants. These structures, however, are quite different internally. The cell walls of the algae are made of cellulose similar to that found in red algae; the outsides of the walls are covered by a gelatinous pectic compound called algin. Bladder wrack contains 18–30% of algin. The algenic acid is a mix of polyuronic acids ($C_6H_8O_6$) that are made of "homogenous blocks" of mannuronic or guluronic acids of 20–30 units. Between the blocks, there are some sequences that are formed by association of the monomers. The well-structured blocks resist to hydrolysis while the badly structured blocks quickly degrade. The rate that D-manuronic acid or guluronic acid changes depends on the season and the place of the harvest. It is mainly used for its thickening, gel forming and stabilizing properties.

Algin

Algin is an extract from brown algae* (seaweed) and used as a stabilizing ingredient that binds oil and water components together. Derivatives may also be listed as potassium alginate and sodium alginate, while alginic acid is a gellike extract that provides thickening and stabilizing properties. Since algin can absorb up to 300 times its own weight in water, it also provides lubricating and moisturizing benefits to body and skin care products.

Alginic acid - see also Algae Extract

The gelling properties of alginic acid, the major polysaccharide in brown seaweeds including fucus, are extensively used in the dairy and baking industries to improve texture, body and smoothness of products.

***Amorphophallus konjac*. Konjac**

It is a food species of Yunnan and Sichuan. It is grown in fields by upland farmers. Its tubers have received attention as a diabetes food. They contain *Konjac glucomannan* which is an excellent dietary fiber. Konjac flour is obtained from the tubers of various species of *Amorphophallus*. It is a soluble dietary fiber that is similar to pectin in structure and function. Konjac flour consists mainly of a hydrocolloidal polysaccharide, glucomannan. Glucomannan is composed of glucose and mannose subunits. It is a slightly branched polysaccharide having a molecular weight of 200,000–2,000,000 Daltons. Acetyl groups along the glucomannan backbone contribute to solubility properties and are located, on average every nine to 19 sugar units. In general, the konjac tuber is ground and milled, and its impurities are separated by either mechanical separation, or washing with water or aqueous ethanol to produce konjac flour. A proprietary mixture of konjac mannan and xanthan gum is available (under the name Glucovis) and is the nearest natural alternative to Carbomer.

Anogeissus latifolia. Ghatti Gum

Gum Ghatti is the amorphous translucent exudate of the *Anogeissus latifolia* tree. The gum is locally known as *Dhavda* and when first exuded, is soft and plastic. The color varies from whitish yellow to amber. Gum Ghatti is a complex polysaccharide of high molecular weight. It occurs in nature as a mixed calcium, magnesium, potassium and sodium salt. It is composed of L-arabinose, D-galactose, D-mannose, D-xylose and D-glucoronic acid in a molar ratio of 10:6:2:1:2 plus traces (<1%) of 6-deoxyhexose. Gum Ghatti disperses in water to form a colloidal dispersion. It does not form a true gel, but forms viscous solutions that exhibit typical non-Newtonian behavior. Gum Ghatti is a moderately viscous gum lying intermediate between Gum Arabic and Gum Karaya. It is quite a good emulsifier. The normal pH of the dispersion is 4.8 and the solutions are sensitive to alkali. The viscosity increases sharply with pH up to a maximum at about pH 8 after which the solutions tend to become stringy. Viscosity values increase with age.

Astragalus gummifer. Tragacanth

Tragacanth contains from 20–30% of water-soluble fraction called tragacanthin (composed of tragacanthic acid and arabinogalactan). It also contains 60–70% of a water-insoluble fraction called bassorin. Tragacanthic acid is composed of D-galacturonic acid, D-xylose, L-fructose, D-galactose and other sugars. Tragacanthin is composed of uronic acid and arabinose and dissolves in water to form a viscous colloidal solution and bassorin swells to form a thick gel. Tragacanth will partially dissolve and partially swell in water to yield a viscous colloid. The maximum viscosity is achieved after 24 hours at room temperature or after heating for eight hours at high temperatures. The viscosity of these solutions is generally considered to be the highest among the plant gums and is heat and pH stable over a wide range of values.

Caesalpinia spinosa. Tara Gum

Tara gum is a neutral galactomannan with a linear main chain of (1-4) linked D-mannose units to which galactose units are bound (1-6) linking. The thickening effect is similar to guar and locust bean gum.

Cellulose Gum

Most cellulose gums are naturally derived as an unwanted part of the wood pulp process to produce paper. Cellulose is the main constituent in plant fiber. Cotton, for example, is 90% cellulose. Used as a thickening agent and emulsifier, it is widely used in cosmetics, hair and skin care because it swells in water.

Ceratonia siliqua. Locust Bean Gum, Carob Seed Gum

Locust bean gum is obtained from the carob seed or *Ceratonia siliqua*. Carob seed gum is another common name for locust bean gum. The pods are nourishing and contain starch, protein and sugar. The seeds are not unlike cocoa in taste and texture when ground. The pharmaceutical industry uses carob in the preparation of cough linctuses. Locust bean gum and its thickening properties date back to the ancient times and the ancient Egyptians used the paste of locust bean gum as an adhesive in mummy binding.

***Chondrus crispus*. Carrageenan**

Carrageenan is extracted from red seaweeds and available in three distinct chemical forms: *Kappa* carrageenans produce strong, brittle gels, *Iota* carrageenans give soft elastic gels, and *Lambda* carrageenans are non-gelling thickeners. These three forms provide the formulator with a large number of possibilities. Carrageenan can be formulated in very low concentrations to form fluid gels that stabilize solid particles and help suspend them and prevent settlement with time, e.g. foundations, scrubs and toothpastes. The formulator can create rheological profiles ranging from free-flowing liquids to thixotropic fluid gels and self-supporting solid gels by choosing the right form of the gellant.

***Cichorium intybus*. Fructan Gum**

This gum is obtained from chicory roots and is described as a vegetable-derived gum with moisturizing and thickening properties. It forms white, opaque, non-tacky, quick-breaking and -flowing gels in water.

***Cochlospermum planchoni*. Karaya Gum**

The gum of this shrub, Karaya gum, is used in India as a laxative and is said to be twice as effective as agar-agar. It is also used as a suspending and binding agent. In the Volta region of Africa it is used in the treatment of fever, conjunctivitis, hemorrhoids, etc. The roots contain a dyestuff, tannins, mucilage and an alkaloid.

***Cyamopsis tetragonolobus* [Syn. *Cyamopsis psoralioides*]. Guar Gum**

Guar gum is obtained from the Indian Cluster bean or *Cyamopsis tetragonolobus*. It has been used for centuries as a thickening agent for foods and pharmaceuticals. It is a dietary fiber obtained from the endosperm of the Indian cluster bean. The endosperm can account for more than 40% of the seed weight and is separated and ground to form commercial guar gum. Guar is a polysaccharide galactomannan that forms a viscous gel when placed in contact with water. Guar gum forms a mucilaginous mass when hydrated.

Diatomaceous Earth

Diatomaceous earth or Kieselguhr is composed of the siliceous shells of fossil diatoms (minute unicellular plants), or of the debris of fossil diatoms. It has similarities and uses to a material known as Fuller's Earth. It is listed in the British Pharmacopoeia and according to the Pharmaceutical Codex of 1923 is used for the preparation of absorbent and emollient dusting powders. Apart from having the ability to absorb essential oils and other active materials, it is also well-known as a filtering agent and can be used as a bulking agent in scrubs and masks. It is used as an absorbent for nitroglycerin, after which it becomes known as dynamite.

Fuller's Earth

Fuller's earth is a variety of kaolin, natural clay. It contains aluminum magnesium silicate, and is often used as a bulking agent to increase the viscosity in scrub formulations and masks.

Gelatin

Gelatin is a protein produced by partial hydrolysis of collagen extracted from the bones, hooves, connective tissues, organs and some intestines of animals such as domesticated cattle, pigs and horses. Gelatin melts to a liquid when heated and solidifies when cooled again. In water, it forms a semisolid colloidal gel that forms a highly viscous solution in water which sets to a gel on cooling. Its chemical composition is similar to that of its parent collagen. The popularity of this excellent gellant has decreased with the consumer preference for nonanimal-derived ingredients and the scare caused by bovine spongiform encephalopathy (BSE) in cattle.

***Gelidium amansii, G. cartilagineum*. Agar-agar**

The word "agar" comes from the Indonesia word agar-agar (meaning jelly). It is also known by the names Bengal Isinglass, Ceylon Isinglass, Chinese Isinglass, China Grass, Gelose, Japan Agar, Japanese Isinglas, Kanten, Layor Carang. The gelling agent is an unbranched polysaccharide obtained from the cell walls of some species of red algae, primarily from the genera *Gelidium* and *Gracilaria*, or seaweed (*Sphaerococcus euchema*). Commercially it is derived primarily from *Gelidium amansii*. Chemically, agar is a polymer made up of subunits of the sugar galactose. Agar polysaccharides serve as the primary structural support for the algae's cell walls. The various species of alga or seaweed from which agar is derived are sometimes called Ceylon moss. It is used as a demulcent, disintegration agent, emulsifying agent, gelating agent for suppositories, laxative, suspension agent, tablet excipient and for plating cultures in microbiology.

Gellan Gum

Gellan gum is a polysaccharide manufactured by microbial fermentation. It is a gelling agent that is effective at extremely low levels, forming solid gels at concentrations as low as 0.1%. This gellant can be used to form fluid gels and will stabilize suspensions without adding viscosity.

***Gracilaria lichenoides*. Agal-agal, Ceylon agar**

Agar is a polysaccharide found in the cell walls of some red algae and is unusual in that it contains sulfated galactose monomers. It is a simple process to extract and purify the plant to produce agar. *Gracilaria* and *Gelidium* spp (see above) are among the best of gelling agents.

Kaolin

Kaolin or China Clay is mainly hydrated aluminium silicate and is a mined mineral. It has been used in pharmaceuticals, cosmetics and skin care for many generations. It is used for its absorbent properties and for its emollient feel on the skin. Kaolin is used in face powders, eye shadows and masks. It is also used as a bulking agent in scrubs and masks.

Laminaria digitata

Laminaria has a number of strange polysaccharides such as laminarin (a storage polysaccharide) and alginic acid (which comes from the cell walls). They are chemically different from agar and have not been widely used.

***Larix occidentalis*. Arabinogalactan**

One of the most recent to hit the commercial market is a branched polysaccharide (arabinogalactan) that has been obtained from larch or *Larix occidentalis*, the properties are said to be similar to guar gum.

***Maranta arundinacea*. Arrowroot Starch**

Arrowroot is also the name of the edible starch from the rhizomes (rootstock) of West Indian arrowroot. The tubers contain about 23% starch. Pure arrowroot, like other pure starches, is a light, white powder, odorless when dry, but emitting a faint, characteristic odor when mixed with boiling water, and swelling on heating into a perfect jelly. In Victorian times it was used (boiled with a little flavor) as an easily digestible food for children and convalescents. Arrowroot is used to make clear, shimmering fruit gels and will prevent ice crystals from forming in homemade ice cream. It can also be used as a thickener under acid pH conditions. Arrowroot thickens at a lower temperature than flour or cornstarch, is not weakened by acidic ingredients and is not affected by freezing. It is recommended that arrowroot is premixed with cold water before adding to a hot mixture. The new blend should be heated only until the mixture thickens then heated no further otherwise the mixture will begin to thin.

Mica

There are specialist micas that can be used as thickeners.

- Mica (Aluminum Fluoro Magnesium Sodium Silicate). In water this produces a non-tacky, opaque, colloidal flowing gel. The mica swells in the water and is transparent on the skin, but it is affected by electrolytes. It can be used to formulate emulsifier free emulsions that have a silky non-greasy skin feel, e.g. Submica E.
- Mica (Aluminum Fluoro Magnesium Potassium Silicate). In oil it swells to form a non-tacky, opaque, flowing gel. It can be used to formulate emulsifier-free emulsions that are non-greasy and non-pearlizing with a silky skin feel, e.g. Submica M.

Montmorillonite

Montmorillonite is a type of refined clay, composed of hydrated aluminium silicate. It often has a natural color that is caused by the presence of trace amounts of minerals such as iron in the structure of the clay. It has similar properties to kaolin.

***Pistachia lentiscus*. Mastic Gum**

Gum mastic (Masticha) is the product of the Gum Mastic tree (*Pistachia lentiscus*) that grows in the island of Chios, Greece. It is a resinous substance obtained by tapping the trunk of the tree, when it is exuded in liquid form and then coagulates. Initially, it is pale green and then becomes yellowish and transparent. Gum mastic has a very old history that can be traced back to classical times. It was well-known in ancient Egypt. In Greek mythology, the virgins dedicated the tree to Artemis who wore branches of the tree as a symbol of purity and youth. Many ancient Greek authors, including Hippocrates, Pliny, Dioscorides, Galen, Theophrastus and others, mentioned it for its healing properties. Mastic of Chios was recommended for the treatment of baldness. Mastic was also suggested in paste for toothache. Roman ladies used toothpicks made from the tree wood, as it would brighten teeth. The gum was used to fill dental cavities. Mastic resin has been proven to reduce bacterial plaque, and so is useful in toothpastes and mouthwashes. Gum mastic is considered to be a color stabilizer.

Sclerotium Gum

Sclerotium gum is a polysaccharide gum similar to Xanthan gum (see later), and is also produced by a fermentation process using the bacterium *Sclerotiium rolfsii*. It increases the viscosity of a product and confers a unique skin feel.

Smectite Clays

These clays are of volcanic origin and are rheological additives for water and emulsions

- ***Bentonite.*** This is Aluminium silicate (with a Magnesium substitution). When hydrated, the platelets of bentonite clays are rather large (1400 angstroms). The positive and negative charges are weak. Therefore, bonding is weaker for bentonite than for hectorite, allowing higher usage levels.

- ***Hectorite.*** This is a Magnesium silicate (with Lithium substitution). When hydrated, the platelets of hectorite clays are very small (800 angstroms). The positive and negative charges are strong, due to the small particle size and high surface area (750 m^2/g). Therefore, bonding and thixotropy are higher for hectorite.

- ***Magnesium Aluminium Silicate.*** This is a blend of the two smectite clays—bentonite and hectorite

The smectite clay builds up a structure (gel) when properly hydrated (temperature, agitation, time of mixing). It suspends pigments and increases viscosity. If agitated (or energy put into the system), the gel breaks down, the mix becomes more fluid and the viscosity drops. The gel builds up again if agitation stops (this is thixotropic behavior).
Thixotropy of smectite clays adds body to the product and so controls the flow and dispensing characteristics of the final formulation.

Smectite clays' colloidal network of particles develops an internal structure capable of providing excellent suspension of fine particles. Smectite clay-water dispersions have high gel strength to viscosity ratio. Organic thickeners can give a synergistic increase in gel strength that gives superior suspension at very low solids levels of clay.

The smectite clays function as a protective colloid that surrounds oil droplets and keeps them dispersed by increasing the viscosity of the external phase. Smectite clays' network of particles insulates the oil and prevents coalescing so providing emulsion stabilization.

When smectite clays are in contact with water, the platelets expand due to hydration of exchangeable cations. This expansion permits water to infiltrate the crystal structure, increasing the overall volume 10–30 times and it is the hydration of smectite clays that develops the classic "house of cards" structure. This stabilizes both O/W and W/O emulsions at low concentrations. Its colloidal structure holds particles separate and suspended, even at high temperatures.

pH: most natural smectite clays contain residual carbonates (mainly $CaCO_3$). A direct consequence is that the pH of a water dispersion of smectite can increase well above 7 after a few hours, and this is sometimes too high (especially for pH-balanced formulations). The excess of carbonate can be neutralized by adding a small amount of boric acid, which should be added just after hydration of the smectite. The use of weak acids (such as citric acid) should be avoided. Once the pH is down to the required level, it will not affect the performances of the smectite. pH values lower than 5 are not recommended for natural smectite.

Sodium Alginate

Alginate gums from various seaweeds.

Starch

Starch is a polysaccharide found widely in nature in such grains as corn, wheat, and rice and such tubers and roots as potatoes and tapioca. Starch is a combination of two glucose-based polymers, amylose and amylopectin. Starch, farina, a whole flour, grated manioc and tapioca are obtained from the manioc tuber. Tapioca is used as a thickener in puddings and soups. Medicinally, the poisonous juice is boiled down to syrup and given as an aperient (Guiana). Fresh rhizome made into a poultice is applied to sores. The flour cooked in grease, the leaf stewed and pulped, and the root decocted as a wash are said to be folk remedies for tumors. Cassava is a folk remedy for abscesses, boils, conjunctivitis, inflammation, sores and swellings.

The size and shape of starch particles depend on the source of starch. The particles can vary in size from about 5 microns for rice starch up to 100 microns for potato starch. The shape of the starch particle will also vary depending on the plant source. Starch can be spherical and oval as with potato starch, truncated and round as with tapioca starch, angular and polygonal as with rice starch, and round and polygonal as with cornstarch.
The size and shape of the starch granules will have an impact on their performance properties in cosmetic applications. For example, rice starch is composed of very fine particles (average about 5 microns), and is reported to have problems with caking and stickiness. The larger corn and

tapioca starch particles (average about 15–20 microns) have improved flow properties. Although similar in size, tapioca starch granules give a different feel (more cushioned, less silky) than cornstarch granules because of its truncated shape.

Sterculia villosa. Karaya or Indian Tragacanth

All parts of the tree exude a soft gum when injured. Karaya gum is produced by scarring the tree trunk and removing a piece of the bark or by drilling holes in the tree. The gum is then collected and washed, then dried.

The use of Karaya gum became widespread as an adulterant for tragacanth gum. However, experience showed that karaya possessed certain physiochemical properties that made it more useful than tragacanth and it was less expensive. Today the gum is used in a variety of products such as cosmetics, hair sprays and lotions. It swells in water (it may absorb up to 100 times its own weight) to produce viscous colloidal solutions. The polysaccharide content of karaya has a high molecular weight and is composed of residues containing galacturonic acid, ß-D-galactose, glucuronic aicd, L-rhamnose and other residues. Because the gum is partially acetylated, upon degrading it may release acetic acid.

In face masks it appears to give more flexibility than tragacanth. It has been sold as denture adhesive. Karaya mucilage will develop a characteristic stringiness when treated with borax.

It has been used in mucilaginous hair setting lotions and dressings usually with a small amount of glycerin added to minimise flaking on the hair shaft. It is claimed that hair preparations made with karaya are superior to those made with tragacanth in that they do not fix the hair with such board like precision, but give a softer set to the hair.

Talc

Talc is a naturally occurring mineral composed of magnesium silicate, which (because of its crystalline shape) forms flat plates that slip over one another to give a sensuous lubricity. It is hard to believe, but the hardness of the skin is marginally greater than talc (Moh scale 1.5 for skin compared to 1 for talc). Talc is used to keep the skin soft, dry and silky and is frequently used not only as talcum powder but also as a component of foundations and other color cosmetics, as well as a bulking agent in scrubs and masks.

Xanthan Gum

This is an exocellular biopolysaccharide obtained from a fermentation of *Xanthomonas campestris*. Xanthan gum is used to modify the texture of personal care products and to stabilize suspensions, oil-in-water emulsions and foams against separation. The high viscosity associated with xanthan gum solutions at low shear rates enables products to keep particles suspended or prevent oil droplets from coalescing. The viscosity drops when shear is applied, so that products can be easily removed, poured or squeezed from their containers. Once the force is removed, the solutions regain their initial viscosity almost immediately.

Zeolite

Zeolite (means "the stone that boils" in Greek) is also known as crystallite. It is a porous claylike mineral derived from silicates, although it can be made synthetically. It has powerful thickening and water-binding properties.

Chapter 17

Scrubs and Abrasives

The following materials are employed as scrubs or abrasives. They are inert and used for their exfoliating and sloughing effects. The list looks a little dull, but might be useful if stuck for new ideas.

Actinidia chinensis. Kiwi seed
Actinidia chinensis. Kiwi seed powder
Agave america. Agave leaf powder
Alumina powder. Corindon
Aluminum Silicate powder. Garnet
Amethyst powder
Amorphyllus konja. Konjac root powder
Ampelopsis japonica. Ampelopsis root powder
Angelica acutiloba. The ground root of Touki
Angelica dahurica. Angelica root powder
Arachis hypogaea. Peanut flour
Argania spinosa. Argan kernel powder
Artemisia carvifolia. Chinese wormwood powder
Ascophyllum nodosum. Knotted wrack (ground seaweed)
Astragalus membranaceus. Astragalus root powder
Astrocaryum murumuru. Murumuru seed powder
Atractyloides macrocephala. Atractyloides macrocephala root/stalk powder
Attapulgite—a variety of Fuller's Earth
Avena sativa. Oat bran
Avena sativa. Oat kernel flour
Avena sativa. Oat kernel meal (fine or medium grades)
Bambusa arundinacea. Bamboo stem powder
Bertholettia excelsa. Brazil nut seed powder
Bletilla striata. Bletilla (Bai Ji) root/stalk powder
Butyrospermum parkii. Shea nut shell powder, Karité
Calcium carbonate. Chalk
Carapa guaianensis. Andiroba seed powder
Carica papaya. Papaya seed
Carum carvi. Caraway seed
Carya illinoensis. Pecan shell powder
Chalk
Charcoal Powder
Chenopodium quinoa. Quinoa seed
Chitin
Chondrus crispus. Irish moss powder
Cinnamomum cassia. Cinnamon bark powder, Cassia
Citrus aurantium amara. Bitter orange peel powder
Citrus aurantium dulcis. Sweet orange peel powder

Citrus grandis. Grapefruit peel
Citrus medica limonum. Lemon peel powder
Citrus nobilis. Mandarin peel powder
Citrus tangerina. Tangerine peel powder
Cnidium monnieri. Cnidium seed powder
Cocos nucifera. Coconut fruit
Cocos nucifera. Coconut shell (granules or powder)
Codonopsis pilosula root powder
Coffea arabica. Coffee seed powder
Coffea robusta. Green coffee seed powder
Coix lacryma-jobi or Job's Tears outer seed coat powder
Colloidal Oatmeal
Conchiolin powder. Pearl
Coptis chinensis root/stalk powder
Coral Powder
Corchorus capsularis. Jute flock
Corylus avellana. Hazelnut seed powder
Corylus avellana. Hazelnut shell granules
Cymbopogon flexuosus. Lemongrass leaf powder
Diamond Powder
Diatomaceous Earth
Dolomite
Egg shell powder. Eggshells ground (roasted)
Eijitsu—*Rosa multiflora* fruit
Elguea Clay
Emerald
Empetrum nigrum. Crowberry powder
Eucalyptus globules. Eucalyptus leaf powder
Euterpe oleracea. Acai pulp powder
Ficus carica. Fig seed
Foeniculum vulgare. Fennel seed
Fuller's Earth
Fragaria vesca. Strawberry seed
Garcinia mangostana. Mangosteen peel powder
Garnet Powder
Glycine soja. Soybean flour
Glycine soja. Soybean seed powder
Helianthus annuus. Sunflower seed
Helianthus annuus. Sunflower seed cake
Helianthus annuus. Sunflower seed flour
Hibiscus sabdariffa. Hibiscus flower powder
Hippophae rhamnoides. Sea buckthorn husk powder
Hippophae rhamnoides. Sea buckthorn seed powder
Honey Powder. Mel (and) Glucose
Hordeum distichion. Barley seed flour
Hordeum vulgare. Barley powder

Hordeum vulgare. Barley seed flour
Hydroxyapatite. A mineral from phosphate rock
Illicium verum. Star Anise fruit powder
Illite. A micalike clay
Ipomoea hederacea. Morning Glory seed powder
Isiatis tinctoria. Woad root powder
Juglans mandshurica. Walnut shell powder
Juglans regia. Walnut shell powder
Kaolin
Kochia scoparia. Belvedere Cypress fruit powder
Kurumi Kaku. Walnut shell powder
Lactose (and) Cellulose. Beads (different colors and contents)
Leonurus artemisia. Leonurus powder
Linum usitatissimum. Linseed, Flax seed powder
Litchi chinensis. Lychee seed powder
Lithothamnium calcarum. Lithothamnium powder
Lithothamnium corallioides. Lithothamnium powder
Loess. A loose mineral sediment formed during the ice age
Luffa cylindrica (and) *Bixa orellana.* Loofah fruit powder and Annatto
Luffa cylindrica. Loofah fruit
Luffa cylindrica. Loofah fruit powder
Lygodium japonicum. Lygodium fern spore
Macadamia integrifolia. Macadamia shell powder
Macadamia ternifolia. Macadamia shell powder
Macadamia ternifolia. Macadamia nut granules (coarse or fine)
Magolia denudate. Magnolia bud powder
Mangifera inica. Mango seed powder
Melaleuca alternifolia. Tea Tree leaf
Melaleuca alternifolia. Tea Tree leaf powder
Maris sal. Dead Sea salt
Mel. Honey powder
Montmorillonite
Moroccan Lava Clay
Mother of Pearl
Myristica fragrans. Nutmeg powder
Nacre Powder. Pearl powder
Nelumbo nucifera. Lotus flower powder
Nelumbo nucifera. Lotus seed powder
Oenothera biennis. Evening primrose seed
Olea europaea. Olive fruit
Olea europaea. Olive husk powder
Olea europaea. Olive seed
Olea europaea. Olive seed powder
Oryza sativa. Rice bran ground
Oryza sativa. Rice seed powder
Oryza sativa. Rice germ powder

Oubaku—Ground bark of *Phellodendron amurense*
Oyster Shell Powder
Paeonia lactiflora. Peony root powder
Paeonia officinalis. Peony flower powder
Papaver somniferum. Blue poppy seed
Passiflora edulis. Passion fruit seed (whole or ground)
Paullinia cupana. Guarana seed powder
Perlite
Persea gratissima . Avocado fruit powder
Persea gratissima. Avocado seed
Phaseolus angularis. Adzuki beans milled
Phaseolus radiatus. Adzuki starch
Phellodendron amurense. Oubaku ground bark
Plantago psyllium. Psyllium husk (whole or powdered)
Platinum Powder
Poria cocos. Fu Ling powder
Porphyra umbilicalis. Nori powder
Prunus amygdalus dulcis. Sweet almond seedcoat powder
Prunus amygdalus dulcis. Sweet almond seed meal
Prunus amygdalus dulcis. Sweet almond shell powder
Prunus armeniaca. Apricot kernel (granules or powder)
Prunus mume. Japanese apricot fruit
Prunus persica. Peach flower powder
Prunus persica. Peach seed powder
Pumice powder. Rhyolite
Pumice. Pumice (sand, granules or powder)
Punica granatum. Pomegranate seed powder
Pyrus malus. Apple fruit powder
Quartz
Quassia amara. Quassia wood powder
Ribes nigrum. Blackcurrant fruit powder
Rosa canina. Rose hip seed
Rosa gallica. Rose flower powder
Rosa multiflora. Eijitsu—fruit
Rosa rugosa. Rose bud powder
Rubus idaeus. Raspberry seed
Rubus idaeus. Raspberry seed powder
Salt Mine Mud—The sediment from salt mines
Salvia hispanica. Chia seed
Sesamum indicum. Sesame seed
Salvia hispanica. Chia seed powder
Sand
Sanguisorba officinalis. Burnet root/stalk powder
Scutellaria baicalensis. Scullcap root powder
Sea Salt. *Maris Sal*
Secale cereale. Rye seed flour

Silica
Simmondsia chinensis. Jojoba meal and seed powder
Smilax lanceaefolia. Smilax root/stalk powder
Sodium Bicarbonate. Baking soda
Solum Diatomeae, Diotomaceous earth
Sophora flavescens. Sophora root powder
Stemona sessifolia. Stemona root powder
Sucrose. Sugar (demerara, brown or white)
Symphitum officinale. Comfrey leaf powder
Synthetic Ruby
Synthetic Ruby Powder
Syringa oblata. Lilac bark/leaf powder
Talc
Theobroma cacao. Cocoa husk
Theobroma cacao. Cocoa seed powder
Theobroma cacao. Cocoa shell powder
Theobroma grandiflorum. Cupuaçu seed powder
Topaz
Touki. This is the ground root of *Angelica acutiloba*
Tribulus terrestris. Puncture Vine fruit powder
Trichosanthes kirilowii. Chinese Cucumber fruit powder
Triticum vulgare. Wheat bran
Triticum vulgare. Wheat germ powder
Triticum vulgare. Wheat kernel flour
Triticum vulgare. Wheat starch
Vaccinium angustifolium. Blueberry seed
Vaccinium macrocarpon. Cranberry seed
Vaccinium macrocarpon. Cranberry seed powder
Vanilla planifolia. Vanilla seed powder
Vanilla tahitensis. Vanilla seed powder
Viola prionantha. Viola powder
Virola sebifera, Acuuba seed powder
Vitis vinifera. Grape seed (whole or ground)
Vitis vinifera. Grape seed powder
Volcanic Ash
Volcanic Rock
Wood Powder. Consists mainly of cellulose, hemicellulose and lignin
Yokuinin. *Coix lacryma-jobi* or Job's Tears outer seed coat powder
Zea mays. Corn cob meal
Zea mays. Corn cob powder
Zea mays. Corn kernel meal
Zea mays. Corn seed flour
Zea mays. Cornstarch
Zingiber officinale. Ginger root powder

Chapter 18

The Legal Challenge

Introduction

The industry is increasingly becoming obsessed with litigation and more companies were targeted for compensation than ever before. In this chapter, some ideas for the fight ahead are offered, and it is suggested that patience and a sense of humor are the keys to retaining sanity.

Litigation and Related Issues

The growth of opportunistic treasure seekers has led to large increases in accident claim solicitors fighting on behalf of clients to seek compensation for irritation received as a result of using a cosmetic or skin care product. It is hard to remember when so much time was spent fighting off these parasites whose claims are at best ill-founded and opportunistic.

Companies should have a strategy and action plan for this event, because it is likely that they will have to pull a team together at a moment's notice with a leader given responsibility for ensuring the matter is dealt with competently and quickly. This chapter hopes to provide a checklist of ideas and procedures to ensure success in the case of a challenge. An industry expert from outside of your team or company can often discourage a case from going to court.

The scenario often starts with a consumer demanding compensation and claiming that they have skin damage resembling a third-degree burn and have been to a physician who says that medical science is unable to help despite having prescribed vast amounts of corticosteroids. Pictures are often available, but only through their solicitors. On no account admit any liability at this stage; by all means express concern that this event has occurred in their lives and without prejudice refund the cost of the product and the postage. A useful paragraph in these circumstances is to confirm that your products completely satisfy The Cosmetic Products (Safety) Regulations 2004, as amended, and that there has been no complaint for this product in its history of sale. Always reply to a consumer and always fully investigate every complaint. (Naturally, if there has been a history of complaints then as a responsible company you should have fully investigated and dealt with every single one and even recalled a suspect product).

In the event that customers write again, then consider the option of offering a patch test with a dermatologist to find out what raw material they may be allergic to or may find irritating. Explain that on very rare occasions this can occur despite every duty of care and that is why every ingredient is listed on the pack to enable consumers to identify those things that may not agree with their skin type.

If this helpful attention fails, then the consumers may go to a specialist solicitor or to the Trading Standards Office (TSO). The legal representative is often aggressive, hateful and not at all understanding of the law and the regulations we follow. The tenant to remember is that, "We are all innocent until proved guilty." The second major point to remember is that you have all the time in the world and are only required to acknowledge the receipt of their letter within a week or so. The TSO is normally most understanding and accustomed to this type of complaint, which rarely makes it to court (if you are properly prepared). These are the action points that any company faced with a claim against them should follow. Review the files to see whether this customer wrote to your company initially and retrieve your reply/replies, which should have been dealt with as described above. Make copies of this correspondence and the solicitor's letter and start a new file. Contact your insurers and copy them on the correspondence in the new file.

Trading Standards and the Product Information File

In the case of the TSO, you must have a full Product Information File (PIF) available in two to three working days. The faster you can supply the information, the more respect it will have for you and the less worried it will be about the competency of your business.

The data should be as follows:

- Product description and codes or formulation code.
- Product formula including percentages, INCI names, trade names, CAS/EINECS numbers, material function, toxicological summary and material suppliers.
- Raw material ingredient specifications and technical data sheets.
- Manufacturing procedure summary or single-page flow chart.
- Summary GMP statement on company-headed paper declaring ISO standards and/or GMP compliance.
- Product safety assessment statement. This should include the calculation for the margin of safety and the total daily exposure for each raw material,
- Undesirable health effects summary.
- Claim substantiation summary with references—proof that pack claims are able to be substantiated.
- Stability summary, with reference to methods—usually a spreadsheet showing the stability at ambient, 30ºC, 40ºC, freeze-thaw etc.
- Specification—viscosity, pH and other test data listed.
- Microbiological challenge test records for products that contain water.
- Proposed pack copy or artwork for each carton and label.
- It is useful to include a picture of the final product in the PIP.
- Perfumes and flavors should have IFRA- and/or RIFM-compliance statements from the perfume house.
- Perfumes should have the 26 potential allergens content list from the perfume house.
- Additional data may include user trials, instrumental data, analytical data and other facts to support the claims and/or safety of the product.

The TSO will prosecute if this PIF and all of its components are not available, because it is mandatory in law. If your company can demonstrate that it is in control of its legal requirements, has shown all fair and reasonable duty of care to its consumers, not been negligent in its surveillance and monitored its product through its entire history (including all complaints), then it is not likely to prosecute.

Claims Solicitor

This is never going to be pleasant if the claims solicitors do not win because then, like their client, they do not usually get paid. They try and prove all manner of ridiculous assertions, such as your company is using forbidden drug materials. There are some materials that are not allowed internally in pharmaceuticals but are perfectly legal in topical products.

It is not uncommon for ludicrous pieces of pharmaceutical law to be thrown back that have no relevance to a cosmetic product or the laws that might govern it. Take care to answer each point with patience and meticulous care regardless of the irrelevance. A phrase guaranteed to gain a modicum of revenge is, "We are very particular when it comes to the law governing cosmetics and the need for accuracy should not be obscured or threatened by an ignorance of that law."

Contest every statement made against your company and provide evidence to support every statement you make. Challenge every nonsensical statement with authority, because many of these solicitors know a lot less than you do and are trying to intimidate with legalistic jargon. Eventually there will come a time when the correspondence cycle is complete, often dozens of letters will have been exchanged between your insurers, the claimants legal team and whoever else has been lucky enough to be sucked into the loop. Hundreds of hours will have been wasted and it is a wonderful scenario for those who are being paid for their time.

After more than a year the attorneys will be called in, which means that the legal costs escalate from around £50 per hour for each player to around £500–1000 per hour for the new teamsters. If your company does not have insurance, then this is the time to go into receivership. Remember that diligence and the willingness to fight every argument leads to many cases failing because they appear too hard to win. If your company seems to have covered all the bases and is eager to fight, then most cases will evaporate.

In all the time that this round of legal banter has been in progress, you have been assembling a file of information that puts your product beyond all reasonable doubt. You have commissioned a repeat insult patch test, conducted user trials and sought dermatological proof that your product is harmless. An eager defendant will have done a full battery of animal alternative tests in preparation for a court appearance.

During this time there will be two files, one is a beautiful electronic version on CD ROM or memory stick that is all perfectly indexed and has all the documents linked through to a hyperlinked spreadsheet (for the defense team) with the really useful documents highlighted. The other version is the paper alternative in at least half a dozen notebook binder files that is for the plaintiff's team. Most legal teams are exceptionally grateful for this seemingly endless supply of paperwork and are always thrilled to receive a crate full of information.

The Legal Portfolio of Evidence

Every document you can lay your hands on has been placed in these files and all the documents are labeled in a way that is fully understood by you and your defense team. The files contain:

- The full PIP, now extensively supplemented as described above (and a lot more).

- The MSDS for every single raw material used in the product.
- The Technical Data sheet for each and every raw material.
- The Folklore and Ethnobotany of a natural product is a really fascinating topic that sadly can run into vast numbers of pages which are essential evidence in proving the safety of an individual raw material.
- Toxicity files (where available) should also be included for each ingredient, and regrettably these can also tend to be quite verbose and lengthy.
- Copies of the inspection reports for every single raw material at the time of production are essential but dull reading. Inspection reports for lots taken before and after the production date is always helpful for comparison to show a normal compliance and that there was nothing out of the ordinary. Regrettably this puts an additional burden on the volume of paper supplied.
- Certificates of Conformance for every raw material used are absolutely vital.
- The batch sheet for the product under dispute and all related documents such as the demineralized water conformance report for every day that month (just in case it is asked for).
- The qualifications and personnel reports for the staff involved in the production of the batch including details of all relevant training received.
- The Quality Control sheet for the proof of compliance to the product's specification and the full protocol for the use of the equipment to make those measurements accurately (the standard operating procedure).
- The provision of helpful information such as published papers should be included for the active ingredients.
- A survey of information from the Internet should be a part of the package showing a history of safe and common usage for the raw materials.
- A copy of the Cosmetic Regulations and all its appendices.
- The Product After Opening (PAO) explanation document.
- The Challenge Test protocol.
- The MHRA Guidance Note 8 is an excellent read and will provide a good read to the attacking solicitor and his team.
- Hazard and Critical Control Points (HACCP) analysis will also have been conducted for this product and should be included in the data.

This information should be sectioned into reasonable categories where you and your defense team can easily locate any document, but which the plaintiff's team will find a paper maze of unfathomable complexity. Your attorney has the headlines of your defense well laid out in two to three concise pages.

The Winning Stroke

In one last notebook binder file is kept the:

- instrumental data;
- clinical data; and
- statements of opinion and other last-minute defense strategies.

This last file is supplied on the morning or on the day before the trial as late evidence and best supplied direct to the plaintiff's solicitor by courier. The first morning of most trials is spent seeing what evidence has been lost or misplaced. The chances of losing are quite slim if you have all the data and if the case has been prepared well. The last consolation is to remember that this vast data can be assembled very easily and will tie up a legal brain for many hundreds of hours, which is why cases can last four years or more.

It is only by making this industry an unattractive target that will discourage the "ambulance chasers" from making a profitable living.

Appendix V
Glossary

A

Absorbents—Increase the absorption of diseased tissue.
Abortifacient—Causing abortion.
Acrid—Caustic, having a hot, biting taste.
Alterant—[L. *alterare*, to change.] Alterative. Causing a favorable change in the disordered functions of the body or in metabolic processes. Promoting repair.
Alterative—Improving the nutrition of the tissues and absorbing diseased tissues, thereby restoring them to their normal condition. Changing the morbid action of the secretions.
Alexeteric—[G. *alexeterios*, able to defend.] Protective, defensive; in reference especially to infectious diseases; antidotal.
Alexipharmic—[G. *alexipharmakos*, preserving against poisoning.] An antidote.
Alexipyretic—[G. *alexo*, I ward off, + *pyretos*, fever.] Febrifuge.
Alexiteric—Thwarting the action of venom.
Alible—[L. *alibilis*, nutritive.] Capable of nourishing, nutritive, nutritious.
Amebicide—[L. *caedere*, to kill.] Any agent that causes the destruction of amoebas.
Anacathartic—Emetic.
Anesthetic—Producing insensibility to pain.
Analeptic—[G. *analeptikos*, restorative.] A strengthening, invigorating remedy.
Analgesic—[G. *an*-priv. + *algos*, pain.] Analetic, causing loss of sensibility to pain, relieving pain. A pain-stilling remedy.
Anatriptic—[G. *anatriptos*, rubbed up.] A remedy to be applied by friction or inunction.
Anatrophic—[G. *ana*, up, + *trophe*, nourishment.] Nourishing. Also, [G. *an*-priv. + *atrophia*.] Preventing atrophy.
Anhidrotic, Anidrotic—Checking perspiration. Having a tendency to arrest or prevent sweating.
Anodyne—[G. *an*—priv. + *odyne*, pain.] An agent that has the power to allay pain, milder in action than the analgesics, generally used to indicate a systemic effect. Mitigating pain, quieting.
Antacid—[G. *anti*, against + L. *acidum*, acid.] Neutralizing the acidity of the gastric juice.
Antalgesic—[G. *anti*, against, + *algos*, pain.] Anodyne.
Antemetic—[G *anti*, against, + *emetikos*, emetic.] A remedy that tends to prevent or control vomiting.
Anthehemorrhagic—Having the power to prevent or arrest hemorrhage.
Anthelmintic or anthelminthic—[G. *anti,* against, + *helmins*, worm.] Having the ability to destroy or expel worms. Vermifuge.
Anthydropic—Antihydropic.
Antiarthritic—Relieving gout.
Antibechic—[G. *anti*, against, + *bex(bech-)*, cough.] A cough remedy.
Antiblennorhagic—Preventive or curative of catarrh or of gonorrhea.
Anticalculous—Antilithic. Preventing the formation of calculi or promoting their solution.
Antibromic—[G. *anti*, + *bromos,* smell.] Deodorizing.
Antibilious—Correcting the bilious secretions.
Anticatarrhal—Preventive or curative of catarrhal inflammation of the mucous membranes.
Anticholagogue—Depressing the hepatic function, opposing the secretion of bile.

Anticoagulant—Anticoagulative, preventing coagulation.
Anticoncipiens—[G. *anti*, against, + L. *concipere*, to conceive.] Preventing contraception.
Anticonvulsive—Preventing or arresting convulsions.
Antidinic—[G. *anti,* against, + *dinos*, dizziness.] Preventing or relieving vertigo.
Antidiuretic—Lessening the secretion of urine.
Antidote—[G. *antidotos* from *anti*, against, + *dotos*, what is given.] An agent that neutralizes a poison or counteracts its effects.
Antidysenteric—Relieving dysentary.
Antidysuric—Relieving strangury or distress in urination.
Antiemetic—[G. *anti*, against, + *emetikos*, nauseated.] Relieving nausea, arresting vomiting.
Antigalactic, antigalactagogue—[G. *anti*, against, + *gala(galakt-)*, milk, + *agogos*, drawing forth.] Diminishing or arresting the secretion of milk.
Antihidrotic—[G. *anti*, against, + *hidrotikos*, sudorific.] Antisudorific. Preventing or arresting the secretion of sweat.
Antihydropic—[G. *anti,* against, + *hydropikos*, dropsical.] An agent that causes dropsical effusions to diasappear.
Antiicteric—Preventing or curing jaundice.
Antileptic—[G. *antileptikos*, able to prevent.] Preventing an attack of disease.
Antilithics—[G. *anti*, against, + *lithos*, stone.]. Preventing the formation of calculus matter in the kidney or bladder.
Antimycotic—[G. *anti*, against, + *mykes*, fungus.] Destructive to fungi.
Antinephritic—Preventing or relieving inflammation of the kidneys.
Antiodontalgic—[G. *anti,* against, + *odous*(*odont-*) tooth, + *algos*, pain.] A toothache remedy.
Antiperiodic—Relieving regular recurrence of a disease or symptom, as in malaria.
Antiphialtic—[G. *anti,* against, + *ephialtes,* nightmare.] Tending to prevent distressing dreams.
Antiphlogistic—[G. *anti*, against, + *phlogistos*, on fire.] Subdues inflammation.
Antiplastic—Preventing cicatrization.
Antipodagric—Antiarthritic.
Antipruritic—Relieving itching.
Antipyogenic—[G. *anti*, against, + *pyon*, pus, + *gennae*, I produce.] Preventing suppuration. Antipyic.
Antipyretic—[G. *anti*, against, + *pyretos*, fever.] Reducing fever, a febrifuge.
Antipyrotic—[G. *anti*, against, + *pyrotikos*, burning, inflammation.] Antiphlogistic. An application for burns.
Antiscorbutic—Useful in scorbutus or scurvy.
Antiseptic—Preventing mortification. Substances that check the growth of microorganisms.
Antisialagogue—[G. *anti*, against, + *sialon*, saliva, + *agogos*, drawing forth.] Antisialic. Diminishing or arresting the flow of saliva.
Antisideric—[G. *anti*, against, + *sideros*, iron.] Chemically incompatible with iron, such as tannin.
Antispasmodic—[G. *anti*, against, + *spasmos,* convulsion] Antispastic. Preventing contractions of muscle and also lessening convulsions. The term is also applied to that which lessens nervousness, because of the tremors of the muscles that often occur in nervous people. Calming spasms, relaxing nervous irritation.
Antitussive—[G. *anti*, against, + L. *tussis*, cough.] A cough remedy. Antibechic.
Antizymotic—Antiseptic, checking the action of germs.

Aperient—[L. *aperire*, to open.] Producing mild movements of the bowels. Gently purgative.
Aperitive—[Fr. *aperitif*, from L. *aperire*, to open.] Aperient.
Arthrifuge—[G. *arthron*, joint, + *-itis*-arthritis + L. *fugare*, to chase away.] A gout remedy.
Ascaricide—[L. *caedere*, to kill.] Destroying pinworms.
Astringent—[L. *astringere*, to contract.] Causing contraction or hardening of tissues. Arresting secretion or hemorrhage. Styptic.

B

Balsamic—Unctuous, mitigating, fragrant, aromatic.
Bibulous—[L. *bibere*, to drink.] Absorbent.
Blennostatic—[G. *blennos*, mucous, + *stasis*.] Diminishing mucous secretion.

C

Calmative—Quieting excitement, a sedative.
Carminative—[L. *carminare*, to cleanse.] Preventing the formation, or causing the expulsion, of gas in the stomach and intestines, relieving flatulence.
Catastaltic—[G. *katastello*, I check.] An inhibitory agent, such as an antispasmodic or an astringent.
Cathartic—[G. *katharsis*, purification.] Causing purgation, movements of the bowels.
Caustic—[G. *kaustikos; kaio*, I burn.] Corrosive, causing a destruction of tissues. Escharotic.
Cauterant—A cauterizing agent.
Cholagogue—[G. *chole*, bile, + *agogos*, leading.] Promoting the flow of bile.
Cicatrize - To heal by forming scar tissue.
Coagulant—An agent that causes clotting.
Convulsant—Producing convulsions.
Corrective—Modifying or changing what is not desirous or is injurious in another substance, as making unpleasant drugs pleasant to the taste.
Corrigent—[L. *corrigere*.] Corrective.
Corroborant—Invigorating, tonic.
Counterirritant—Acting on the skin, causing redness and irritation, thus relieving deep-seated inflammation in remote organs or tissues. By acting on the nerve endings in the skin, they also relieve pain in remote organs, producing vesication, even pustulation.
Culicide—[L. *culex(culic-)*, mosquito, + *coedere*, to kill.) An agent that destroys mosquitos.
Culicifuge—[L. *culex(culic-)*, mosquito, + *fugare*, to drive away.] An agent that drives away gnats and mosquitos, or keeps them from biting.

D

Decalvant—[L. *decalvare*, to make bald.] Depilatory. Removing the hair.
Decongestant—Reducing congestion.
Delirifacient—[L. *delirium* + *facere,* to make.] Deliriant. Increasing the activity of the brain.
Demulcent—[L. *demulcere*, to smooth down.] Soothing agent, usually of an oily or mucilaginous nature, used to coat, protect and lubricate a mucous membrane or surface of the body. Lubricating, softening, relieving irritation.
Deobstruent—[L. *de*—priv. + *obstruere*, to obstruct.] Having power to resolve or remove obstruction, as to secretion or excretion. Resolvent.
Deodorants—[L. *de*—priv. + *oderare*, to smell.] Remedies that destroy unpleasant odors.

Deoppilative—[L. *de*—priv. + *oppilare*, to stop up.] Removing obstructions, deobstruent.
Depilatories—[L. *depilare*, to deprive of hair.] Substances used to remove hair.
Depressant—[L. *depressus*, *deprimere*, to press down.] Lowering vital tone, lowering nervous or functional activity. Sedative.
Depurants—[L. *depurare*, to purify.] Increasing the excretions of the body, removing waste products, cleansing and purifying the system, particularly the blood. Also Depurative.
Dessicant—[L. *desiccare*, to dry up.] Substance that absorbs moisture, especially from skin or mucosa. Also Dessicative or Desiccator.
Desquamative—[L. *de*, from, + *squama,* scale.] Causing the shedding of the outer layer of the skin.
Detergents—[L. *detergere*, to wipe off.] Cleansing agent. Detersive.
Diaphoretic—[G. *dia*, through + *phoreo,* I carry.] Promoting or increasing perspiration. Sudorific.
Diapyetic—Causing suppuration.
Digestives or Digestants—Aiding the digestion of food.
Dinical—[G. *dinos*, dizziness.] Relieving vertigo.
Disinfectants—Destroying germs.
Discutient—[L. *discutere,* to shake apart.] Dissolving, or dispersing a pathological accumulation.
Discussive—Discutient.
Diuretic—[G. *dia*, intensive, + *ouresis*, urination.] Increasing the excretion of urine.

E

Ecbolics—[G. *ekbolos*, abortive.] Accelerating childbirth, producing abortion or assisting labor. Oxytocic.
Eccritic—[G. *separation.*] Promoting excretion, expulsion of waste.
Edulcorant—[L. *dulcedo*, sweetness; *edulcorare*, to sweeten.] An agent that sweetens, takes away acridity. Used in the sense of sweetening tissues.
Eliminant—[L. *eliminans*; *eliminare*, to turn out of doors.] Increasing excretion of wastes.
Emetics—Causing vomiting.
Emictory—Diuretic.
Emmenagogue—[G. *emmenos*, monthly, + *agogos*, leading.] Promoting or increasing menstruation.
Emollient—[L. *emolliens*; *emollire*, to soften.] Softening, soothing to skin and mucosa.
Emunctory—[L. *emungere*, to blow the nose.] Causing the removal of an excretion.
Epilatory—Depilatory.
Epispastic—[G. *epi*, upon, + *spao*, I draw.] A vesicating agent, causing blisters.
Errhine—[G. *en*, in, + *rhis(rhin-)*, nose.] Sternutatory, provoking mucous discharge from the nose.
Errifine—Increasing the nasal secretions and producing sneezing.
Epispastic—[G. *epi*, upon, + *spao*, I draw.] Producing blisters on the skin.
Escharotic—[G. *eschara*, scab] Caustic, destroying the area of skin over which they are applied.
Expectorant—[L. *ex*, out, + *pectus*, chest.] Increasing bronchial secretions and augmenting and promoting the discharge from the lungs.
Exsiccant—[L. *exsiccare*, to dry out.] Drying, absorbing a discharge.
Evacuant—[L. *evacuare*, to empty.] Promoting excretion, especially of the bowels.

F

Febrifuge—[L. *febris*, fever, + *fugare*, to put to flight.] Reducing fever, also called Febricant, Febricide, Febrifacient. Antipyretic.

G

Galactogogues—[G. *gala(galakt-)*, milk, + *agogos*, leading.] Increasing the secretion of milk. Galactapoetic. Galactapoietic.

H

Helminthagogue—[G. *helmins(helminth-)*, worm + *emesis*, vomiting.] Anthelmintic. Vermifuge, Helminthic.
Hemafacient—[G. *haima*, blood, + L. *facere*, to make.] Hematopoietic. Hematogenic. Hematoplastic. Promoting the formation of blood.
Hemagogue—[L. *haima*, blood, + *agogas*, leading.] Promoting a flow of blood, usually refers to menstrual. Hemagogic. Emmenagogue.
Hematic—A remedy for anemia.
Hematinic, Hematonic—Stimulating the formation of blood cells and hemoglobin.
Hemostatic—Arresting the flow of blood. Styptic.
Hepatic—Affecting the liver.
Herpetic—Curing cutaneous diseases.
Hydragogue—[G. *hydor*, water, + *agogos*, drawing forth.] A cathartic that produces frequent watery movements of the bowels.
Hypnotic—[G. *hypnotikos*, causing one to sleep.] Producing sleep. Hypnic. Hypnagogic.

I

Irritant—[L. *irritare*, to excite.] An irritating agent, stimulant.
Ischogalactic—[G. *ischo*, I keep back, + *gala(galakt-)*, milk.] Antigalactic. Lactifuge. Supressing the secretion of milk.
Ischomenic—[G. *ischo*, I keep back, + *men*, monthly.] Supressing menses.

J

K

Keratolytic—[G. *keras(kerat-)*, horn, + *lysis*, solution.] Desquamative. Causing shedding of the horny layer of the epidermis.

L

Lactifuge—[L. *lac(lact-)*, milk, + *fugare*, to drive away.] Arresting the secretion of milk.
Lapactic—[G. *lapaktikos*.] Purgative laxative.
Liquifacient—[L. *liquire*, to be fluid, + *facere*, to make.] Causing the resolution of a solid tumor by liquifying its contents. Resolvent. Liquifactive.
Lithagogue—[G. *lithos*, stone, + *agogos,* drawing forth.] Causing dislodgment or expulsion of calculi.
Lithontryptic, Lithotryptic—[G. *lithos*, stone, + *tripsis*, a rubbing.] An agent that dissolves calculi.

Laxative—producing mild movements of the bowels.

M

Malactic—[G. *malaktikos*, softening.] Emollient.
Mydriatic—Dilating the pupil of the eye. Cyclopegic.
Myotic—Contracting the pupil of the eye. Miotic.

N

Narcotic—[G. *narkosis*, a benumbing.] In mild doses, relieving pain. In larger amounts, producing profound sleep, stupor or general anesthesia.
Nervine—[L. *nervus*, nerve.] A therapeutic agent, especially a sedative, whose action is specific to nerve or nerve function. Strengthening the nerves, or lessening irritability.
Neurotic—[G. *neuron*, nerve.] Subduing nervous erethism, a nervine.
Neurotonic—[G. *neuron*, nerve, + *tonus*, tension.] Improving the tone or force of the nervous system.

O

Obstruent—[L. *ob*, before, + *struere*, to build.] Obstructing or preventing a normal discharge, usually refers to bowels.
Obtundent—[L. *obtundere, ob*, against, + *tundre*, to pound.] Blunting sensibility, especially to pain.
Odinagogue—[G. *odis(odin-)*, labor-pains, + *agogos*, drawing forth.] Oxytocic.
Oppilative—[L. *oppilare*, to stop up.] Obstructive to any secretion, but commonly used in reference to constipating.
Oxytocic—Increasing contractions of the uterus.

P

Parasiticide—[L. *parasitus*, + *caedere*, to kill.] Destructive to parasites.
Paratriptic—[G. *para*, beside, + *tripsis*, rubbing.] Retarding tissue waste (catabolism).
Parturifacient—[L. *parturire*, to be in labor + *facere*, to make.] Oxytocic. Facilitating, inducing or accelerating labor.
Pectoral—Beneficial in diseases of the lung.
Pellant—[L. *pellans; pellare*, to drive.] Causing the removal of "peccant humors." Depurative.
Phygogalactic—[G. *phyge*, flight, + *gala(galakt-)*, milk.] Checking secretion of milk. Lactifuge. Galactophygous. Ischogalactic.
Phylacagogic—[G. *phylaxis (phylak-)*, protector, + *agogos*, leading.] Stimulating the production of antibodies.
Prophylactic—Preventing the development of a disease.
Psilothron—[G. *psilosis*, a stripping.] A depilatory. Psilotic.
Ptarmic—[G. *ptarmikos*, causing to sneeze.] A sternutatory.
Ptalagogue—[G. *ptyalon*, saliva, + *agogos*, leading.] Sialagogue. Producing or increasing the flow of saliva.
Purgative—[L. *purgare*, to cleanse.] Producing moderately active and frequent movements of the bowels. Cathartic.
Pyostatic—[G. *pyon*, pus, + *statikos*, causing to stand.] Arresting the formation of pus.

R

Refrigerant—[L. *refrigerare.*] Relieving thirst and cooling the patient in fever.
Relaxant—Relieving tension.
Repellent—[L. *repellare*, to drive back.] Astringent or other agent that reduces swelling.
Restorative—[L. *restaurare*, to restore, repair.] Renewing health or strength.
Revulsant—[L. *revulsus, revellare*, to pull back.] Drawing blood from the deeper parts to the surface. Also Revulsent. Also Revulsive. Also Revellent. A counterirritant.
Roborant—[L. *roborare*, to strengthen.] A tonic.
Rubefacient—Reddening the skin by widening the capillaries. Producing inflammation and redness of the skin.

S

Salivant—[L. *saliva*] Increasing the flow of saliva. Salivator.
Saxifragant—[L. *saxum*, stone + *frangere*, to break .] Lithotritic. Having the power to dissolve calculi.
Sedative—[L. *sedare*, to allay.] Lessening the activity of an organ of the body. They are usually drugs that lessen nervous excitation. The sedatives are named according to the system they are specific for.
Sialogogue—[G. *sialon*, saliva, + *agogos*, drawing forth.] Increasing the flow of saliva. Ptyalogogue.
Siccative—[L. *siccans; siccare*, to dry.] Drying.
Somnifacient—[L. *somnus*, sleep, + *facere*, to make.] Soporific. Somnific. Hypnotic. Producing sleep. Relieving pain, stupefying.
Soporific—[L. *sopor*, deep sleep, + *facere,* to make.] Somnifacient. Soporiferous. Causing sopor.
Sorbefacient—[L. *sorbere*, to suck up, + *facere*, to make.] Facilitating absorption.
Specific—[L. *species*, a kind or sort, + *facere*, to make.] Having specific action in specific disease, having definite curative action, usually by destroying, or combining with, the causative agent.
Stimulant—[L. *stimulans*, *stimulare*, to urge on, to incite.] An agent that strengthens activity, increasing action. Stimulants are classified according to the parts they act upon.
Stomachic—[G. *stomakos*, L. *stomachus*.] Increasing the activity of the stomach and intestines. They increase the appetite and aid digestion.
Stupefacient—[L. *stupor* + *facere*, to make.] Stupefactive. Narcotic. Causing unconsciousness.
Styptic—[G. *styptikos*.] Astringent. Hemostatic. Arresting bleeding.
Sudorific—[L. *sudor*, sweat, + *facere*, to make.] Producing sweating. Diaphoretic. Sudoriferous. Sudoriparous.

T

Taenicide—[G. *taenia*, band, tape, + *caedere*, to kill.] Destroying, or causing the expulsion of, tape worms. Also Taenifuge. Also Teniacide. Also Teniafuge. Also Tenifuge, Tenicide.
Tonic—[G. *tonikos*; *tonos*, tone.] Increasing physical strength, restoring function and promoting vigor. Tonics are classified according to the system acted upon.
Torpent—[L. *torpere*, to be numb.] Benumbing agent.

U

Ulotic—[G. *oule*, scar.] Cicatricial.
Uragogue—[G. *ouran*, urine, + *agogos*, drawing forth.] Diuretic.
Uretic—[G. *ouretikos*, relating to urine.] Promoting the flow and excretion of urine. Diuretic.

V

Vasoconstrictor—[L. *vas*, vessel] Contracting blood vessels.
Vasodilator—Dilating blood vessels.
Vermicide—[L. *vermis*, worm, + *fugare*, to chase away.] Destroying worms.
Vermifuge—[L. *vermis*, worm, + *caedere*, to kill.] Expelling worms.
Vesicatory or **Vesicant**—[L. *vesicae*, bladder.] Producing blisters.
Vulnerary—[L. *vulnus,(vulner-)*, wound.] Promoting the healing of wounds.

17201593R00173

Printed in Great Britain
by Amazon